Advances in Experimental Medicine and Biology

More information about this series at http://www.springer.com/series/5584

Monica Gotta • Patrick Meraldi
Editors

Cell Division Machinery and Disease

Springer

Editors
Monica Gotta
Department of Cell Physiology
and Metabolism
Faculty of Medicine, Geneva University
Geneva, Switzerland

Patrick Meraldi
Department of Cell Physiology
and Metabolism
Faculty of Medicine, Geneva University
Geneva, Switzerland

ISSN 0065-2598 ISSN 2214-8019 (electronic)
Advances in Experimental Medicine and Biology
ISBN 978-3-319-57125-6 ISBN 978-3-319-57127-0 (eBook)
DOI 10.1007/978-3-319-57127-0

Library of Congress Control Number: 2017941190

Printed on acid-free paper

This Springer imprint is published by Springer Nature
The registered company is Springer International Publishing AG
The registered company address is: Gewerbestrasse 11, 6330 Cham, Switzerland

Preface

Cells are the fundamental units of living organisms that arise from the division of pre-existing cells. Cell division is not only crucial during development, it is also essential in adult organisms for tissue homeostasis and regeneration and to produce gametes for sexual reproduction. The goal of every cell division is to equally distribute the genetic material onto the two daughter cells by building up a mitotic spindle that will segregate the sister chromatids. Already in 1914 Boveri suggested a link between abnormal number of centrosomes, the organelle that form the poles of the mitotic spindle, defective cell divisions, and one prominent disease, cancer. By now a combination of genetics, cell biology, biochemistry, and the study of a wide range of model organisms led to the discovery of the principles governing cell division and to the identification of thousands of components participating in this process. Moreover, it highlighted how erroneous cell divisions or more generally mutations in genes encoding components of the cell division machinery are linked to further pathologies, such as malformation of the central nervous system, aging, and infertility.

Despite a vast knowledge about the molecular processes governing cell division, the exact physiological contribution of the cell division machinery in a disease setting is still under investigation and, in some cases, controversial. In some pathologies, such as microcephaly or lissencephaly, the genetics are well established; however, the precise mechanisms leading to the disease are not fully understood. In other cases, such as cancer, it is still controversial whether cell division defects are only a symptom of the disease or a contributing factor. The focus of this book is to review the current knowledge, to evaluate the causal link between the cell division machinery and diseases, and to identify the key open questions in the field and how to address them. In the different chapters, the authors present the evidence for and against a causal link between cell division processes and some diseases.

The opening chapter, from **El Yakoubi** and **Wassmann**, describes the process of meiosis, a specialized cell division that results in the formation of gametes. Meiosis in humans, which occurs in the absence of centrosomes, is particularly error prone. This can result in severe consequences going from infertility to severe defects in children. The review presents the differences between female and male meiosis and the potential molecular basis for the higher error rate in female meiosis.

Dysfunction of centrosomes in a mitotic division has severe consequences for cell division and has been associated to multiple diseases. In the second chapter, **Nano** and **Basto** provide a comprehensive overview of centrosomes' function in dividing and non-dividing cells and discuss how mutations that affect centrosomes' function can result in different forms of microcephaly, a developmental disorder leading to small brains. The chapter describes lessons learned from invertebrates and vertebrates, highlights the controversy of the field and the possible reasons for such controversies, and suggests possible ways to reconcile the findings from the different model systems.

In the third chapter, **Rhys** and **Godinho** review the importance of regulating the centrosome number in the context of cancer. The centrosome duplicates only once per cell cycle, to give rise to two centrosomes. The presence of two centrosomes dictates the formation of a bipolar spindle; cancer cells, however, often have supernumerary centrosomes, which favors the formation of multipolar spindles. Having more centrosomes comes, however, with a cost and can only be tolerated, and be advantageous for cancer cells, in specific genetic backgrounds. The authors discuss the mechanisms that allow the maintenance of supernumerary centrosomes in cancer cells and how a precise knowledge of these mechanisms can help to design anticancer therapies.

The mitotic spindle, the beautiful structure responsible for chromosome segregation, consists of dynamic microtubules. Chromosome's capture by these microtubules and their segregation depend on two important factors: the proper assembly of the kinetochores, a multiprotein complex bound to specialized regions of the chromosomes that binds to spindle microtubules, and the proper regulation of spindle microtubule dynamics. **Wolf** and **Kops** review the structure and the role of kinetochores during mitosis and the congenital diseases associated with mutations in genes coding for kinetochore proteins. In the following chapter, **Cirillo et al.** discuss the link between the regulation of microtubule dynamics, chromosome instability, and cancer. The authors highlight that the role that microtubule dynamics play in cancer is still elusive (the elephant in the room) and propose ways to address this issue in future studies.

Microtubule-targeting drugs used in chemotherapy have long been known to impair cell division. **Olziersky** and **Labidy-Galy** review the effect of these drugs in mitotic and non-mitotic cancer cells and discuss how the discovery of key mitotic kinases or microtubule-motor proteins has led to the development of more specific mitotic inhibitors. They review why this novel generation of antimitotic cancer drugs has so far not proven as effective as classical microtubule-targeting drugs and stress the importance of understanding the mechanisms causing resistance to anticancer drugs, to propose better combinations of chemotherapies.

The next two chapters focus on the link between cell division and aging. On one side, **Macedo et al.** discuss a yet unexplored field, how cell division is affected by organismal aging. On the other side, **Polymenis** and **Kennedy** review the mechanisms leading to cell senescence and how mutations in mitotic genes can result in accelerated cellular senescence.

Finally, the last chapter goes beyond cell division, as we discover how components of the cell division machinery can be associated to diseases through non-mitotic functions. Indeed, **Tadenev** and **Tarchini** highlight the additional roles of proteins regulating mitotic spindle orientation in postmitotic cells and their link to different diseases.

Altogether, the chapters of this book review some of the instances in which mutation of genes regulating mitosis or altered cell division processes are linked to human diseases. This book will be of interest to medical doctors and MD-PhD students interested in learning the molecular mechanisms underlying the discussed diseases and to PhD students and scientists interested in how mutations in mitotic genes can lead to human diseases.

Geneva, Switzerland Monica Gotta
Geneva, Switzerland Patrick Meraldi

Acknowledgments

M.G. and P.M. thank members of the Gotta and Meraldi laboratories for the discussion and critical input, colleagues in the field for suggesting author names, and all the authors for their contribution.

Acknowledgements

Contents

Contributors

Renata Basto Institut Curie, PSL Research University, CNRS UMR144, Paris, France

Luca Cirillo Department of Cell Physiology and Metabolism, Faculty of Medicine, Geneva University, Geneva, Switzerland

Bas de Wolf Hubrecht Institute – KNAW (Royal Netherlands Academy of Arts and Sciences), Utrecht, The Netherlands

Warif El Yakoubi Sorbonne Universités, UPMC Univ Paris 06, Institut de Biologie Paris Seine (IBPS), UMR7622, Paris, France

CNRS, IBPS, UMR7622 Developmental Biology Lab, Paris, France

Susana A. Godinho Centre for Molecular Oncology, Barts Cancer Institute, Queen Mary University of London, London, UK

Monica Gotta Department of Cell Physiology and Metabolism, Faculty of Medicine, Geneva University, Geneva, Switzerland

Brian K. Kennedy The Buck Institute for Research on Aging, Novato, CA, USA

Geert J.P.L. Kops Hubrecht Institute – KNAW (Royal Netherlands Academy of Arts and Sciences), Utrecht, The Netherlands

Cancer Genomics Netherlands, University Medical Center Utrecht, Utrecht, The Netherlands

Center for Molecular Medicine, University Medical Center Utrecht, Utrecht, The Netherlands

S. Intidhar Labidi-Galy Department of Oncology, Geneva University Hospitals, Geneva, Switzerland

Elsa Logarinho Aging and Aneuploidy Laboratory, Instituto de Biologia Molecular e Celular, Instituto de Investigação e Inovação em Saúde – i3S, Universidade do Porto, Porto, Portugal

Cell Division Unit, Department of Experimental Biology, Faculdade de Medicina, Universidade do Porto, Alameda Prof. Hernâni Monteiro, Porto, Portugal

Joana Catarina Macedo Aging and Aneuploidy Laboratory, Instituto de Biologia Molecular e Celular, Instituto de Investigação e Inovação em Saúde – i3S, Universidade do Porto, Porto, Portugal

Patrick Meraldi Department of Cell Physiology and Metabolism, Faculty of Medicine, Geneva University, Geneva, Switzerland

Maddalena Nano Institut Curie, PSL Research University, CNRS UMR144, Paris, France

Anna-Maria Olziersky Department of Cell Physiology and Metabolism, Faculty of Medicine, Geneva University, Geneva, Switzerland

Michael Polymenis Department of Biochemistry and Biophysics, Texas A&M University, College Station, TX, USA

Alexander D. Rhys Centre for Molecular Oncology, Barts Cancer Institute, Queen Mary University of London, London, UK

Abigail L.D. Tadenev The Jackson Laboratory, Bar Harbor, ME, USA

Basile Tarchini The Jackson Laboratory, Bar Harbor, ME, USA

Sara Vaz Aging and Aneuploidy Laboratory, Instituto de Biologia Molecular e Celular, Instituto de Investigação e Inovação em Saúde – i3S, Universidade do Porto, Porto, Portugal

Katja Wassmann Sorbonne Universités, UPMC Univ Paris 06, Institut de Biologie Paris Seine (IBPS), UMR7622, Paris, France

CNRS, IBPS, UMR7622 Developmental Biology Lab, Paris, France

Chapter 1
Meiotic Divisions: No Place for Gender Equality

Warif El Yakoubi and Katja Wassmann

Abstract In multicellular organisms the fusion of two gametes with a haploid set of chromosomes leads to the formation of the zygote, the first cell of the embryo. Accurate execution of the meiotic cell division to generate a female and a male gamete is required for the generation of healthy offspring harboring the correct number of chromosomes. Unfortunately, meiosis is error prone. This has severe consequences for fertility and under certain circumstances, health of the offspring. In humans, female meiosis is extremely error prone. In this chapter we will compare male and female meiosis in humans to illustrate why and at which frequency errors occur, and describe how this affects pregnancy outcome and health of the individual. We will first introduce key notions of cell division in meiosis and how they differ from mitosis, followed by a detailed description of the events that are prone to errors during the meiotic divisions.

Keywords Meiosis • Oocytes • Spermatogenesis • Aneuploidy • Fertility • Trisomy • Maternal and paternal age • Spindle assembly checkpoint • Cohesin • Chiasmata

Abbreviations

APC/C	Anaphase promoting complex/cyclosome
CK1	Casein kinase 1
CPC	Chromosome passenger complex
DSB	Double strand break
MCC	Mitotic checkpoint complex
PB	Polar body
Plk1	Polo-like kinase 1
SAC	Spindle assembly checkpoint

W. El Yakoubi • K. Wassmann (✉)
Sorbonne Universités, UPMC Univ Paris 06, Institut de Biologie Paris Seine (IBPS), UMR7622, Paris 75252, France

CNRS, IBPS, UMR7622 Developmental Biology Lab, Paris 75252, France
e-mail: katja.wassmann@upmc.fr

M. Gotta, P. Meraldi (eds.), *Cell Division Machinery and Disease*, Advances in Experimental Medicine and Biology 1002, DOI 10.1007/978-3-319-57127-0_1

1.1 Meiotic Cell Division

Diploid organisms contain two copies of the genome, originating from the two parents and organized as thread-like structures named chromosomes. In S-phase, which precedes cell division in somatic and germ cells, paternal and maternal genomes are replicated, the cell harbors therefore four copies of the genome. As cells enter mitosis, DNA condensation takes place, and chromatids, each corresponding to one copy of the genome, become visible. Two sister chromatids of the same parental origin are paired and segregated to generate two identical daughter cells. In meiosis on the other hand, four cells each harboring only one copy of the genome are produced to form the haploid gametes (Fig. 1.1). In female meiosis, only one of these four cells develops into an oocyte, the remaining cells degenerate as Polar Bodies (PBs). The generation of haploid gametes is possible due to the specialized segregation pattern of the meiotic cell division which ensures that each gamete receives one chromatid of each chromosome pair [1–3]. Contrary to mitosis, the aim of meiosis is not to generate identical daughter cells, but to recombine the maternal and paternal genome to generate new genetic variations in the offspring. This requires the pairing of homologous chromosomes from different parental origin, and DNA recombination, which take place in meiotic prophase, before the meiotic divisions. When germ cells enter the meiotic cell division, recombination events are finalized. But homologous chromosomes are still held together by chiasmata, which are the remaining physical connections of meiotic recombination. To generate four haploid cells, two divisions without intervening S-phase take place in meiosis. In the first division (meiosis I), homologous chromosomes are segregated, and in the second (meiosis II), sister chromatids [1–3]. To bring about this specific segregation pattern, key events of the mitotic cell division have been adapted or changed for meiosis, and they are outlined in detail below.

1.2 Monopolar Attachment in Meiosis I

Chromosome segregation in mitosis and meiosis depends on their correct attachment to a macromolecular structure named the bipolar spindle [4]. The spindle consists of microtubules that attach through a search-and-capture mechanism to chromosomes, and move them to the opposite poles of the mother cell, to form two daughter cells. Microtubules need to form tension bearing attachments to remain stably attached [5]. Each of the two sister chromatids of a chromosome harbors an attachment site which is named kinetochore and consists of more than 100 proteins that are found at a specific chromosome region, the centromere [6]. In mitosis, the kinetochores of the two sister chromatids get attached to the opposite poles of the spindle in prometaphase, and at the metaphase-to-anaphase transition the physical connections holding the sister chromatids together (cohesins, see below), are removed so that each sister can glide to the opposite pole (Fig. 1.2a). As kinetochores

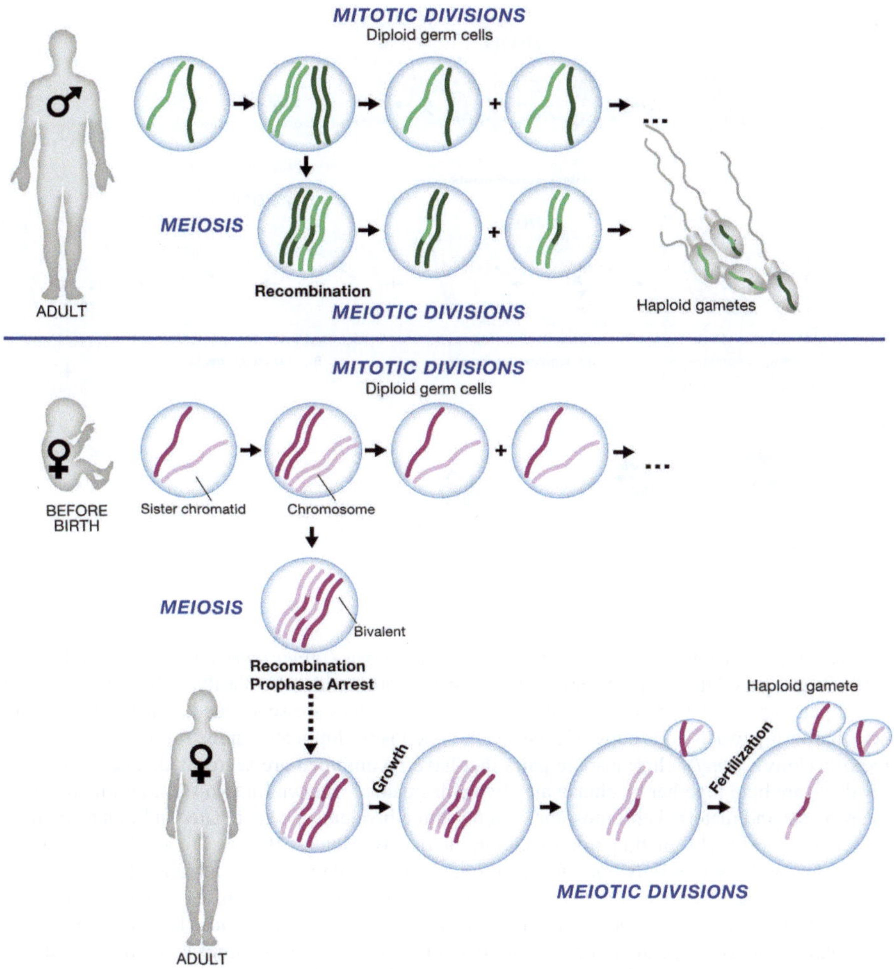

Fig. 1.1 Key differences in the developmental timeline of male and female meiosis in mammals. In this figure we illustrate when the mitotic divisions giving rise to germ cells, and meiosis take place in individuals of different sex. For simplicity, only one chromosome pair (bivalent) out of 22 autosomes and the sex chromosomes in humans is shown. The different shades of *green* (male) or *pink* (female) correspond to the different parental origins. In mitosis, replicated sister chromatids are segregated, and chromosomes of different parental origin never pair. The *upper panel* illustrates that mitotic divisions to generate male germ cells take place in the testis of the adult, as well as all stages of meiosis. In the *lower panel* we illustrate that mitotic divisions and the following meiotic recombination occur before birth in the female, whereas meiosis I and II take place in the adult. In oocytes, meiotic divisions are highly asymmetric, giving rise to one very large oocyte, and two small polar bodies that degenerate

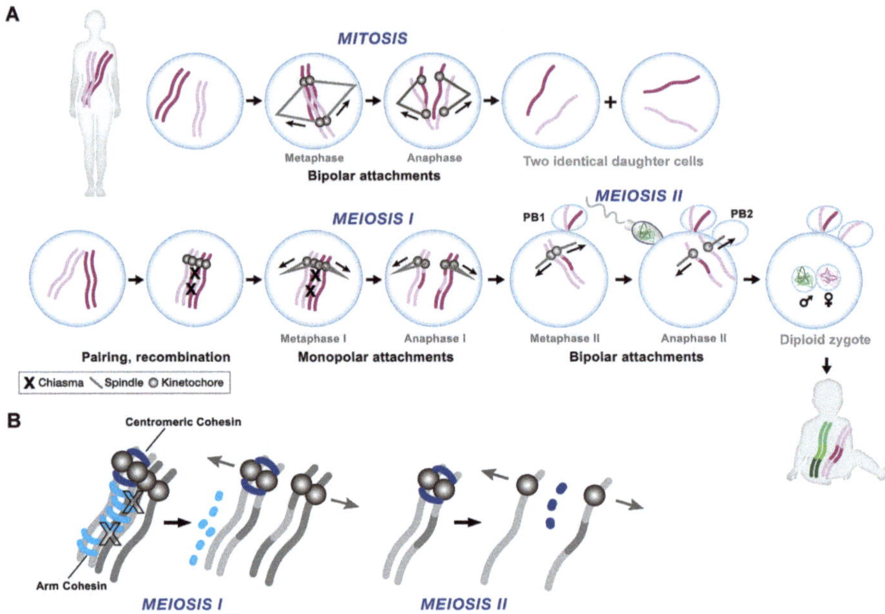

Fig. 1.2 Cell division in mitosis and meiosis. (**a**) In mitosis (*upper panel*), two identical daughter cells are generated through the segregation of sister chromatids. For clarity, only two sister chromatid pairs are shown. In humans, 46 sister chromatid pairs are segregated, generating daughter cells harboring exactly one copy of each sister, which is re-duplicated in the following S-phase. In meiosis (*lower panel*), chromosome pairs that have recombined are segregated. These so-named bivalents are held together by chiasmata. Two chiasmata are shown for two sister chromatids from the paternal and maternal chromosome. Twenty-three bivalents are segregated in humans in meiosis I, and 23 sister chromatid pairs (dyads) in meiosis II. Meiosis II comes to a halt in metaphase in oocytes, and resumes only upon fertilization, whereas male meiosis is already finalized prior to fertilization. Once a female pronucleus is formed, the male pronucleus fuses with the female pronucleus to form a diploid zygote, which again harbors 46 sister chromatids. The different meiotic segregation patterns depend on monopolar attachment of sister kinetochores in meiosis I, and step-wise cohesin removal. (**b**) Step-wise cohesin removal from chromosome arms in meiosis I, and from the centromere region in meiosis II is required for correct segregation. For simplicity, cohesins are visualized only for one of the two chromosomes shown, and recombination (the position of the chiasmata is shown as a transparent X) for only one sister of each chromosome. Arm cohesin is shown in *light blue*, centromeric cohesin in *dark blue*. Cohesin removal from arms is necessary to permit segregation of sister chromatids that have recombined. In the centromere region, no recombination occurs, therefore centromeric cohesin can be maintained throughout meiosis I to allow correct alignment of paired sister chromatids in meiosis II. Centromeric cohesin is then removed at the metaphase-to-anaphase transition of meiosis II

are attached in a back-to-back configuration this is named bipolar attachment. The separation of sister chromatids and their poleward movement take place in a completely synchronized manner for all chromosomes [4, 6].

In meiosis I—in contrast to mitosis—two chromosomes each consisting of two sister chromatids are paired and segregated into two daughter cells. Sister kinetochores are therefore attached in a monopolar fashion, to the same pole [1, 3] (Fig. 1.2b).

Chromosomes are maintained together through chiasmata, structures that are visible on condensed chromosomes. In mice, monopolar orientation is brought about by initial orientation of the two sister chromatids to the same pole due to the presence of chiasmata, and proteins such as Polo-like kinase 1 (Plk1) and the meiosis-specific kinetochore protein Meikin [7, 8]. In the following meiosis II, attachments are bipolar as sister chromatids are paired and segregated, similar to a mitotic division [1] (Fig. 1.2a). Failures to attach sister kinetochores in a monopolar fashion in meiosis I will result in missegregation events. If sister chromatids are attached in a bipolar manner, they may be segregated in meiosis I in a way similar to a mitotic division. These sister chromatids cannot biorient in the second meiotic division and will segregate at random and therefore with a high error rate [9].

1.3 Chiasmata

In mitosis, maternal and paternal chromosomes never pair, this event is specific to meiosis I. To segregate chromosomes correctly in meiosis I, chromosome pairs (bivalents) have to be maintained together through a structure named chiasmata, to keep them aligned at the metaphase plate and attached to the bipolar spindle, until anaphase I onset takes place (Fig. 1.2a). Each chromosome pair needs at least one chiasma, and usually, one to three chiasmata are found per bivalent. The formation of chiasmata depends on recombination events that take place before entry into meiosis I. Chiasmata formation depends on how double strand breaks (DSB) that are induced by Spo11 are repaired (cross-over which leads to the formation of a chiasma, or non-cross over) [10, 11]. Meiotic DSB repair occurs preferentially through the non-sister chromatid, to mix maternal and paternal genomes—this is named non-sister bias. Cross-over interference ensures that only certain DSBs are repaired as cross-overs, and therefore only a limited number of chiasmata is formed per chromosome pair [12]. Obviously, the longer a given chromosome, the more chiasmata are found, and smaller chromosomes often harbor only one chiasma [9].

1.4 Cohesin Removal

In mitosis, the correct and tension-bearing stable attachment of sister chromatids to the spindle requires that they are physically maintained together. The "glue" holding sister chromatids together is a protein complex forming a ring-like structure around the paired chromatids. This complex is named cohesin, and consists of several subunits conserved from yeast to man [13, 14]. The complex contains Smc (stability of mini chromosomes) heterodimers that form the ring and is closed by a protein belonging to the kleisin family, which are proteins interacting with Smc proteins [15, 16]. Variants of yeast Scc3 such as SA2 and SA3 additionally bind the kleisin subunit and are essential for cohesion. Cohesin on chromosome arms is removed by the so-named prophase pathway, which depends on Wapl, the kinases Plk1 and Aurora B, and phosphorylation of the SA2 cohesin subunit, in early mitosis [16]. But residual cohesion in the centromere region is not removed and maintains sisters together until metaphase. To open up the remaining cohesin rings for the separation of sister chromatids, a protease named Separase is activated at the metaphase-to-anaphase transition. Separase cleaves the kleisin subunit Scc1 of the cohesin complex, allowing the two DNA strands to be liberated and torn to the opposite poles of the spindle [13, 14, 17].

In meiosis, sister chromatids are held together by cohesins as well (Fig. 1.2b). Furthermore, cohesion stabilizes chiasmata on chromosome arms, and this maintains the paired chromosomes, each consisting of two sister chromatids, together. For correct chromosome segregation in meiosis I, sister chromatid cohesion is maintained until metaphase I. It is unknown whether some prophase pathway dependent removal of cohesin takes place at this time, but no separation of chromatid arms indicating significant prophase pathway dependent cohesin removal can be observed in male or female prometaphase I. At the metaphase-to-anaphase transition of meiosis I, Separase becomes active and cleaves the meiosis-specific kleisin Rec8, which substitutes mitotic Scc1. But contrary to mitosis, not all cohesion is removed in metaphase of meiosis I. Cohesin in the centromere region is maintained until meiosis II, and this is essential to correctly attach the meiosis II spindle to sister kinetochores in the second meiotic division. Therefore, a key feature of cohesin removal in meiosis is the fact that it has to occur in a step-wise manner during the two divisions [1, 3]: First from chromosome arms in meiosis I, and then from the centromere region in meiosis II. Removal of cohesin from arms allows separation of chromosomes that have recombined and exchanged parts of their DNA, as these regions are still maintained together by cohesin with the original homologous sister chromatid, and chiasmata. Recombination occurs only along the chromosome arms, therefore cohesin can be maintained in the centromere region throughout meiosis I and this does not prevent chromosome segregation. On the contrary, maintaining cohesion in the centromere region is essential to biorient sister chromatids on the meiosis II spindle and establish tension-bearing attachments (Fig. 1.2b).

Cohesin from chromosome arms and the centromere region is removed by Separase. But cohesin in the centromere region is protected from cleavage by Separase throughout meiosis I and it is only in meiosis II that the centromeric fraction of Rec8 can be cleaved by Separase. For this step-wise cohesin removal, the kleisin subunit Scc1 is substituted by a meiosis-specific subunit, Rec8. To be cleaved by Separase, Rec8 on arms has to be phosphorylated by the kinases Casein Kinase 1 (CK1) and Cdc7/Dbf4 in budding yeast [18, 19], but whether this is also the case in mammalian meiosis is still an open question. Maintaining Rec8 in its unphosphorylated state is thought to protect it from cleavage by Separase also in mammals, and this is achieved by localising the phosphatase PP2A-B56 to the centromere through the adaptor protein Sgo2 [20, 21]. In meiosis II, cohesin at the centromere has to be "deprotected", and the exact mechanisms of deprotection are currently unknown, but seem to involve removal of Sgo2 (and therefore PP2A) from the vicinity of Rec8 [21, 22] and co-localization of the histone chaperone and PP2A inhibitory protein I2PP2A/Set [23, 24]. Obviously, correct chromosome segregation in meiosis I and II depends on rigorous control of cohesin protection and deprotection during both divisions.

1.5 Spindle Assembly Checkpoint

During cell division the correct attachment of kinetochores is verified by a checkpoint named the spindle assembly checkpoint or SAC. Kinetochores without or with wrongly attached microtubules maintain this checkpoint active, which leads to a transient cell cycle arrest in metaphase. Essential core checkpoint proteins such as Mad1, Mad2, Bub1, BubR1, Bub3, and Mps1 are localized to incorrectly attached kinetochores, and lead to the formation of the Mitotic checkpoint complex or MCC, consisting of BubR1, Mad2, Bub3 and Cdc20 [25, 26]. This MCC prevents degradation of the Separase inhibitors Securin and Cyclin B, by inhibiting the E3 ubiquitin ligase Anaphase Promoting Complex/Cyclosome (APC/C). Upon bipolar attachment of all kinetochores in mitosis, the SAC is satisfied, Securin and Cyclin B are ubiquitinated and degraded, Separase becomes active and metaphase-to-anaphase transition takes place [27]. Interestingly, the SAC not only recognizes whether attachments are present at individual kinetochores, but it also recognizes the quality of attachments in terms of stability. When kinetochores are not attached to both poles, they lack tension resulting in a destabilization of attachments which is detected by the SAC [28]. Bipolar tension itself is thought to be recognized through a tension-sensing distortion of the kinetochore, the so-named intra-kinetochore stretch, which would silence the SAC and permit metaphase-to-anaphase transition [29]. Recently though the necessity of an established intra-kinetochore stretch for SAC activation has been put into question by two studies showing that stabilizing attachments satisfies the SAC even if there is no bipolar tension and intra-kinetochore stretch [30, 31]. The SAC's role is interwoven with another signalling cascade, namely error correction, implicating the Chromosome Passenger Complex (CPC) proteins Incenp, Borealin, Survivin, and Aurora B [32]. The Aurora B kinase, required for detaching wrongly attached microtubule fibers that are not under tension, is an integral part of the mitotic SAC because it controls the checkpoint kinase Mps1 [33, 34].

In female meiosis attachments also underlie checkpoint control, even though attachments are monopolar [35]. (No data exist so far on SAC functionality in mammalian male meiosis I). But it seems that the checkpoint is much more permissive in oocyte meiosis I than in mitosis, and a few unattached kinetochores escape detection by the SAC [36, 37]. It is far from clear whether the checkpoint in meiosis can distinguish monopolar from bipolar attachments, and whether the quality of attachments such as presence of tension is recognized. Core checkpoint proteins are required for SAC control in meiosis, but also for correct timing of the extremely long prometaphase I in oocytes. In the absence of SAC control, gross missegregations take place in meiosis I and lead to sterility in mice [38–41].

1.6 Age-Related Differences in Male and Female Gametogenesis

As we will outline below, advanced age is the most important risk factor for female reproductive health in humans [9, 42]. To understand why this is the case for female but by far not as much for male meiosis, it is important to introduce the crucial sex-specific differences in mammalian gametogenesis.

Primordial germ cells are set aside early in embryo development in both sexes. They migrate during early development to the place of the future gonads. Here, the fate of female and male germ cells is quite different: male germ cells remain dormant until hormonally stimulated to further divide in the sexually mature adult. The spermatogonial stem cell population divides mitotically to maintain the stem cell population, and to generate progenitor cells that are committed to undergoing meiosis. In the seminiferous tubules these progenitor cells will differentiate to primary and secondary spermatocytes, spermatides and finally, spermatozoa. During these different stages they undergo meiotic recombination, followed by meiosis I and II. Spermatogonial stem cells divide continuously in the adult male, giving rise to more stem cells, or alternatively, to spermatozoa ([43], and references therein).

In the female on the other hand germinal cells form the primordial follicle, which consists of pregranulosa cells surrounding an oocyte in prophase, before birth. Oocytes of the primordial follicle have already finished meiotic recombination and await growth and entry into the first meiotic division upon hormonal stimulation in the sexually mature female. Importantly, homologous chromosomes that are paired for recombination, have to be maintained together by cohesins and chiasmata until entry into the first meiotic division, which can be several decades later, such as in humans ([44], and references therein). Therefore, oocytes which are generated only in the embryo (there is no convincing evidence that oocytes in significant numbers could be generated in the adult [45]) are under much more temporal stress than spermatozoa, which are produced continuously and throughout adult life (Fig. 1.1).

1.7 How Meiotic Segregation Errors Occur

1.7.1 Errors in Male Meiosis

Chromosomes and sister chromatids must be maintained together for correct alignment of chromosomes at the metaphase plate in meiosis I. All steps of male meiosis take place in the adult in a continual manner, starting from somatic cells that enter the meiotic program. In humans, it takes about a month from the start of meiotic recombination to generate spermatids ([43], and references therein) as pairing of chromosomes and recombination is immediately followed by entry into meiosis I and chromosome segregation. The cohesin complex and chiasmata have to be maintained for a time frame that is rather short compared to female meiosis, to ensure correct segregation in meiosis I and II. In humans, paternal age seems therefore not to be a very important factor for fidelity of chromosome segregation, and an increase in chromosome missegregations is observed in men above 50 only according to some studies, whereas other studies have not found an increase [46]. The overall amount of aneuploid sperm in healthy men lies within a range of 4–5%, according to [47]. When individual chromosomes are analyzed for their probability to missegregate, an important challenge for male meiosis comes to light, namely the segregation of sex chromosomes. As male meiosis requires pairing and segregation of an X and a Y chromosome, and because the region of homology of the X and Y does not span the whole chromosome, precocious separation of those two sex chromosomes occurs at a higher frequency than that of autosomes. As a consequence, the male contribution to numerical abnormalities of the sex chromosome in the offspring is more important than the female, because in oocytes two homologous X chromosomes are paired and segregated, which does not constitute a challenge. Approximately 80% of sex chromosome aneuploidies are therefore thought to be of paternal origin [9, 47]. Survival rates of individuals with aneuploidies of a sex chromosome are high, and we will detail below how health of the individual is affected in the different cases.

1.7.2 Errors in Female Meiosis

Female meiosis is much more error prone than male meiosis. It is estimated that 25–30% of oocytes in a healthy, young woman are aneuploid after the meiotic divisions. This high error rate furthermore increases drastically with the age of the mother, and it has been shown that women closer to menopause have a risk of more than 30% of being pregnant with a trisomic baby [42]. Why especially in humans female meiosis is so error prone is still a matter of debate. The reason lies most likely at least in part in the different timeline of male and female gametogenesis. Even in oocytes of a woman at the start of her reproductive life, chromosomes have been paired for nearly two decennia, putting enormous strain on sister chromatid cohesion and chiasmata maintenance. It has been shown in the mouse that there is no renewal or turnover of cohesin proteins during meiotic prophase arrest, at least during the growing phase of the oocyte [48, 49]. Therefore, it is attractive to speculate that deterioration of cohesin complexes with age contributes to increased missegregation rates in females.

Bipolar spindle assembly occurs without centrosomes and takes extremely long (in human oocytes around 16 h), with frequent attachments and detachments [50]. It has been observed that sister kinetochores instead of presumably acting as a sole unit, frequently split in human oocytes, indicating that they are at least transiently attached in a bipolar and not monopolar manner in meiosis I [51]. Additionally, SAC control seems to be leaky, at least in mouse oocytes [36, 37, 52]. One or two not correctly or not at all attached kinetochores may escape SAC surveillance, further contributing significantly to the high error rate.

The most important factor negatively influencing fidelity of chromosome segregation in female meiosis is maternal age. Studies from mouse and human oocytes suggest that cohesins diminish with age, which concomitantly leads to the destabilization and eventually, loss of chiasmata [53–57]. It is interesting to note that in human oocytes the frequency of chromosomes that show sister kinetochore splitting in meiosis I increases with age, and this may be due to weaker cohesion between sister chromatids because of cohesin loss [51]. Weakening of the cohesin complex and loss of chiasmata will result in a failure to properly attach chromosome pairs to the meiotic spindle in meiosis I. Achiasmatic chromosomes may instead attach in a manner similar to mitosis, with sister chromatids oriented in a bipolar, and not monopolar manner. This can lead to precocious separation of sister chromatids in meiosis I, followed by their random segregation in meiosis II, because those single sister chromatids will fail to correctly align at the meiosis II metaphase plate.

Some chromosomes are more affected by precocious meiosis I separation than others, which may be explained by the fact that for example smaller chromosomes have less chiasmata to start with. Also the position of chiasmata (close to the centromere or the telomere) and resulting orientation on the meiotic spindle influence how likely a bivalent may missegregate in meiosis I, and whether the chiasma will be lost upon weakening of cohesin with age [9, 42, 44].

Additionally, the abundance of other proteins required for correct meiotic cell division diminishes with maternal age, these are for example the SAC protein BubR1 [58, 59], which is also required for stable microtubule-kinetochore attachments [41, 60, 61], and Sgo2 [54, 62] which is required for centromeric cohesin protection. The age-related increase in chromosome missegregations in oocytes is therefore considered as depending on a combination of multiple factors. Loss of cohesin or of Sgo2 in combination with weaker SAC control is expected to be detrimental for oocyte maturation because chromosome segregation depends on it. Indeed, reproductive age in humans seems to be restrained by the fidelity of meiotic chromosome segregation in oocytes [42, 44, 63].

1.8 Genetic Predispositions to Error Prone Chromosome Segregation in Meiosis

Mutations that affect proper segregation of the genetic material during cell division are expected to be essential and to prevent the generation of a viable organism. Indeed, loss of spindle checkpoint components Mad2, Bub1 and BubR1 for example leads to early emrbyonic lethality, at least in mice [64–66]. Therefore, only mutations in genes that are required for meiotic and not mitotic chromosome segregation are expected to correlate with a predisposition to erroneous chromosome segregation in oocytes and during spermatogenesis. One such example is the Aurora C kinase, which is partly redundant with Aurora B and which is a component of the Chromosomal Passenger Complex [67–69]. Naturally occurring mutations in Aurora

C have been reported in humans, and they are associated with polyploid or multiflagellar spermatozoa, and therefore, male infertility [70, 71]. This is quite likely due to the function of Aurora C in cytokinesis. On the other hand, female meiosis is not affected by these mutations, probably because Aurora B can substitute for some of Aurora C's functions [72]. Functional assays in mouse oocytes indicate that the observed male phenotype is most likely due to failures to exit meiosis I [73].

Another example for a genetic predisposition affecting meiotic chromosome segregation is the cohesin subunit Stag3, which is -in addition to the previously mentioned Rec8-specific to meiosis. A frameshift mutation in Stag3 was identified in a family with a history of premature ovarian failure (POF). Whereas the molecular reasons for this mutation to cause POF in humans have not been elucidated, results from the mouse suggest that POF is due to partial cohesin loss and the fact that double strand break-repair is initiated, but not finalized and oocytes are lost in prophase [74, 75].

1.9 Other Genetic Reasons Why Meiosis May Go Awry

Other factors that influence oogenesis and spermatogenesis can be of a wide variety of origins. They may be environmental (exposure to toxic substances), hormonal, or related to improper oocyte development or spermatogenesis. Genetic factors associated with these conditions are beyond the scope of this chapter and are for example listed in [76].

1.10 Consequences of Meiotic Missegregations

Here we will outline the fate of chromosome missegregations in meiosis for pregnancy outcome in humans and consequences for the individual. Aneuploid gametes can still be functional and upon fertilization lead to the formation of an (aneuploid) embryo, but depending on the chromosome affected, survival may be compromised. Indeed, a third of all spontaneous abortions in the first trimester concern embryos with the wrong number of chromosomes. Importantly, this takes into account only recognized pregnancies, as certain abortions occur so early during pregnancy (e.g. before implantation) that they go undetected. This makes aneuploidy the most frequent reason for early pregnancy loss [42].

Aneuploidies that arise from missegregations in meiosis affect all the cells of the embryo, in contrary to so-named mosaic trisomies, which will not be discussed here and which are due to missegregations during the mitotic divisions in the embryo, affecting only a subset of cells. Monosomies (one copy) of autosomes are not viable in humans, whereas a small number of trisomies (three copies) are. Individuals with trisomy 21, which gives rise to the Down syndrome, usually survive to adulthood and reach a median age of 60 years [77], but individuals with trisomy 13 (Patau syndrome) and 18 (Edwards syndrome) usually die during the first weeks of life [78]. Trisomies of other autosomes are incompatible with survival up to birth.

Newborns with trisomy 13 and 18 have severe developmental defects of the heart, the nervous system, and to a varying degree of muscles, of the members, and of brain. Trisomy 21 is by far the most frequent, affecting 1 birth out of 1000 [79]. Survival of children affected has steadily increased from around 50% in the 1950s to around 95% today, in industrialized nations [77]. This is mainly due to early intervention for health problems associated with Down syndrome, such as cardiac malformations and respiratory problems. Trisomy 21 is compatible with life into adulthood, but individuals show intellectual disabilities, and often early onset of Alzheimer disease, and disorders of the respiratory, cardiovascular and endocrine system, hearing and vision disorders, and the typical cranofacial disorders [80]. Children with trisomy 21 are predisposed to the development of certain cancers, such as testicular germ cell tumors, myeloid leukemias, and acute lymphoblastic leukemia [81]. Overall number of births has strongly decreased since the development of more reliable prenatal screening methods and ultrasound exams. Often, diagnosis of trisomy 21 is followed by voluntary termination of the pregnancy, explaining why less children with trisomy 21 than expected are born today [82], even though trisomy 21 pregnancies have increased by 70% within 20 years, due to overall increased maternal age [83].

Viable sex chromosome aneuploidies are monosomy of the X chromosome; or three sex chromosomes, with two copies of the X or the Y chromosome in addition to the X chromosome. Even though these aneuplodies are associated with characteristic disabilities, individuals are often not diagnosed. Below, we will shortly describe each of the syndromes associated with a specific sex-chromosome aneuploidy.

Individuals harboring only one sex chromosome, namely the X chromosome (monosomy of the Y chromosome is not compatible with life), are suffering from Turner's syndrome. One out of 2000 females is affected by this disorder [84]. In 70–80% of cases the remaining X chromosome is maternally derived, meaning that the segregation error had occurred in male meiosis [42]. Women with Turner's Syndrome suffer from the consequences of haploinsufficiency of regions of the X chromosome that are normally not inactivated on the second X chromosome. Ovarian failure is observed during foetal development, and leads to decreased production of sex hormones and infertility. Certain cognitive capacities are affected, and shorter stature, renal and cardiovascular problems are observed [84, 85]. Currently treatment consists of cognitive rehabilitation and hormone-replacement therapy.

One out of 600 male live births harbors two X and one Y chromosome, which gives rise to Klinefelter's syndrome. This aneuploidy is in equal proportions due to missegregation in male and female meiosis. Again, certain neurological features and cognitive abilities such as learning abilities are affected (but less so than with Turner's syndrome), testosterone production stops at puberty and individuals are infertile, and an increased risk for usually very rare mediastinal germ-cell tumors is observed. Some health problems such as early onset of osteoporosis are probably due to deficiencies in testosterone levels [84].

Another aneuploidy of sex chromosomes leads to the so-named XYY syndrome, where male individuals have a supplementary Y chromosome. Obviously, only errors in male meiosis lead to this aneuploidy, and it is estimated that one out of 1000 male livebirths is affected. XYY syndrome is characterized mainly by tall stature, impaired neurological features and cognitive abilities with learning disorders comparable to Klinefelter's syndrome. Increased impulsivity has been described, but earlier reports indicating that crime rate is higher among individuals with XYY syndrome have not been confirmed [86]. In contrary to Turner's and Klinefelter's syndrome, man with XYY syndrome do not have sex hormone deficiencies, and are fertile [84].

The presence of three X chromosomes is due to errors in female meiosis and touches one out of 1000 female live births, which are consequently suffering from the triple X syndrome. Cognitive performance seems to be affected in a manner similar to individuals with Klinefelter 's and Turner's syndrome, and behavioral abnormalities are equally observed [84].

1.11 Conclusion

Correct execution of the meiotic program is essential to obtain male and female gametes with the correct number of chromosomes, a prerequisite for the generation of healthy offspring. Female meiosis is much more error prone than male meiosis, due at least in part to the fact that meiotic recombination takes place already decennia before the meiotic divisions in oocytes. Furthermore, meiosis is strongly affected by maternal age in humans. Aneuploidies in the embryo severely affect pregnancy outcome and health of the individual. Gaining insights into the molecular mechanisms regulating meiotic divisions in male and female gametes is important to understand how aneuploidies with such severe consequences can occur.

Acknowledgements We thank lab members for discussion, and Sophie Gournet (UMR7622) for help with figures. W.E.Y. is supported by a postdoctoral fellowship from ARC (Association de la Recherche Contre le Cancer), and the group of K. Wassmann by an ANR grant (ANR-12-BSV2-0005-01), a grant by the Fondation de la Recherche Médicale (EquipeDEQ20160334921), the UPMC and CNRS.

References

1. Petronczki M, Siomos MF, Nasmyth K (2003) Un menage a quatre: the molecular biology of chromosome segregation in meiosis. Cell 112:423–440
2. MacLennan M, Crichton JH, Playfoot CJ, Adams IR (2015) Oocyte development, meiosis and aneuploidy. Semin Cell Dev Biol 45:68–76
3. Page SL, Hawley RS (2003) Chromosome choreography: the meiotic ballet. Science 301:785–789

4. Heald R, Khodjakov A (2015) Thirty years of search and capture: the complex simplicity of mitotic spindle assembly. J Cell Biol 211:1103–1111
5. Lampson MA, Cheeseman IM (2011) Sensing centromere tension: Aurora B and the regulation of kinetochore function. Trends Cell Biol 21:133–140
6. Cheeseman IM (2014) The kinetochore. Cold Spring Harb Perspect Biol 6:a015826
7. Watanabe Y (2012) Geometry and force behind kinetochore orientation: lessons from meiosis. Nat Rev Mol Cell Biol 13:370–382
8. Kim J, Ishiguro K, Nambu A, Akiyoshi B, Yokobayashi S et al (2015) Meikin is a conserved regulator of meiosis-I-specific kinetochore function. Nature 517:466–471
9. Nagaoka SI, Hassold TJ, Hunt PA (2012) Human aneuploidy: mechanisms and new insights into an age-old problem. Nat Rev Genet 13:493–504
10. Lam I, Keeney S (2015) Mechanism and regulation of meiotic recombination initiation. Cold Spring Harb Perspect Biol 7:a016634
11. Bascom-Slack CA, Ross LO, Dawson DS (1997) Chiasmata, crossovers, and meiotic chromosome segregation. Adv Genet 35:253–284
12. Zickler D, Kleckner N (2016) A few of our favorite things: pairing, the bouquet, crossover interference and evolution of meiosis. Semin Cell Dev Biol 54:135–148
13. Jeppsson K, Kanno T, Shirahige K, Sjogren C (2014) The maintenance of chromosome structure: positioning and functioning of SMC complexes. Nat Rev Mol Cell Biol 15:601–614
14. Peters JM, Nishiyama T (2012) Sister chromatid cohesion. Cold Spring Harb Perspect Biol 4
15. Nasmyth K, Haering CH (2005) The structure and function of SMC and kleisin complexes. Annu Rev Biochem 74:595–648
16. Zheng G, Yu H (2015) Regulation of sister chromatid cohesion during the mitotic cell cycle. Sci China Life Sci 58:1089–1098
17. Peters JM, Tedeschi A, Schmitz J (2008) The cohesin complex and its roles in chromosome biology. Genes Dev 22:3089–3114
18. Katis VL, Lipp JJ, Imre R, Bogdanova A, Okaz E et al (2010) Rec8 phosphorylation by casein kinase 1 and Cdc7-Dbf4 kinase regulates cohesin cleavage by separase during meiosis. Dev Cell 18:397–409
19. Ishiguro T, Tanaka K, Sakuno T, Watanabe Y (2010) Shugoshin-PP2A counteracts casein-kinase-1-dependent cleavage of Rec8 by separase. Nat Cell Biol 12:500–506
20. Llano E, Gomez R, Gutierrez-Caballero C, Herran Y, Sanchez-Martin M et al (2008) Shugoshin-2 is essential for the completion of meiosis but not for mitotic cell division in mice. Genes Dev 22:2400–2413
21. Lee J, Kitajima TS, Tanno Y, Yoshida K, Morita T et al (2008) Unified mode of centromeric protection by shugoshin in mammalian oocytes and somatic cells. Nat Cell Biol 10:42–52
22. Gomez R, Valdeolmillos A, Parra MT, Viera A, Carreiro C et al (2007) Mammalian SGO2 appears at the inner centromere domain and redistributes depending on tension across centromeres during meiosis II and mitosis. EMBO Rep 8:173–180
23. Chambon JP, Touati AS, Berneau S, Hebras C, Groeme R et al (2013) The PP2A inhibitor I2PP2A is essential for sister chromatid segregation in meiosis II. Curr Biol 23:485–490
24. Wassmann K (2013) Sister chromatid segregation in meiosis II: deprotection through phosphorylation. Cell Cycle 12:1352–1359
25. Sacristan C, Kops GJ (2015) Joined at the hip: kinetochores, microtubules, and spindle assembly checkpoint signaling. Trends Cell Biol 25:21–28
26. Musacchio A (2015) The molecular biology of spindle assembly checkpoint signaling dynamics. Curr Biol 25:R1002–R1018
27. Lara-Gonzalez P, Westhorpe FG, Taylor SS (2012) The spindle assembly checkpoint. Curr Biol 22:R966–R980
28. Jia LY, Kim S, Yu HT (2013) Tracking spindle checkpoint signals from kinetochores to APC/C. Trends Biochem Sci 38:302–311
29. Maresca TJ, Salmon ED (2010) Welcome to a new kind of tension: translating kinetochore mechanics into a wait-anaphase signal. J Cell Sci 123:825–835

30. Etemad B, Kuijt TE, Kops GJ (2015) Kinetochore-microtubule attachment is sufficient to satisfy the human spindle assembly checkpoint. Nat Commun 6:8987
31. Tauchman EC, Boehm FJ, DeLuca JG (2015) Stable kinetochore-microtubule attachment is sufficient to silence the spindle assembly checkpoint in human cells. Nat Commun 6:10036
32. van der Horst A, Lens SM (2014) Cell division: control of the chromosomal passenger complex in time and space. Chromosoma 123:25–42
33. Santaguida S, Vernieri C, Villa F, Ciliberto A, Musacchio A (2011) Evidence that Aurora B is implicated in spindle checkpoint signalling independently of error correction. EMBO J 30:1508–1519
34. Saurin AT, van der Waal MS, Medema RH, Lens SM, Kops GJ (2011) Aurora B potentiates Mps1 activation to ensure rapid checkpoint establishment at the onset of mitosis. Nat Commun 2:316
35. Touati SA, Wassmann K (2016) How oocytes try to get it right: spindle checkpoint control in meiosis. Chromosoma 125:321–335
36. Gui L, Homer H (2012) Spindle assembly checkpoint signalling is uncoupled from chromosomal position in mouse oocytes. Development 139:1941–1946
37. Lane SI, Yun Y, Jones KT (2012) Timing of anaphase-promoting complex activation in mouse oocytes is predicted by microtubule-kinetochore attachment but not by bivalent alignment or tension. Development 139:1947–1955
38. Wassmann K, Niault T, Maro B (2003) Metaphase I arrest upon activation of the Mad2-dependent spindle checkpoint in mouse oocytes. Curr Biol 13:1596–1608
39. McGuinness BE, Anger M, Kouznetsova A, Gil-Bernabe AM, Helmhart W et al (2009) Regulation of APC/C activity in oocytes by a Bub1-dependent spindle assembly checkpoint. Curr Biol 19:369–380
40. Hached K, Xie SZ, Buffin E, Cladiere D, Rachez C et al (2011) Mps1 at kinetochores is essential for female mouse meiosis I. Development 138:2261–2271
41. Touati SA, Buffin E, Cladiere D, Hached K, Rachez C et al (2015) Mouse oocytes depend on BubR1 for proper chromosome segregation but not for prophase I arrest. Nat Commun 6:6946
42. Hassold T, Hunt P (2001) To err (meiotically) is human: the genesis of human aneuploidy. Nat Rev Genet 2:280–291
43. Griswold MD (2016) Spermatogenesis: the commitment to meiosis. Physiol Rev 96:1–17
44. Herbert M, Kalleas D, Cooney D, Lamb M, Lister L (2015) Meiosis and maternal aging: insights from aneuploid oocytes and trisomy births. Cold Spring Harb Perspect Biol 7
45. Telfer EE, Gosden RG, Byskov AG, Spears N, Albertini D et al (2005) On regenerating the ovary and generating controversy. Cell 122:821–822
46. Fonseka KG, Griffin DK (2011) Is there a paternal age effect for aneuploidy? Cytogenet Genome Res 133:280–291
47. Templado C, Uroz L, Estop A (2013) New insights on the origin and relevance of aneuploidy in human spermatozoa. Mol Hum Reprod 19:634–643
48. Tachibana-Konwalski K, Godwin J, van der Weyden L, Champion L, Kudo NR et al (2010) Rec8-containing cohesin maintains bivalents without turnover during the growing phase of mouse oocytes. Genes Dev 24:2505–2516
49. Revenkova E, Herrmann K, Adelfalk C, Jessberger R (2010) Oocyte cohesin expression restricted to predictyate stages provides full fertility and prevents aneuploidy. Curr Biol 20:1529–1533
50. Holubcova Z, Blayney M, Elder K, Schuh M (2015) Human oocytes. Error-prone chromosome-mediated spindle assembly favors chromosome segregation defects in human oocytes. Science 348:1143–1147
51. Zielinska AP, Holubcova Z, Blayney M, Elder K, Schuh M (2015) Sister kinetochore splitting and precocious disintegration of bivalents could explain the maternal age effect. Elife 4
52. Sebestova J, Danylevska A, Novakova L, Kubelka M, Anger M (2012) Lack of response to unaligned chromosomes in mammalian female gametes. Cell Cycle 11:3011–3018
53. Liu L, Keefe DL (2008) Defective cohesin is associated with age-dependent misaligned chromosomes in oocytes. Reprod Biomed Online 16:103–112

54. Lister LM, Kouznetsova A, Hyslop LA, Kalleas D, Pace SL et al (2010) Age-related meiotic segregation errors in mammalian oocytes are preceded by depletion of cohesin and Sgo2. Curr Biol 20:1511–1521
55. Tsutsumi M, Fujiwara R, Nishizawa H, Ito M, Kogo H et al (2014) Age-related decrease of meiotic cohesins in human oocytes. PLoS One 9:e96710
56. Chiang T, Duncan FE, Schindler K, Schultz RM, Lampson MA (2010) Evidence that weakened centromere cohesion is a leading cause of age-related aneuploidy in oocytes. Curr Biol 20:1522–1528
57. Jessberger R (2012) Age-related aneuploidy through cohesion exhaustion. EMBO Rep 13:539–546
58. Riris S, Webster P, Homer H (2014) Digital multiplexed mRNA analysis of functionally important genes in single human oocytes and correlation of changes in transcript levels with oocyte protein expression. Fertil Steril 101:857–864
59. Baker DJ, Jeganathan KB, Cameron JD, Thompson M, Juneja S et al (2004) BubR1 insufficiency causes early onset of aging-associated phenotypes and infertility in mice. Nat Genet 36:744–749
60. Suijkerbuijk SJ, Vleugel M, Teixeira A, Kops GJ (2012) Integration of kinase and phosphatase activities by BUBR1 ensures formation of stable kinetochore-microtubule attachments. Dev Cell 23:745–755
61. Xu P, Raetz EA, Kitagawa M, Virshup DM, Lee SH (2013) BUBR1 recruits PP2A via the B56 family of targeting subunits to promote chromosome congression. Biol Open 2:479–486
62. Yun Y, Lane SI, Jones KT (2014) Premature dyad separation in meiosis II is the major segregation error with maternal age in mouse oocytes. Development 141:199–208
63. Hassold T, Hunt P (2009) Maternal age and chromosomally abnormal pregnancies: what we know and what we wish we knew. Curr Opin Pediatr 21:703–708
64. Dobles M, Liberal V, Scott ML, Benezra R, Sorger PK (2000) Chromosome missegregation and apoptosis in mice lacking the mitotic checkpoint protein Mad2. Cell 101:635–645
65. Taylor SS, McKeon F (1997) Kinetochore localization of murine Bub1 is required for normal mitotic timing and checkpoint response to spindle damage. Cell 89:727–735
66. Wang Q, Liu T, Fang Y, Xie S, Huang X et al (2004) BUBR1 deficiency results in abnormal megakaryopoiesis. Blood 103:1278–1285
67. Balboula AZ, Schindler K (2014) Selective disruption of Aurora C kinase reveals distinct functions from Aurora B kinase during meiosis in mouse oocytes. PLoS Genet 10:e1004194
68. Schindler K, Davydenko O, Fram B, Lampson MA, Schultz RM (2012) Maternally recruited Aurora C kinase is more stable than Aurora B to support mouse oocyte maturation and early development. Proc Natl Acad Sci U S A 109:E2215–E2222
69. Tang CJ, Lin CY, Tang TK (2006) Dynamic localization and functional implications of Aurora-C kinase during male mouse meiosis. Dev Biol 290:398–410
70. Dieterich K, Soto Rifo R, Faure AK, Hennebicq S, Ben Amar B et al (2007) Homozygous mutation of AURKC yields large-headed polyploid spermatozoa and causes male infertility. Nat Genet 39:661–665
71. Dieterich K, Zouari R, Harbuz R, Vialard F, Martinez D et al (2009) The Aurora kinase C c.144delC mutation causes meiosis I arrest in men and is frequent in the north African population. Hum Mol Genet 18:1301–1309
72. Yang KT, Tang CJ, Tang TK (2015) Possible role of Aurora-C in meiosis. Front Oncol 5:178
73. Fellmeth JE, Ghanaim EM, Schindler K (2016) Characterization of macrozoospermia-associated AURKC mutations in a mammalian meiotic system. Hum Mol Genet 25:2698–2711
74. Llano E, Gomez HL, Garcia-Tunon I, Sanchez-Martin M, Caburet S et al (2014) STAG3 is a strong candidate gene for male infertility. Hum Mol Genet 23:3421–3431
75. Caburet S, Arboleda VA, Llano E, Overbeek PA, Barbero JL et al (2014) Mutant cohesin in premature ovarian failure. N Engl J Med 370:943–949

76. Butler MG, Rafi SK, McGuire A, Manzardo AM (2016) Currently recognized clinically relevant and known genes for human reproduction and related infertility with representation on high-resolution chromosome ideograms. Gene 575:149–159
77. Englund A, Jonsson B, Zander CS, Gustafsson J, Anneren G (2013) Changes in mortality and causes of death in the Swedish down syndrome population. Am J Med Genet A 161A:642–649
78. Wu J, Springett A, Morris JK (2013) Survival of trisomy 18 (Edwards syndrome) and trisomy 13 (Patau syndrome) in England and Wales: 2004-2011. Am J Med Genet A 161A:2512–2518
79. Fitzgerald P, Leonard H, Pikora TJ, Bourke J, Hammond G (2013) Hospital admissions in children with down syndrome: experience of a population-based cohort followed from birth. PLoS One 8:e70401
80. Arumugam A, Raja K, Venugopalan M, Chandrasekaran B, Sampath KK et al (2016) Down syndrome – a narrative review. Clin Anat 29:568–577
81. Ganmore I, Smooha G, Izraeli S (2009) Constitutional aneuploidy and cancer predisposition. Hum Mol Genet 18:R84–R93
82. de Graaf G, Haveman M, Hochstenbach R, Engelen J, Gerssen-Schoorl K et al (2011) Changes in yearly birth prevalence rates of children with down syndrome in the period 1986-2007 in The Netherlands. J Intellect Disabil Res 55:462–473
83. De Souza E, Alberman E, Morris JK (2009) Down syndrome and paternal age, a new analysis of case-control data collected in the 1960s. Am J Med Genet A 149A:1205–1208
84. Hong DS, Reiss AL (2014) Cognitive and neurological aspects of sex chromosome aneuploidies. Lancet Neurol 13:306–318
85. Sybert VP, McCauley E (2004) Turner's syndrome. N Engl J Med 351:1227–1238
86. Stochholm K, Bojesen A, Jensen AS, Juul S, Gravholt CH (2012) Criminality in men with Klinefelter's syndrome and XYY syndrome: a cohort study. BMJ Open 2:e000650

Chapter 2
Consequences of Centrosome Dysfunction During Brain Development

Maddalena Nano and Renata Basto

Abstract Development requires cell proliferation, differentiation and spatial organization of daughter cells to occur in a highly controlled manner. The mode of cell division, the extent of proliferation and the spatial distribution of mitosis allow the formation of tissues of the right size and with the correct structural organization. All these aspects depend on cell cycle duration, correct chromosome segregation and spindle orientation. The centrosome, which is the main microtubule-organizing centre (MTOC) of animal cells, contributes to all these processes. As one of the most structurally complex organs in our body, the brain is particularly susceptible to centrosome dysfunction. Autosomal recessive primary microcephaly (MCPH), primordial dwarfism disease Seckel syndrome (SCKS) and microcephalic osteodysplastic primordial dwarfism type II (MOPD-II) are often connected to mutations in centrosomal genes. In this chapter, we discuss the consequences of centrosome dysfunction during development and how they can contribute to the etiology of human diseases.

Keywords Centrosome • Microcephaly • Animal models of microcephaly • Autosomal recessive primary microcephaly (MCPH) • Seckel syndrome (SCKS) • Microcephalic osteodysplastic primordial dwarfism type II (MOPD-II)

2.1 The Centrosome Duplication Cycle

The centrosome is a non-membranous organelle composed of a pair of orthogonally organized centrioles, which during mitosis organize the pericentriolar material (PCM) [1]. The two centrioles are composed of nine sets of microtubules (MTs), polarized filaments of tubulin. The PCM surrounds the centrioles to support mitotic

M. Nano • R. Basto (✉)
Institut Curie, PSL Research University, CNRS UMR144,
12 rue Lhomond, 75005 Paris, France
e-mail: renata.basto@curie.fr

M. Gotta, P. Meraldi (eds.), *Cell Division Machinery and Disease*, Advances in Experimental Medicine and Biology 1002, DOI 10.1007/978-3-319-57127-0_2

spindle assembly and consists of a highly organized matrix of more than one hundred proteins [2–5].

The centrosome is the main MT-organizing centre of animal cells: two centrosomes organize the mitotic spindle during mitosis and—most of the time—it is required to build the MT-cytoskeleton in interphase [6, 7]. In addition, in interphase, the centriole functions as basal body to template the assembly of cilia and flagella [8]. MT nucleation from the centrosome depends on the presence of γ-tubulin containing complexes [9]. The minus-ends of MTs are embedded at the centrosome, while the plus-ends extend in the cytoplasm, forming a polarized network that sustains chromosome (or molecular cargoes) movement. MTs possess an intrinsic dynamic instability and are in general built by 13 polar protofilaments, each being composed by heterodimers of α and β-tubulin [10, 11].

At the end of cell division, each daughter cell contains a single centrosome composed of a pair of orthogonally organized centrioles [12]. This allows the assembly an organized MT network in interphase [13]. To form a bipolar spindle at the next mitosis, the centrosome is duplicated only once per cell cycle in a tightly regulated process (Fig. Fig. 2.1) [15]. Centrosome duplication is licensed by centriole disengagement, which occurs during G1 when centrioles lose their orthogonal configuration [15]. Centriole disengagement allows the assembly—during S-phase—of one daughter procentriole next to each mother. In preparation for mitosis, the centrosomes starts recruiting PCM and the two centrosomes separate and nucleate MTs in order to assemble the mitotic spindle. The very last step of cell division, cytokinesis, will then separate the two centrosomes in two distinct cells.

2.1.1 Proteins Required for Centrosome Duplication

Proteins involved in the centrosome duplication cycle have been initially identified through genome wide screens in *C. elegans* and they present functional homologs in the fruit fly *Drosophila melanogaster* (D) and in humans (*Homo sapiens*, Hs). Their recruitment to the centrosome and their activity are sequential [18–20]. Throughout evolution the three major steps required for centriole duplication have been conserved [16, 17] (Fig. Fig. 2.2). They consist of:

1. Recruitment of kinase activity to the centrosome (proteins involved: SPD-2/HsCEP192/DSPD-2 or HsCEP152/DAsl and ZYG-1/HsPLK4/DSak)
2. Formation of a procentriole-primordium (proteins involved: SAS-6/HsSAS6/DSas-6 and SAS-5/HsSTIL/DAna2)
3. Incorporation of MTs at the newly formed procentriole (proteins involved: SAS-4/HsCPAP/DSas4, HsCP110, Hsγ-tubulin, HsCEP135/DBld10). Of note, other members of the tubulin superfamily (zeta-, epsilon- and delta-tubulin) are required to form centrioles in certain cells (for a review, see [25])

For simplicity, throughout this chapter, we will refer to these genes with their Hs name.

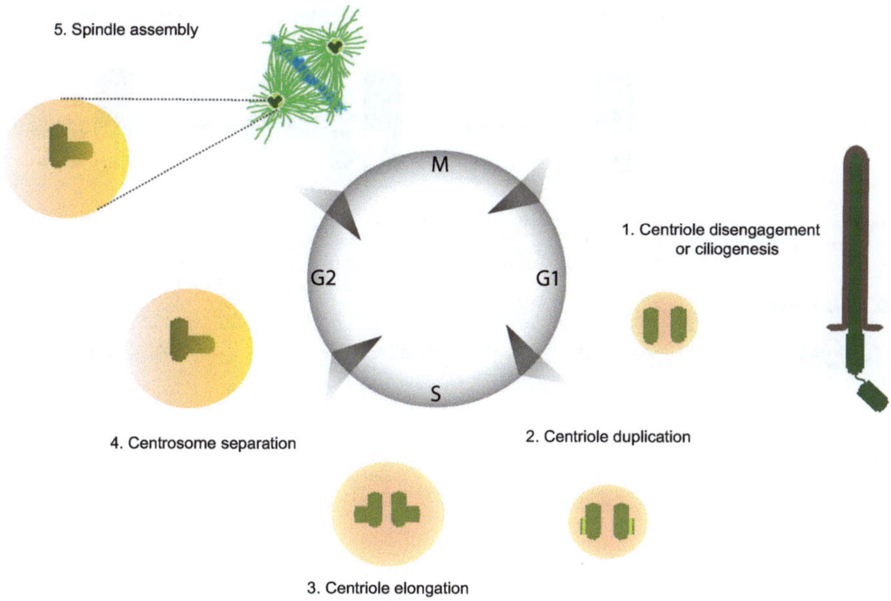

Fig. 2.1 The centrosome duplication cycle. The centrosome is composed of a pair of centrioles (*green cylinders*) [12] surrounded by pericentriolar material (PCM) (*orange*) [14]. It normally duplicates once, in coordination with the cell cycle (composed by sequential G1-, S-, G2- and M-phase) [15]. (*1*) Centriole disengagement or ciliogenesis. At the end of cell division, each daughter cell contains a single centrosome composed by a pair of orthogonally organized centrioles. This configuration is lost in G1 in a process called "disengagement", which license the centrosome duplication cycle [15]. Alternatively, the mother centriole can dock at the membrane and form the basal body that templates the assembly of cilia and flagella [8]. (*2*) Centriole duplication. During S-phase a number of proteins are timely recruited on the disengaged centrioles and trigger the assembly of a new centriole next to each parental one (see Fig. 2.2 for a more detailed description of the process) [18–20]. (*3*) Centriole elongation. The newly formed centrioles continue to elongate during S-phase and the rest of the cell cycle [21]. From G2 onwards, the centrosomes reinforce the recruitment of PCM material, which will serve to organize the mitotic spindle in M-phase [22]. (*4*) Centrosome separation. At the beginning of M-Phase, centrosomes separate and migrate at opposite side of the cell [23]. Each centrosome will be composed of two centrioles that organize the PCM: an older, parental centriole and a daughter one. (*5*) Spindle assembly. In mitosis, the centrosomes nucleate MTs and organize the mitotic spindle, a network that contacts the chromosomes and allows their segregation at opposite pole of the cells [24]

2.2 The Centrosome and the Mitotic Spindle

The mitotic spindle supports chromosome separation during mitosis. It consists of a bipolar, antiparallel array of MTs [11, 26, 27]. In animal cells it is composed by MTs nucleated mainly by the centrosome. Astral MTs are typical of centrosomal-spindles: they emanate from the centrosome into the cytosol and—contacting the cortex—play important roles in spindle orientation [28–31]. The mitotic spindle is composed by different populations of coexisting MTs. These can be distinguished

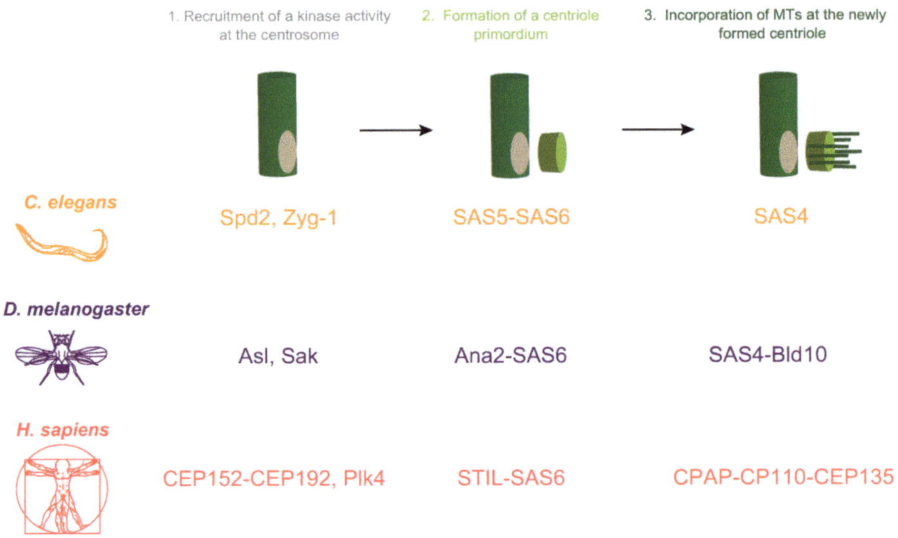

Fig. 2.2 Proteins required for centrosome duplication. The core machinery required for centrosome duplication is well conserved [16, 17]. The proteins sequentially required for centriole duplication in *C. elegans*, *D. melanogaster* and *H. sapiens* are listed. They act timely to induce recruitment of kinase activity to the centrosome (*1*), formation of a centriole-primordium (*2*) and incorporation of MTs at the newly formed centriole (*3*)

based on their orientation, function and stability. Kinetochore MTs connect the chromosomes to the spindle machinery while MTs that emanate from opposite poles and interact in an antiparallel fashion are named interpolar MTs [27] (Fig. 2.1).

MTs can also be assembled by the Augmin complex through branching from pre-existing MTs [32]. Additionally, MTs are nucleated at the level of the chromatin by the establishment of a RanGTP gradient after nuclear envelope breakdown [33–38]. Chromatin-dependent spindle assembly occurs physiologically in different systems. For instance, this occurs frequently during female meiosis, even in human oocytes [39]. However, this process can also support spindle assembly when centrosomes are experimentally removed (e.g. [40–43]).

Importantly, while bipolar mitotic spindle assembly can occur in the absence of centrosomes, these organelles are normally required to ensure correct spindle orientation. Spindle orientation determines the position of daughter cells at the end of mitosis and contributes to the differential inheritance of cytoplasmic and cortical factors [40, 44–48].

2.3 The Centrosome and the Cilium

One important role of the centrosome in interphase is to function as a basal body to template the assembly of cilia and flagella. In light of human brain anatomy and function, two kinds of cilia need to be described: motile cilia and primary cilia.

Motile cilia normally present 9 MT doublets plus a central pair of MTs and are required to mediate fluids movement in the human body. In the brain, ependymal cells are responsible for the flux of the cerebrospinal fluid (CSF). These cells are multiciliated: they present hundreds of cilia, which are formed by sequential, multiple non-canonical centriole duplication cycles [49, 50].

Primary cilia slightly differ from motile cilia in their structure, as they lack the central pair of MTs [51]. These cilia, also known as non-motile cilia, mainly function as signalling hubs. In the brain, they are important to sense signalling molecules transported in the CSF, including growth factors, sonic hedgehog (Shh) and Wnt ligands [52–58].

2.4 General Principles Governing the Effects of Centrosome Dysfunction

Centrosomes play important roles in determining the outcome of cell division, both in terms of cell fate and cell survival. We conceive that centrosome defects can impair this process by affecting at least four different mechanisms: orientation of the mitotic spindle [28, 30, 40, 45, 48, 59, 60], correct chromosome segregation [61–64] and assembly of a primary cilium [53]. In addition, centrosome loss can affect cell proliferation: it has been recently shown that centrosome removal triggers p53 activation and arrests vertebrate cell in G1 [65, 66].

The position of the spindle defines where daughter cells will be positioned at the end of mitosis as well as the cortical/cytoplasmic inheritance they will receive. During symmetric cell division, the factors inherited by daughter cells are equivalent and the daughter cells will have similar fates. During asymmetric cell division, daughters are unequal and will differ in fate [46]. In *Drosophila* neural stem cells, cell fate-determinants are anchored to the membrane and directly transmitted to daughter cells by cortical inheritance, while in mammalian NSCs structural elements such as junctions, the apical membrane and the basal lamina are proposed to be responsible for cell fate determination [46, 57, 67–69]. Interestingly, the centrosome itself can be inherited asymmetrically by stem cells and differentiating cells [70–73]. The role of centrosome asymmetry during development or in maintaining tissue integrity is still not understood. A plausible explanation comes from a study performed in vertebrate cells in culture where it has been shown that, after mitosis, primary cilia grow asynchronously in the two daughter cells [74]. The cell that inherits the older centrosome will initiates ciliogenesis earlier than its sister, putting forward the concept that centrosome asymmetry might influence the capacity to sense environmental signals [74].

Centrosome defects can generate abnormal cilia, both in terms of number or structure. For instance, the nucleation of extra primary cilia can impact on the ability of a cell to transduce Sonic Hedgehog (Shh) and Wnt signaling and might lead to alterations in cell fate [53, 58].

In addition, defects in centrosome number (mainly centrosome amplification) and function can impair spindle activity during chromosome segregation. This can lead to alterations in the karyotype of daughter cells, a condition known as aneuploidy [75]. The pathological implications of an abnormal number of chromosomes are extremely broad and can be associated with both over proliferation (e.g. cancer) [76] or defective growth (e.g. microcephaly) [62], which will be further discussed in the following paragraphs.

Aneuploidy *per se* can be at the basis of premature differentiation or cell death [62, 77–80]. Additionally, lengthening of the G1 phase of the cell cycle has been shown to cause exhaustion of proliferative divisions and favour differentiation, probably by allowing the cell extra time to sense differentiation signals [81–83].

2.5 Neocortex Development: Evolutionary Insights

Drosophila is an invertebrate organism commonly used to understand the genetic bases of developmental processes. Brain development in *Drosophila* is quite stereotyped and the deep knowledge of its anatomy and of the cell types composing it render it an ideal model to explore the role of different factors in neurogenesis. During embryonic development, a population of neural stem cells called neuroblasts (NBs) delaminates from the neuroectoderm and give rise to the larval brain, composed by two lobes and a ventral nerve cord [84]. After a period of dormancy, proliferation of the larval NBs resumes. NBs divide in an asymmetric fashion and generate two daughter cells with distinct fates: a new NB, which retains the stem-cell potential, and a ganglion mother cell, which undergoes a single additional division. At the end, each NB gives rise to a reasonably invariant set of neurons and glial cells [85].

The vertebrate brain (including Zebrafish, mouse and human brain) can roughly be subdivided in three parts: the forebrain, the midbrain, and the hindbrain [86].

The mammalian brain is characterized by the development of the neocortex, composed of six layers of neurons. It is the part of cerebral cortex which underwent the biggest and most recent phylogenetic expansion, mainly by growth in the lateral and radial dimension [57, 87, 88]. It originates from the divisions of neuroepithelial cells, which give rise to a set of intermediate progenitors that will undergo additional divisions. Taverna et al. have proposed a classification of these populations based on the localization of progenitor mitosis with respect to the ventricular zone (VZ) (apical—AP, basal—BP and subapical progenitors—SAP), the extent of cell polarity and their proliferative capacity [57].

The evolution of different types of progenitor cells in the primate and human brain has contributed to the expansion of the cerebral cortex [88]. In particular, a novel type of non-epithelial progenitors has been described in the outer subventricular zone (OSVZ), which are proposed to have contributed to the evolutionary expansion of the human brain [89].

The human brain strongly differs from that of other mammals by its degree of corticalization, which accompanies an increase in cognitive functions [90, 91]. This impressive growth in brain size has been accommodated in the skull thanks to gyrification. Interestingly, it has been reported that the formation of gyri in the otherwise unfolded mouse cortex (lissencephalic) can be induced by modifying the expression pattern of a single protein or by forcing the expression of a human gene [92, 93].

2.5.1 Centrosome and Brain Development: Lessons from Drosophila

In order to explain the effects of centrosomal defects on brain development, we will start by presenting two extreme cases: what happen when centrosomes are absent and when they are present in excess (while for others—more specific—models, we address the readers to the corresponding paragraphs). Both scenarios can be obtained by mutating or overexpressing proteins involved in the centrosome duplication cycle. For instance, mutations in Plk4 result in centrosome loss, while Plk4 overexpression causes centrosome amplification [59, 94].

In the absence of centrosomes, mitotic spindles can be assembled from the vicinity of chromatin or from pre-existing MTs [32, 36, 37, 95]. However, centrioles are indispensible to nucleate sensory cilia and sperm flagella [96–98]. Flies without centrioles can develop in viable adults, but they die shortly after eclosion because their sensory neurons lack cilia, affecting vital function such as movement, smell and prioception [59, 94, 99–101]. Acentrosomal mitoses do not generate aneuploidy in dividing NSCs. However, due to lack of astral MTs, spindle orientation and asymmetric cell division are perturbed, leading to an expansion of the stem cell pool, which is tumorigenic in allogeneic transplantations [40, 44, 47, 102].

When extra centrosomes are present, their efficient clustering at the spindles poles allows the formation of bipolar spindles and ensures correct chromosome segregation. Flies with centrosome amplification do not present gross cilia defects. However, asymmetric cell division is again perturbed, and the brain holds tumorigenic potential [59].

Overall, we think that centrosome defects mostly affect spindle positioning in *Drosophila* and that this is the main route by which cell proliferation is affected. However, the mammalian brain can respond differently to the same kind of perturbations (see paragraph, see below for further details).

Of note, not all tissues have the same way of responding to centrosome loss and amplification. While centrosomes number in the brain is not a variable influencing faithful chromosome segregation, it can generate aneuploidy in the wing disc [63, 103]. These results suggests that, when centrosomes are perturbed, mitotic fidelity will rely on the strength of alternative mechanisms for spindle assembly or on the capacity to achieve centrosome clustering.

2.5.2 Centrosome and Brain Development: Lessons from Zebrafish

Small head size phenotypes were obtained by Novorol and colleagues in Zebrafish after knockdown (KD) of four centrosomal genes: *stil*, *aspm wdr62* and *odf2* [104] (see also [105]), generating abnormal centrosome number and localization. In these mutants the microcephalic brain results mainly from mitotic defects - namely prometaphase delay [104], while the contribution of p53-dependent apoptosis is minor. In addition, the work from Pfaff, K.L. and colleagues [106], also reported a strong mitotic phenotype with highly disorganized spindles in Zebrafish in the absence of *Stil*. An additional Zebrafish model with non-functional centrosomes was obtained by targeting NEDD1, which is required to recruit γ-tubulin at the centrosome [107]. NEDD1 knockdown (KD) resulted in mitotic arrest and apoptosis. Depending on the intensity of the KD, the phenotype was spanning from embryonic lethality to severe defects in the brain [107].

Depletion of Plk4 causes a strong size reduction in Zebrafish, mainly due to abnormal spindles and mitotic defects, including a substantial delay in mitotic progression [108]. Moreover, the impairment of centriole duplication due to Plk4 knockdown results in a dilution of basal bodies and causes a dose-dependent ciliary phenotype [108]: while a mild reduction in Plk4 levels mainly affect mitosis by reducing the number of centrioles and thus altering the bipolar configuration of the spindle, a stronger KD resulting in complete centriole/basal body loss impairs the ability of cells of growing cilia [108].

The consequences of centrosome amplification in the Zebrafish brain have been studied recently [109]. In this system, extra centrosomes do not cluster and induce the formation of multipolar spindles. Multipolar divisions can lead to the presence of multiple nuclei in one of the daughter cells. When occurring in the neuroepithelial progenitors, this leads to apoptosis, tissue degeneration and death, affecting mostly retinal neuronal layering [109].

2.5.3 Centrosome and Brain Development: Lessons from Mouse

Similarly to *Drosophila*, centrosome removal and amplification in mouse can be achieved genetically by manipulating genes involved in the centrosome duplication cycle [62, 110].

Since most vertebrate cells are ciliated, lack of centrioles would be expected to cause lethality due to the absence of cilia. However, CPAP mutant mouse embryos die earlier than mutants lacking cilia. In rodents, centriole presence becomes essential from embryonic day 9 [110–114] and experimental removal of centrosomes causes mitotic delay and p53-dependent cell death in the embryo [110, 115, 116].

Bazzi and Anderson recently showed that a null mutation in CPAP results in embryonic lethality at midgestation. They also observed a prometaphase delay and demonstrated its involvement in p53 activation and p53-dependent apoptosis [110].

When progressive loss of centrioles is taking place in the neuronal precursors, mice develop microcephaly [45, 110]. p53 removal rescues cell death and the reduced brain size phenotype. However, it does not rescue the defects in tissue architecture due to abnormal spindle orientation that leads to misplacement of neural progenitors [45]. Very recently it has been shown that—in addition to randomizing spindle orientation—CPAP silencing in post-mitotic neurons leads to abnormal morphology and slower neuronal migration [117]. The defective neuronal migration described here is ascribable to the function that CPAP exert on interphase rather than mitotic MTs, and open new possible roles for centrosomal genes in contributing to growth disorders. Supporting this view, Gabriel, E. and colleagues have shown that CPAP promotes neural progenitors fate by promoting cilia disassembly, rather than by a centrosomal function [118].

As centrosome loss, centrosome amplification in the mouse central nervous system causes microcephaly [62]. Compared to *Drosophila*, mouse neural stem cells have less efficient clustering mechanisms. Multipolar spindles cause errors in chromosome segregation that leads to aneuploidy and p53-dependent cell death. Interestingly, p53 inhibition rescues cell death, but prompts premature differentiation of progenitor cells, mirroring results obtained in *Drosophila* aneuploid brains [119]. However, while in flies premature differentiation is a primary response to aneuploidy, in mouse it is probably a secondary mechanism taking over only when the p53 primary response is not efficient. These observations seem paradoxical, since aneuploidy has long been regarded uniquely as conferring proliferative advantage. However, recent studies showing a negative effect of aneuploidy on proliferation put forward the novel concept that chromosome imbalance mostly hinders proliferation, and only specific gain or losses might favor malignant transformation (which might not occur through proliferative advantage—e.g. see [120–122]).

2.6 The Centrosome and Its Role in Primordial Microcephalic Disorders

Primordial microcephalic disorders include a spectrum of diseases characterized by severe growth retardation. Autosomal recessive primary microcephaly (MCPH) (2.6.1), microcephalic osteodysplastic primordial dwarfism type II (MOPD-II) (2.6.2) and primordial dwarfism disease Seckel syndrome (SCKS) (2.6.3) are all primordial microcephalic disorders that share phenotypic and genetic traits. In the following paragraphs, we will present a summary of clinical description and molecular insights for each of these syndromes. These disorders can be linked to premature exhaustion of proliferative division of stem cells, due to premature differentiation or cell death. The causes underlying cell death/differentiation can be multiple and include defects in spindle robustness [123], in spindle positioning/ orientation [124], in cell cycle progression or DNA damage repair [125] or in chromosome segregation [62].

2.6.1 Autosomal Recessive Primary Microcephaly (MCPH: Microcephaly Primary Hereditary)

MCPH is characterized by a reduction of the occipito-frontal circumference, which can be nearly normal at birth (−2) but is inevitably worsen in the first year of life (−3) [126]. It is a rare genetic disease, found in about 100 families [127]. Brain size reduction is proportionate, albeit affecting particularly the cerebral cortex [128]. Other clinical features are mental retardation, mild seizures and particular neuronal migration defects [126–129].

MCPH causal mutations are found in a large number of genes and can all cause premature differentiation, cell death and displacement of neural progenitors. Thirteen MCPH loci (1–13) have so far been identified in human patients, encoding for 13 different genes. MCPH genes can be grouped in 3 (partially overlapping) categories:

1. genes with a role in centrosome and spindle function (CEP152, CEP63, SAS-6, STIL, CPAP/CENPJ, CEP135, CDK5RAP2, CDK6, ASPM, WDR62 and STIL; see next paragraphs for a detailed description);
2. genes with a role in chromosome dynamics. This include the kinetochore gene CASC5, encoding for KNL1 [130] and CENPE. CENPE is encoded in an MCPH locus, but generates a more severe phenotype, similar to MOPD-II [131];
3. genes with a role in DNA-damage related pathways. In this category we can include Microcephalin, PHC1, ZNF335 [132–135]. Of note, CDK6, ASPM, STIL are involved in both cell cycle regulation and centrosome function.

In light of recent results an additional, fourth class of genes involved in cell cycle regulation (Microcephalin and ASPM), should be considered [81, 133].

For the purposes of this chapter, we will focus on the genes with a role in centrosome and spindle function (category 1) and we will describe their identification in MCPH, MOPD-II and SCKS. For the other genes, we address the reader to [127, 136, 137]. However, a small digression should be made on Microcephalin (Mcph1). Microcephalin is mostly known for its role in DNA-related processes (DNA damage response, chromosome condensation) [129, 138, 139] but was also involved in the etiology of microcephaly through its role in coupling the centrosome cycle with mitosis [140]. Indeed, mutations in *Drosophila* MCPH1 are early embryonic lethal. Mutant embryos exhibit asynchronous nuclear and centrosome cycle, abnormal centrosomes and spindles, chromatin bridging due to premature chromosome condensation and mitotic arrest [141, 142].

Genes with a role in centrosome and spindle function that are at the origin of MCPH were previously classified in centriole duplication genes, genes encoding PCM proteins and genes that encode spindle-pole associated proteins [143].

2.6.1.1 Centriole Duplication Genes: CEP152, STIL, CEP135, CEP63, SAS-6 and CPAP/CENPJ

CEP152 underwent positive selection in humans [144]. A non-conservative amino acid change in CEP152 has been identified by SNP genotyping in three MCPH families. Only one of the cases reported was heterozygous for the missense mutation, with the second allele characterized by a premature stop codon [144].

Mutations in STIL that results in truncation of the protein has been reported in 4 MCPH families [145]. In 2014, Arquint and Nigg [146] described two truncating mutations in STIL found in MCPH patients which perturbs its ubiquitination, thus compromising its degradation and causing centrosome amplification.

MCPH mutations in CPAP/CENPJ have fist been described by Bond and colleagues, [147]. The authors found a homozygous single-base deletion and a substitution resulting in an amino acid change in a very conserved residue of the protein.

A truncated form of the protein encoded by CEP135 has been described in MCPH patients by Hussain et al. [148] and Farooq et al. [149]. So far, reported MCPH mutations in CEP135 are a single base deletion in exon 8 and a splice site mutation leading to complete skipping of exon 11 with loss of the C-terminus domain of CEP135 necessary for the interaction with SAS-6 [149].

The protein encoded by CEP63 interacts with CEP152 and plays an important role in regulating centrosome number. Homozygous mutations in this gene generating a premature stop codon have been found to cause MCPH [150]. Further analysis demonstrated that the protein is normally localized in a discrete ring around the parental centriole and that this localization is lost in patient-derived cells. Centrosome maturation and separation were not found to be perturbed, however, the presence of abnormal spindles—probably due to delayed procentriole assembly and erroneous centriole engagement—highlighted possible defects in centrosome duplication [150]. Surprisingly, the mitotic phenotype observed in CEP63KD cells was rescued by exogenous targeting of CEP152 to the centrosome, suggesting that the microcephalic phenotype could be ascribed to the role of CEP63 in recruiting CEP152 in rapidly proliferating cells. Interestingly, CEP63 deficient mice recapitulate SCKS. In this model, mitotic errors due to centrosomal defects cause p53-dependent cell death, resulting in a reduced brain size. In addition, CEP63 loss impairs male meiotic recombination [151].

A missense mutation within a highly conserved region of SAS-6 has been found by homozygosity mapping in individuals from a Pakistani consanguineous family. Tissue culture analysis of the Ile62Thr SAS-6 mutant revealed that this form of the protein is less efficient in sustaining centriole formation. Further analysis by protein KD revealed the presence of monopolar spindles [152].

2.6.1.2 Genes Encoding PCM Proteins: CDK5RAP2 and CDK6

The formation of a functional mitotic centrosome requires the expansion of PCM around the centrioles, in a process known as centrosome maturation [153, 154].

CDK5RAP2 is a pericentriolar protein required for γ-tubulin recruitment and MT nucleation at the centrosome [155, 156]. It is a MCPH protein highly expressed in the neuronal progenitor pool. Its loss causes depletion of apical progenitors due to cell cycle exit and premature differentiation [157]. CDK5RAP2 mutations were initially identified in two MCPH families. The mutations found were in coding and non-coding regions of the gene respectively, causing an amino acid substitution in the first case and aberrant splicing (generating a truncated protein) in the second [147]. Strikingly, mutations in CDK5RAP2 gene in mice (153) or in human pluripotent stem cell-derived 3D organoid culture system (cerebral organoids) can recapitulate microcephaly, probably due to premature neurogenic non-proliferative divisions [159]. Interestingly, CDK5RAP2 mutations most likely contribute to MCPH onset through defects in several processes. Characterization of a Hertwig's anemia mutant mouse model revealed the presence of multipolar spindles, spindle mispositioning and increased apoptosis due to an inversion in CDK5RAP2 [158]. Cerebral organoids derived from reprogramming of patient skin fibroblasts, mainly displayed misoriented spindles [159]. Finally, in the avian B cell line DT40, CDK5RAP2 function has been linked to the cohesion between centrioles and has been shown to promote cell cycle arrest in cells that underwent DNA-damage [160]. In flies, CDK5RAP2 plays several roles: for instance, it is required for centrosome maturation [161], it regulates centrosome size [162], establishes centrosome asymmetry [163] and represses dendrite branching [164], sustaining the view that mutations in this gene might contribute to MCPH in different ways.

CDK6 is a cyclin dependent kinase required for cell cycle progression and it was found to be mutated in a Pakistani family with MCPH [165]. The mutation—Alanine to Threonine conversion—occurs at the level of a very conserved residue. In the same study, the protein was found to localize at the mitotic centrosomes, but this localization was lost in patient primary fibroblast, leading to the conclusion that CDK6 could play a role in organizing centrosomal-MTs and in centrosome positioning.

2.6.1.3 Genes Encoding Spindle-Pole Associated Proteins: ASPM and WDR62

ASPM is a protein required for spindle integrity, as it plays a role in focusing the poles of the mitotic spindle [166]. It is normally down regulated during the switch from proliferative to neurogenic division of neural progenitors and is required to maintain spindle orientation [167]. Mutations in ASPM are the most common cause of MCPH [168] and strong evidence favor a role for ASPM in neurogenesis: Pulvers and colleagues [169] showed that in mice, ASPM mutations similar to those causing microcephaly in humans, cause abnormal protein localization during mitosis. They also report the appearance of mild microcephaly that can be rescued by the human transgene. A very recent paper proposed a role for ASPM in the etiology of MCPH by regulating cell cycle progression through G1 instead of spindle orientation. Capecchi and Pozner [81] generated a new mouse model and demonstrated that ASPM can tune cyclin E ubiquitination and—thus—mitotic progression through G1. Interestingly, mutations in the *Drosophila* functional homolog of ASPM can also cause severe defects in brain size and neuroepithelium morphogenesis. The absence of ASPM results in abnormal mitosis and increased apoptosis. In addition, ASPM mutants present abnormal spindle positioning and abnormal interkinetic nuclear migration, which compromises tissue architecture [170].

Missense and frame-shifting mutations in WDR62 have been identified in seven MCPH families, making this gene the second most commonly mutated in MCPH after ASPM. WDR62 is specifically expressed in neuronal precursors undergoing mitosis and, similarly to what has been observed for CDK6, the centrosomal localization of the protein is lost in the mutant form [171]. More recently, the study of a WDR62 mouse model established that mutant progenitor cells show spindle instability (including multipolar spindle formation) leading to mitotic arrest, cell death and microcephaly [172]. Interestingly, mutations in the *Drosophila* functional homolog of WDR62 affect the asymmetry that normally characterizes apical and basal MTOCs during interphase. This is due to a lack of Plk1/Polo recruitment on the apical centrosome, which mediates MTOC activity [173].

2.6.2 Majewski/Microcephalic Osteodysplastic Primordial Dwarfism Type II (MOPD-II)

MOPD-II patients are characterized by severe pre-natal and postnatal growth failure with proportionate microcephaly at birth, that evolves in disproportionate microcephaly [174–176]. MOPD can be clinically distinguished from Seckel syndrome mainly from the radiologic finding of skeletal dysplasia [175], as well as by less severe mental retardation but more pronounced growth defects [177]. Mutations in the PCNT gene (encoding a PCM protein) were described to be at the origin of both SCKS and MOPD-II, with the diagnosis often revised when evidence of skeletal dysplasia appear [178]. Absence of PCNT causes defects in spindle structure, which leads to chromosome missegregation [179]. In [177] MOPDII was defined as a "genetically homogeneous condition due to loss of function of pericentrin". Analyzing the current knowledge, Delaval A. and Doxsey J. proposed three mechanisms to explain the implication of pericentrin in a phenotype of reduced growth: (1) through its role as a DNA-damage checkpoint protein, (2) because of its function in MT nucleation and (3) in light of its role in spindle orientation and organization [180].

1. Role of pericentrin as a checkpoint protein. Mutations in PCNT were reported in individuals with SCKS and cells from these patients displayed defects in the ATR-dependent checkpoint signaling for DNA damage [181]. In addition, peri-

centrin plays a role in anchoring Chk1 at the centrosome, thus regulating the activation of centrosomal cyclin B-Cdk1. Its mutation would then be responsible for premature mitotic entry, even in the presence of DNA damage [182]
2. Function of pericentrin in MT nucleation. Localization of Pericentrin at the centrosome is required to sustain MT nucleation during mitosis [183] and thus to ensure mitotic spindle organization [123]. Interestingly, pericentrin loss phenocopies CDK5RAP2 loss in mice and causes a reduced recruitment of CDK5RAP2 at the centrosome [157].
3. Role of pericentrin in spindle organization. PCNT depletion in human cells causes γ-tubulin loss at the centrosome and disrupts astral MT nucleation, [123], spindle positioning [184] and organization, impairing chromosome segregation [210].

Of note, mutations in mouse PCNT and in the *Drosophila* functional homolog cause ciliary phenotypes. A mouse model with hypomorphic PCNT mutation displayed malformed cilia in the olfactory chemosensory neurons [185]. Similarly, adult flies present defects in sensory neuron cilia and sperm flagella function. However, in this system, PCNT is dispensable for mitosis and spindle formation [101].

2.6.3 Seckel Syndrome (SCKS)

Seckel syndrome was described by Seckel in 1960 as a severe form of dwarfism accompanied by microcephaly, a distinctive *facies* and mental retardation [186]. It is a rare and heterogeneous type of primordial dwarfism, very similar to MOPD and firstly distinguished from it by Majewski and Goecke [175]. Mutated genes identified so far in SCKS play a role in centriole duplication and are CPAP/CENPJ and CEP152 [144, 187–189], initially associated with MCPH (see MCPH section for further details on CEP152), and PCNT (see MOPD-II section for further details). Interestingly, in mouse, a hypomorphic allele of CPAP/CENPJ recapitulates several features of Seckel syndrome. Those arise from defective spindle formation and genetic defects such as polyploidy, aneuploidy and apoptosis [190].

2.7 Centrosomal Genes and Other Growth Syndromes

Genes with a role in centrosome and spindle function that are at the origin of other forms of primordial microcephalic disorders other than MCPH, SKCS and MOPD-II include CPAP, CEP152, PCNT (see previous sections for detailed descriptions), the centriole duplication gene PLK4 and the PCM gene TUBGCP4.

Plk4 PLK4 is the master regulator of centriole duplication [94, 191]. In 2014, two different groups identified mutation in the Plk4 genes in individuals with microcephalic primordial dwarfism [108, 192]. Martin and colleagues analyzed a Zebrafish model for Plk4 loss of function and found that centriole biogenesis is compromised, causing both longer, abnormal mitosis—which lead to impaired growth—and ciliary phenotypes. Interestingly, the growth and ciliary phenotypes were dependent on Plk4 dosage. Indeed, cilia loss correlates with complete absence of basal bodies, while mitosis can proceed and being perturbed even when centrioles are present, but their number is reduced. This suggests that—in this system—mitosis is more sensitive to centriole depletion than ciliogenesis [108].

TUBGCP4 TUBGCP4 (tubulin gamma complex associated protein 4) is a component of the γ-tubulin ring complex. Compound heterozygous mutations in TUBGCP4 have recently been reported in individuals with autosomal recessive microcephaly and patient-derived fibroblasts presented abnormal MT organization and aneuploidy [193]. Moreover, striking nuclear defects were reported: enlarged nuclei of abnormal shapes, chromatin bridges and multinucleation were detected in patient-derived fibroblasts [193].

In addition, mutations in POC1 centriolar protein A (POC1A) have been reported in patients affected by primordial dwarfism. Patient's fibroblasts displayed abnormal spindles and impaired ciliogenesis [194] and have been shown to contain centrosome amplification [195].

2.8 Other Centrosome-Related Developmental Syndromes

Oral-Facial-Digital Syndrome Type I (OFD1) OFD1 is a complex syndrome, characterized by polycystic kidney disease and malformations of the mouth, face, brain and digits. It represents an interesting case related to defective primary cilium signaling, although it results from mutations in a basal body gene rather than a ciliary gene. The syndrome is mainly caused by caused by mutations in the CXORF5/OFD1 [196], which colocalizes with γ-tubulin, suggesting an association not only with the cilium—but also with the centrosome [197]. Mutations in CXORF5/OFD1 cause dysfunctions of the primary cilium, abnormal proliferation, abnormal Hedgehog and Wnt signalling and defects in planar cell polarity [198]. In normal human embryos, the gene product localizes in the organs affected by the syndrome, including the brain.

Meier Gorlin Syndrome Meier-Gorlin syndrome is a form of microcephalic primordial dwarfism often due to mutations in DNA-replication proteins, like Orc1 [199]. Orc1 depletion in Zebrafish was proposed to cause reduced body size due to an impairment of replication licensing and cell cycle lengthening [199]. However, in addition to controlling DNA replication by interacting with Cyclin A–CDK2, a different domain of Orc1 controls Cyclin E–CDK2-dependent centriole and centrosome copy number [200, 201]. Analysis of Meier-Gorlin causing mutations in Orc1, revealed that they can cause centrosome amplification [201], putting forward the idea that extra centrosomes might contribute to the growth-defective phenotype observed.

2.9 Concluding Remarks and Current Opinions on Microcephaly

The genetic background of animal models used to study human diseases can dramatically influence the phenotypes observed. A representative example can be found in CDK5RAP2 mouse models. While the one used in [158] could recapitulate microcephaly, the system used in [202] did not reveal significant defects in brain size. The high degree of human brain corticalization, its complexity and the presence of specific progenitor cell populations are unique features, difficult to recapitulate with rodent model systems. Thus, the introduction of cerebral organoids represents a great opportunity to study the etiology of human diseases.

Centrosome defects are very often at the origin of microcephaly and reduced growth. However, other genetic or environmental phenomena can generate similar effects. This is for example the case of fetal alcohol syndrome [203]. In addition, a recent outbreak of microcephaly in Brazil with the number of cases increased of 20-fold in the last months [204, 205], indeed suggests that the developing human brain is more vulnerable in terms of size than other organs. Why does the Zika virus, a flavivirus that in adults causes relatively mild syndromes and is transmitted by mosquitoes [206], affects specifically the developing brain remains to be understood.

Zika tropism to the brain was reported in two studies in 1952 and 1971 [207, 208], but only recently it was shown to infect human neural progenitor cells, causing cell death and cell-cycle alterations [209] and providing a possible explanation for the role of Zika in establishing brain size reduction.

Based on the current knowledge, we propose that the main mechanisms by which abnormal centrosomal components can induce defective brain growth are aneuploidy and alteration of cell cycle length. These can cause cell cycle exit or cell death, leading to a premature exhaustion of neural progenitors. The recent Zika outbreak raised awareness on microcephaly, which is normally a rare condition. Despite great advances in elucidating its causes, the reasons why the development of a normal-sized, well-organized brain is more susceptible to centrosome defects than the rest of the body in vertebrates and mammals remains to be understood.

Importantly, the zygotic centrosome mutations found in MCPH mostly generate architecturally normal but smaller brains, without affecting body size. Why is the brain so vulnerable to centrosome mutations is an important question that remains unanswered. One possibility is that neural progenitors are particularly susceptible to centrosome mutations when compared to other progenitors in the body. Alternatively, establishment of brain size might rely more than other organs on cell divisions that occur during developmental stages. However, other possibilities should not be discarded and this remains—in our view—the major open question in the field.

Acknowledgements We thank V. Marthiens for helpful comments on the manuscript. Work in our lab is supported by an ANR (ANR DIVACEN), an INCa (AAP INCa 2015/118) grants, the CNRS, I. Curie, La Ligue contre le Cancer and FRM (M.N.) Our lab is a member of the CelTisPhyBio labex.

References

1. Bornens M (2002) Centrosome composition and microtubule anchoring mechanisms. Curr Opin Cell Biol 14:25–34
2. Bobinnec Y, Khodjakov A, Mir LM, Rieder CL, Edde B, Bornens M (1998) Centriole disassembly in vivo and its effect on centrosome structure and function in vertebrate cells. J Cell Biol 143:1575–1589
3. Mennella V, Keszthelyi B, McDonald KL, Chhun B, Kan F, Rogers GC, Huang B, Agard DA (2012) Subdiffraction-resolution fluorescence microscopy reveals a domain of the centrosome critical for pericentriolar material organization. Nat Cell Biol 14:1159–1168
4. Nigg EA, Raff JW (2009) Centrioles, centrosomes, and cilia in health and disease. Cell 139:663–678
5. Sonnen KF, Schermelleh L, Leonhardt H, Nigg EA (2012) 3D-structured illumination microscopy provides novel insight into architecture of human centrosomes. Biol Open 1:965–976
6. Karsenti E, Newport J, Hubble R, Kirschner M (1984b) Interconversion of metaphase and interphase microtubule arrays, as studied by the injection of centrosomes and nuclei into Xenopus eggs. J Cell Biol 98:1730–1745
7. Kellogg DR, Moritz M, Alberts BM (1994) The centrosome and cellular organization. Annu Rev Biochem 63:639–674
8. Ishikawa H, Marshall WF (2011) Ciliogenesis: building the cell's antenna. Nat Rev Mol Cell Biol 12:222–234
9. Moritz M, Braunfeld MB, Guenebaut V, Heuser J, Agard DA (2000) Structure of the gamma-tubulin ring complex: a template for microtubule nucleation. Nat Cell Biol 2:365–370
10. Desai A, Mitchison TJ (1997) Microtubule polymerization dynamics. Annu Rev Cell Dev Biol 13:83–117
11. Mitchison TJ (1993) Localization of an exchangeable GTP binding site at the plus end of microtubules. Science 261:1044–1047
12. Anderson RG (1972) The three-dimensional structure of the basal body from the rhesus monkey oviduct. J Cell Biol 54:246–265
13. Karsenti E, Kobayashi S, Mitchison T, Kirschner M (1984a) Role of the centrosome in organizing the interphase microtubule array: properties of cytoplasts containing or lacking centrosomes. J Cell Biol 98:1763–1776
14. Stearns T, Evans L, Kirschner M (1991) Gamma-tubulin is a highly conserved component of the centrosome. Cell 65(5):825–836
15. Tsou MF, Stearns T (2006) Mechanism limiting centrosome duplication to once per cell cycle. Nature 442:947–951
16. Carvalho-Santos Z, Machado P, Branco P, Tavares-Cadete F, Rodrigues-Martins A, Pereira-Leal JB, Bettencourt-Dias M (2010) Stepwise evolution of the centriole-assembly pathway. J Cell Sci 123:1414–1426
17. Hodges ME, Scheumann N, Wickstead B, Langdale JA, Gull K (2010) Reconstructing the evolutionary history of the centriole from protein components. J Cell Sci 123:1407–1413
18. Delattre M, Canard C, Gonczy P (2006) Sequential protein recruitment in *C. elegans* centriole formation. Curr Biol 16:1844–1849
19. Kleylein-Sohn J, Westendorf J, Le Clech M, Habedanck R, Stierhof YD, Nigg EA (2007) Plk4-induced centriole biogenesis in human cells. Dev Cell 13:190–202
20. Pelletier L, O'Toole E, Schwager A, Hyman AA, Muller-Reichert T (2006) Centriole assembly in *Caenorhabditis elegans*. Nature 444:619–623
21. Kohlmaier G et al (2009) Overly long centrioles and defective cell division upon excess of the SAS-4-related protein CPAP. Curr Biol 19(12):1012–1018
22. Dobbelaere J et al (2008) A genome-wide RNAi screen to dissect centriole duplication and centrosome maturation in Drosophila. PLoS Biol 6(9):e224
23. Faragher AJ, Fry AM (2003) Nek2A kinase stimulates centrosome disjunction and is required for formation of bipolar mitotic spindles. Mol Biol Cell 14(7):2876–2889

24. Glover DM et al (1995) Mutations in aurora prevent centrosome separation leading to the formation of monopolar spindles. Cell 81(1):95–105
25. McKean PG, Vaughan S, Gull K (2001) The extended tubulin superfamily. J Cell Sci 114:2723–2733
26. Reber S, Hyman AA (2015) Emergent properties of the metaphase spindle. Cold Spring Harb Perspect Biol 7:a015784
27. Wittmann T, Hyman A, Desai A (2001) The spindle: a dynamic assembly of microtubules and motors. Nat Cell Biol 3:E28–E34
28. Gonzalez C (2007) Spindle orientation, asymmetric division and tumour suppression in Drosophila stem cells. Nat Rev Genet 8:462–472
29. Laan L, Pavin N, Husson J, Romet-Lemonne G, van Duijn M, Lopez MP, Vale RD, Julicher F, Reck-Peterson SL, Dogterom M (2012) Cortical dynein controls microtubule dynamics to generate pulling forces that position microtubule asters. Cell 148:502–514
30. Morin X, Bellaiche Y (2011) Mitotic spindle orientation in asymmetric and symmetric cell divisions during animal development. Dev Cell 21:102–119
31. Siller KH, Doe CQ (2009) Spindle orientation during asymmetric cell division. Nat Cell Biol 11:365–374
32. Goshima G, Mayer M, Zhang N, Stuurman N, Vale RD (2008) Augmin: a protein complex required for centrosome-independent microtubule generation within the spindle. J Cell Biol 181:421–429
33. Carazo-Salas RE, Guarguaglini G, Gruss OJ, Segref A, Karsenti E, Mattaj IW (1999) Generation of GTP-bound ran by RCC1 is required for chromatin-induced mitotic spindle formation. Nature 400:178–181
34. Kalab P, Pu RT, Dasso M (1999) The ran GTPase regulates mitotic spindle assembly. Curr Biol 9:481–484
35. Khodjakov A, Cole RW, Oakley BR, Rieder CL (2000) Centrosome-independent mitotic spindle formation in vertebrates. Curr Biol 10:59–67
36. Khodjakov A, Copenagle L, Gordon MB, Compton DA, Kapoor TM (2003) Minus-end capture of preformed kinetochore fibers contributes to spindle morphogenesis. J Cell Biol 160:671–683
37. Maiato H, Rieder CL, Khodjakov A (2004) Kinetochore-driven formation of kinetochore fibers contributes to spindle assembly during animal mitosis. J Cell Biol 167:831–840
38. Ohba T, Nakamura M, Nishitani H, Nishimoto T (1999) Self-organization of microtubule asters induced in Xenopus egg extracts by GTP-bound ran. Science 284:1356–1358
39. Holubcova Z, Blayney M, Elder K, Schuh M (2015) Human oocytes. Error-prone chromosome-mediated spindle assembly favors chromosome segregation defects in human oocytes. Science 348:1143–1147
40. Basto R, Lau J, Vinogradova T, Gardiol A, Woods CG, Khodjakov A, Raff JW (2006) Flies without centrioles. Cell 125:1375–1386
41. Heald R, Tournebize R, Habermann A, Karsenti E, Hyman A (1997) Spindle assembly in Xenopus egg extracts: respective roles of centrosomes and microtubule self-organization. J Cell Biol 138:615–628
42. Nahaboo W, Zouak M, Askjaer P, Delattre M (2015) Chromatids segregate without centrosomes during *Caenorhabditis elegans* Mitosis in a ran- and CLASP-dependent manner. Mol Biol Cell 26:2020–2029
43. Toya M, Terasawa M, Nagata K, Iida Y, Sugimoto A (2011) A kinase-independent role for aurora a in the assembly of mitotic spindle microtubules in *Caenorhabditis elegans* embryos. Nat Cell Biol 13:708–714
44. Giansanti MG, Gatti M, Bonaccorsi S (2001) The role of centrosomes and astral microtubules during asymmetric division of Drosophila neuroblasts. Development 128:1137–1145
45. Insolera R, Bazzi H, Shao W, Anderson KV, Shi SH (2014) Cortical neurogenesis in the absence of centrioles. Nat Neurosci 17:1528–1535
46. Knoblich JA (2010) Asymmetric cell division: recent developments and their implications for tumour biology. Nat Rev Mol Cell Biol 11:849–860

47. Megraw TL, Kao LR, Kaufman TC (2001) Zygotic development without functional mitotic centrosomes. Curr Biol 11:116–120
48. Mora-Bermudez F, Matsuzaki F, Huttner WB (2014) Specific polar subpopulations of astral microtubules control spindle orientation and symmetric neural stem cell division. eLife 3:e02875
49. Al Jord A, Lemaitre AI, Delgehyr N, Faucourt M, Spassky N, Meunier A (2014) Centriole amplification by mother and daughter centrioles differs in multiciliated cells. Nature 516:104–107
50. Klos Dehring DA, Vladar EK, Werner ME, Mitchell JW, Hwang P, Mitchell BJ (2013) Deuterosome-mediated centriole biogenesis. Dev Cell 27:103–112
51. Satir P, Christensen ST (2007) Overview of structure and function of mammalian cilia. Annu Rev Physiol 69:377–400
52. Corbit KC, Aanstad P, Singla V, Norman AR, Stainier DY, Reiter JF (2005) Vertebrate smoothened functions at the primary cilium. Nature 437:1018–1021
53. Mahjoub MR, Stearns T (2012) Supernumerary centrosomes nucleate extra cilia and compromise primary cilium signaling. Curr Biol 22:1628–1634
54. Rohatgi R, Milenkovic L, Scott MP (2007) Patched1 regulates hedgehog signaling at the primary cilium. Science 317:372–376
55. Shimogori T, Banuchi V, Ng HY, Strauss JB, Grove EA (2004) Embryonic signaling centers expressing BMP, WNT and FGF proteins interact to pattern the cerebral cortex. Development 131:5639–5647
56. Simons M, Gloy J, Ganner A, Bullerkotte A, Bashkurov M, Kronig C, Schermer B, Benzing T, Cabello OA, Jenny A et al (2005) Inversin, the gene product mutated in nephronophthisis type II, functions as a molecular switch between wnt signaling pathways. Nat Genet 37:537–543
57. Taverna E, Gotz M, Huttner WB (2014) The cell biology of neurogenesis: toward an understanding of the development and evolution of the neocortex. Annu Rev Cell Dev Biol 30:465–502
58. Wallingford JB, Mitchell B (2011) Strange as it may seem: the many links between wnt signaling, planar cell polarity, and cilia. Genes Dev 25:201–213
59. Basto R, Brunk K, Vinadogrova T, Peel N, Franz A, Khodjakov A, Raff JW (2008) Centrosome amplification can initiate tumorigenesis in flies. Cell 133:1032–1042
60. Yamashita YM, Jones DL, Fuller MT (2003) Orientation of asymmetric stem cell division by the APC tumor suppressor and centrosome. Science 301:1547–1550
61. Ganem NJ, Godinho SA, Pellman D (2009) A mechanism linking extra centrosomes to chromosomal instability. Nature 460:278–282
62. Marthiens V, Rujano MA, Pennetier C, Tessier S, Paul-Gilloteaux P, Basto R (2013) Centrosome amplification causes microcephaly. Nat Cell Biol 15:731–740
63. Sabino D, Gogendeau D, Gambarotto D, Nano M, Pennetier C, Dingli F, Arras G, Loew D, Basto R (2015) Moesin is a major regulator of centrosome behavior in epithelial cells with extra centrosomes. Curr Biol 25:879–889
64. Silkworth WT, Nardi IK, Scholl LM, Cimini D (2009) Multipolar spindle pole coalescence is a major source of kinetochore mis-attachment and chromosome mis-segregation in cancer cells. PLoS One 4:e6564
65. Lambrus BG, Uetake Y, Clutario KM, Daggubati V, Snyder M, Sluder G, Holland AJ (2015) p53 protects against genome instability following centriole duplication failure. J Cell Biol 210:63–77
66. Wong YL, Anzola JV, Davis RL, Yoon M, Motamedi A, Kroll A, Seo CP, Hsia JE, Kim SK, Mitchell JW et al (2015) Cell biology. Reversible centriole depletion with an inhibitor of polo-like kinase 4. Science 348:1155–1160
67. Jan YN, Jan LY (1998) Asymmetric cell division. Nature 392:775–778
68. Liu X, Hashimoto-Torii K, Torii M, Ding C, Rakic P (2010) Gap junctions/hemichannels modulate interkinetic nuclear migration in the forebrain precursors. J Neurosci Off J Soc Neurosci 30:4197–4209

69. Rhyu MS, Jan LY, Jan YN (1994) Asymmetric distribution of numb protein during division of the sensory organ precursor cell confers distinct fates to daughter cells. Cell 76:477–491
70. Paridaen JT, Wilsch-Brauninger M, Huttner WB (2013) Asymmetric inheritance of centrosome-associated primary cilium membrane directs ciliogenesis after cell division. Cell 155:333–344
71. Rebollo E, Sampaio P, Januschke J, Llamazares S, Varmark H, Gonzalez C (2007) Functionally unequal centrosomes drive spindle orientation in asymmetrically dividing Drosophila neural stem cells. Dev Cell 12:467–474
72. Wang X, Tsai JW, Imai JH, Lian WN, Vallee RB, Shi SH (2009) Asymmetric centrosome inheritance maintains neural progenitors in the neocortex. Nature 461:947–955
73. Yamashita YM, Mahowald AP, Perlin JR, Fuller MT (2007) Asymmetric inheritance of mother versus daughter centrosome in stem cell division. Science 315:518–521
74. Anderson CT, Stearns T (2009) Centriole age underlies asynchronous primary cilium growth in mammalian cells. Curr Biol 19:1498–1502
75. Ricke RM, van Deursen JM (2013) Aneuploidy in health, disease, and aging. J Cell Biol 201:11–21
76. Gordon DJ, Resio B, Pellman D (2012) Causes and consequences of aneuploidy in cancer. Nat Rev Genet 13:189–203
77. Gogendeau D, Siudeja K, Gambarotto D, Pennetier C, Bardin AJ, Basto R (2015) Aneuploidy causes premature differentiation of neural and intestinal stem cells. Nat Commun 6:8894
78. Li M, Fang X, Baker DJ, Guo L, Gao X, Wei Z, Han S, van Deursen JM, Zhang P (2010) The ATM-p53 pathway suppresses aneuploidy-induced tumorigenesis. Proc Natl Acad Sci U S A 107:14188–14193
79. McNamee LM, Brodsky MH (2009) p53-independent apoptosis limits DNA damage-induced aneuploidy. Genetics 182:423–435
80. Ohashi A, Ohori M, Iwai K, Nakayama Y, Nambu T, Morishita D, Kawamoto T, Miyamoto M, Hirayama T, Okaniwa M et al (2015) Aneuploidy generates proteotoxic stress and DNA damage concurrently with p53-mediated post-mitotic apoptosis in SAC-impaired cells. Nat Commun 6:7668
81. Capecchi MR, Pozner A (2015) ASPM regulates symmetric stem cell division by tuning cyclin E ubiquitination. Nat Commun 6:8763
82. Fluckiger AC, Marcy G, Marchand M, Negre D, Cosset FL, Mitalipov S, Wolf D, Savatier P, Dehay C (2006) Cell cycle features of primate embryonic stem cells. Stem Cells 24:547–556
83. Stead E, White J, Faast R, Conn S, Goldstone S, Rathjen J, Dhingra U, Rathjen P, Walker D, Dalton S (2002) Pluripotent cell division cycles are driven by ectopic Cdk2, cyclin a/E and E2F activities. Oncogene 21:8320–8333
84. Hartenstein V, Spindler S, Pereanu W, Fung S (2008) The development of the Drosophila larval brain. Adv Exp Med Biol 628:1–31
85. Homem CC, Knoblich JA (2012) Drosophila neuroblasts: a model for stem cell biology. Development 139:4297–4310
86. Northcutt RG 2002 Understanding vertebrate brain evolution. From the symposium recent advances in neurobiology presented at the annual meeting of the society for integrative and comparative biology, 2–6 January 2002, at Anaheim, California
87. Borrell V, Reillo I (2012) Emerging roles of neural stem cells in cerebral cortex development and evolution. Dev Neurobiol 72:955–971
88. Fish JL, Dehay C, Kennedy H, Huttner WB (2008) Making bigger brains-the evolution of neural-progenitor-cell division. J Cell Sci 121:2783–2793
89. Hansen DV, Lui JH, Parker PR, Kriegstein AR (2010) Neurogenic radial glia in the outer subventricular zone of human neocortex. Nature 464:554–561
90. Pollard KS, Salama SR, Lambert N, Lambot MA, Coppens S, Pedersen JS, Katzman S, King B, Onodera C, Siepel A et al (2006) An RNA gene expressed during cortical development evolved rapidly in humans. Nature 443:167–172

91. Roth G, Dicke U (2005) Evolution of the brain and intelligence. Trends Cogn Sci 9:250–257
92. Florio M, Albert M, Taverna E, Namba T, Brandl H, Lewitus E, Haffner C, Sykes A, Wong FK, Peters J et al (2015) Human-specific gene ARHGAP11B promotes basal progenitor amplification and neocortex expansion. Science 347:1465–1470
93. Stahl R, Walcher T, De Juan RC, Pilz GA, Cappello S, Irmler M, Sanz-Aquela JM, Beckers J, Blum R, Borrell V et al (2013) Trnp1 regulates expansion and folding of the mammalian cerebral cortex by control of radial glial fate. Cell 153:535–549
94. Bettencourt-Dias M, Rodrigues-Martins A, Carpenter L, Riparbelli M, Lehmann L, Gatt MK, Carmo N, Balloux F, Callaini G, Glover DM (2005) SAK/PLK4 is required for centriole duplication and flagella development. Curr Biol 15:2199–2207
95. Dinarina A, Pugieux C, Corral MM, Loose M, Spatz J, Karsenti E, Nedelec F (2009) Chromatin shapes the mitotic spindle. Cell 138:502–513
96. Avidor-Reiss T, Maer AM, Koundakjian E, Polyanovsky A, Keil T, Subramaniam S, Zuker CS (2004) Decoding cilia function: defining specialized genes required for compartmentalized cilia biogenesis. Cell 117:527–539
97. Bate M, Martinez-Arias A (1993) The development of *Drosophila melanogaster*. Cold Spring Harbor Laboratory Press, New York
98. Riparbelli MG, Callaini G (2011) Male gametogenesis without centrioles. Dev Biol 349:427–439
99. Baker DJ, Jeganathan KB, Cameron JD, Thompson M, Juneja S, Kopecka A, Kumar R, Jenkins RB, de Groen PC, Roche P et al (2004) BubR1 insufficiency causes early onset of aging-associated phenotypes and infertility in mice. Nat Genet 36:744–749
100. Blachon S, Gopalakrishnan J, Omori Y, Polyanovsky A, Church A, Nicastro D, Malicki J, Avidor-Reiss T (2008) Drosophila asterless and vertebrate Cep152 are orthologs essential for centriole duplication. Genetics 180:2081–2094
101. Martinez-Campos M, Basto R, Baker J, Kernan M, Raff JW (2004) The Drosophila pericentrin-like protein is essential for cilia/flagella function, but appears to be dispensable for mitosis. J Cell Biol 165:673–683
102. Castellanos E, Dominguez P, Gonzalez C (2008) Centrosome dysfunction in Drosophila neural stem cells causes tumors that are not due to genome instability. Curr Biol 18:1209–1214
103. Poulton JS, Cuningham JC, Peifer M (2014) Acentrosomal Drosophila epithelial cells exhibit abnormal cell division, leading to cell death and compensatory proliferation. Dev Cell 30:731–745
104. Novorol C, Burkhardt J, Wood KJ, Iqbal A, Roque C, Coutts N, Almeida AD, He J, Wilkinson CJ, Harris WA (2013) Microcephaly models in the developing zebrafish retinal neuroepithelium point to an underlying defect in metaphase progression. Open Biol 3:130065
105. Kim HT, Lee MS, Choi JH, Jung JY, Ahn DG, Yeo SY, Choi DK, Kim CH (2011) The microcephaly gene aspm is involved in brain development in zebrafish. Biochem Biophys Res Commun 409:640–644
106. Pfaff KL, Straub CT, Chiang K, Bear DM, Zhou Y, Zon LI (2007) The zebra fish cassiopeia mutant reveals that SIL is required for mitotic spindle organization. Mol Cell Biol 27:5887–5897
107. Manning JA, Lewis M, Koblar SA, Kumar S (2010) An essential function for the centrosomal protein NEDD1 in zebrafish development. Cell Death Differ 17:1302–1314
108. Martin CA, Ahmad I, Klingseisen A, Hussain MS, Bicknell LS, Leitch A, Nurnberg G, Toliat MR, Murray JE, Hunt D et al (2014) Mutations in PLK4, encoding a master regulator of centriole biogenesis, cause microcephaly, growth failure and retinopathy. Nat Genet 46:1283–1292
109. Dzafic E, Strzyz PJ, Wilsch-Brauninger M, Norden C (2015) Centriole amplification in zebrafish affects proliferation and survival but not differentiation of neural progenitor cells. Cell Rep 13:168–182
110. Bazzi H, Anderson KV (2014) Acentriolar mitosis activates a p53-dependent apoptosis pathway in the mouse embryo. Proc Natl Acad Sci U S A 111:E1491–E1500

111. Calarco-Gillam PD, Siebert MC, Hubble R, Mitchison T, Kirschner M (1983) Centrosome development in early mouse embryos as defined by an autoantibody against pericentriolar material. Cell 35:621–629
112. David A, Liu F, Tibelius A, Vulprecht J, Wald D, Rothermel U, Ohana R, Seitel A, Metzger J, Ashery-Padan R et al (2014) Lack of centrioles and primary cilia in STIL(−/−) mouse embryos. Cell Cycle 13:2859–2868
113. Gueth-Hallonet C, Antony C, Aghion J, Santa-Maria A, Lajoie-Mazenc I, Wright M, Maro B (1993) Gamma-tubulin is present in acentriolar MTOCs during early mouse development. J Cell Sci 105(Pt 1):157–166
114. Szollosi D, Calarco P, Donahue RP (1972) Absence of centrioles in the first and second meiotic spindles of mouse oocytes. J Cell Sci 11:521–541
115. Hudson JW, Kozarova A, Cheung P, Macmillan JC, Swallow CJ, Cross JC, Dennis JW (2001) Late mitotic failure in mice lacking sak, a polo-like kinase. Curr Biol 11:441–446
116. Izraeli S, Lowe LA, Bertness VL, Good DJ, Dorward DW, Kirsch IR, Kuehn MR (1999) The SIL gene is required for mouse embryonic axial development and left-right specification. Nature 399:691–694
117. Garcez PP, Diaz-Alonso J, Crespo-Enriquez I, Castro D, Bell D, Guillemot F (2015) Cenpj/CPAP regulates progenitor divisions and neuronal migration in the cerebral cortex downstream of Ascl1. Nat Commun 6:6474
118. Gabriel E, Wason A, Ramani A, Gooi LM, Keller P, Pozniakovsky A, Poser I, Noack F, Telugu NS, Calegari F et al (2016) CPAP promotes timely cilium disassembly to maintain neural progenitor pool. EMBO J 35:803–819
119. Gogendeau D, Siudeja K, Gambarotto D, Pennetier C, Bardin AJ, Basto R (2015) Aneuploidy causes premature differentiation of neural and intestinal stem cells. Nat Commun 6:8894
120. Pfau SJ, Silberman RE, Knouse KA, Amon A (2016) Aneuploidy impairs hematopoietic stem cell fitness and is selected against in regenerating tissues in vivo. Genes Dev 30:1395–1408
121. Stingele S, Stoehr G, Peplowska K, Cox J, Mann M, Storchova Z (2012) Global analysis of genome, transcriptome and proteome reveals the response to aneuploidy in human cells. Mol Syst Biol 8:608
122. Thorburn RR, Gonzalez C, Brar GA, Christen S, Carlile TM, Ingolia NT, Sauer U, Weissman JS, Amon A (2013) Aneuploid yeast strains exhibit defects in cell growth and passage through START. Mol Biol Cell 24:1274–1289
123. Zimmerman WC, Sillibourne J, Rosa J, Doxsey SJ (2004) Mitosis-specific anchoring of gamma tubulin complexes by pericentrin controls spindle organization and mitotic entry. Mol Biol Cell 15:3642–3657
124. Kitagawa D, Kohlmaier G, Keller D, Strnad P, Balestra FR, Fluckiger I, Gonczy P (2011) Spindle positioning in human cells relies on proper centriole formation and on the microcephaly proteins CPAP and STIL. J Cell Sci 124:3884–3893
125. Klingseisen A, Jackson AP (2011) Mechanisms and pathways of growth failure in primordial dwarfism. Genes Dev 25:2011–2024
126. Passemard S, Titomanlio L, Elmaleh M, Afenjar A, Alessandri JL, Andria G, de Villemeur TB, Boespflug-Tanguy O, Burglen L, Del Giudice E et al (2009) Expanding the clinical and neuroradiologic phenotype of primary microcephaly due to ASPM mutations. Neurology 73:962–969
127. Kaindl AM, Passemard S, Kumar P, Kraemer N, Issa L, Zwirner A, Gerard B, Verloes A, Mani S, Gressens P (2010) Many roads lead to primary autosomal recessive microcephaly. Prog Neurobiol 90:363–383
128. Woods CG, Bond J, Enard W (2005) Autosomal recessive primary microcephaly (MCPH): a review of clinical, molecular, and evolutionary findings. Am J Hum Genet 76:717–728
129. Trimborn M, Bell SM, Felix C, Rashid Y, Jafri H, Griffiths PD, Neumann LM, Krebs A, Reis A, Sperling K et al (2004) Mutations in microcephalin cause aberrant regulation of chromosome condensation. Am J Hum Genet 75:261–266
130. Genin A, Desir J, Lambert N, Biervliet M, Van Der Aa N, Pierquin G, Killian A, Tosi M, Urbina M, Lefort A et al (2012) Kinetochore KMN network gene CASC5 mutated in primary microcephaly. Hum Mol Genet 21:5306–5317

131. Mirzaa GM, Vitre B, Carpenter G, Abramowicz I, Gleeson JG, Paciorkowski AR, Cleveland DW, Dobyns WB, O'Driscoll M (2014) Mutations in CENPE define a novel kinetochore-centromeric mechanism for microcephalic primordial dwarfism. Hum Genet 133:1023–1039
132. Awad S, Al-Dosari MS, Al-Yacoub N, Colak D, Salih MA, Alkuraya FS, Poizat C (2013) Mutation in PHC1 implicates chromatin remodeling in primary microcephaly pathogenesis. Hum Mol Genet 22:2200–2213
133. Jackson AP, Eastwood H, Bell SM, Adu J, Toomes C, Carr IM, Roberts E, Hampshire DJ, Crow YJ, Mighell AJ et al (2002) Identification of microcephalin, a protein implicated in determining the size of the human brain. Am J Hum Genet 71:136–142
134. Kiyomitsu T, Obuse C, Yanagida M (2007) Human Blinkin/AF15q14 is required for chromosome alignment and the mitotic checkpoint through direct interaction with Bub1 and BubR1. Dev Cell 13:663–676
135. Yang YJ, Baltus AE, Mathew RS, Murphy EA, Evrony GD, Gonzalez DM, Wang EP, Marshall-Walker CA, Barry BJ, Murn J et al (2012) Microcephaly gene links trithorax and REST/NRSF to control neural stem cell proliferation and differentiation. Cell 151:1097–1112
136. Faheem M, Naseer MI, Rasool M, Chaudhary AG, Kumosani TA, Ilyas AM, Pushparaj P, Ahmed F, Algahtani HA, Al-Qahtani MH et al 2015 Molecular genetics of human primary microcephaly: an overview. BMC Med Genomics 8(Suppl 1):S4
137. Mahmood S, Ahmad W, Hassan MJ (2011) Autosomal recessive primary microcephaly (MCPH): clinical manifestations, genetic heterogeneity and mutation continuum. Orphanet J Rare Dis 6:39
138. Trimborn M, Ghani M, Walther DJ, Dopatka M, Dutrannoy V, Busche A, Meyer F, Nowak S, Nowak J, Zabel C et al (2010) Establishment of a mouse model with misregulated chromosome condensation due to defective Mcph1 function. PLoS One 5:e9242
139. Zhou ZW, Tapias A, Bruhn C, Gruber R, Sukchev M, Wang ZQ (2013) DNA damage response in microcephaly development of MCPH1 mouse model. DNA Repair 12:645–655
140. Gruber R, Zhou Z, Sukchev M, Joerss T, Frappart PO, Wang ZQ (2011) MCPH1 regulates the neuroprogenitor division mode by coupling the centrosomal cycle with mitotic entry through the Chk1-Cdc25 pathway. Nat Cell Biol 13:1325–1334
141. Brunk K, Vernay B, Griffith E, Reynolds NL, Strutt D, Ingham PW, Jackson AP (2007) Microcephalin coordinates mitosis in the syncytial Drosophila embryo. J Cell Sci 120:3578–3588
142. Rickmyre JL, Dasgupta S, Ooi DL, Keel J, Lee E, Kirschner MW, Waddell S, Lee LA (2007) The Drosophila homolog of MCPH1, a human microcephaly gene, is required for genomic stability in the early embryo. J Cell Sci 120:3565–3577
143. Nano M, Basto R (2015) The Janus soul of centrosomes: a paradoxical role in disease? Chromosome Res 24(1):127–144
144. Guernsey DL, Jiang H, Hussin J, Arnold M, Bouyakdan K, Perry S, Babineau-Sturk T, Beis J, Dumas N, Evans SC et al (2010) Mutations in centrosomal protein CEP152 in primary microcephaly families linked to MCPH4. Am J Hum Genet 87:40–51
145. Kumar A, Girimaji SC, Duvvari MR, Blanton SH (2009) Mutations in STIL, encoding a pericentriolar and centrosomal protein, cause primary microcephaly. Am J Hum Genet 84:286–290
146. Arquint C, Nigg EA (2014) STIL microcephaly mutations interfere with APC/C-mediated degradation and cause centriole amplification. Curr Biol 24:351–360
147. Bond J, Roberts E, Springell K, Lizarraga SB, Scott S, Higgins J, Hampshire DJ, Morrison EE, Leal GF, Silva EO et al (2005) A centrosomal mechanism involving CDK5RAP2 and CENPJ controls brain size. Nat Genet 37:353–355
148. Hussain MS, Baig SM, Neumann S, Nurnberg G, Farooq M, Ahmad I, Alef T, Hennies HC, Technau M, Altmuller J et al (2012) A truncating mutation of CEP135 causes primary microcephaly and disturbed centrosomal function. Am J Hum Genet 90:871–878
149. Farooq M, Fatima A, Mang Y, Hansen L, Kjaer KW, Baig SM, Larsen LA, Tommerup N (2016) A novel splice site mutation in CEP135 is associated with primary microcephaly in a Pakistani family. J Hum Genet 61:271–273

150. Sir JH, Barr AR, Nicholas AK, Carvalho OP, Khurshid M, Sossick A, Reichelt S, D'Santos C, Woods CG, Gergely F (2011) A primary microcephaly protein complex forms a ring around parental centrioles. Nat Genet 43:1147–1153
151. Marjanovic M, Sanchez-Huertas C, Terre B, Gomez R, Scheel JF, Pacheco S, Knobel PA, Martinez-Marchal A, Aivio S, Palenzuela L et al (2015) CEP63 deficiency promotes p53-dependent microcephaly and reveals a role for the centrosome in meiotic recombination. Nat Commun 6:7676
152. Khan MA, Rupp VM, Orpinell M, Hussain MS, Altmuller J, Steinmetz MO, Enzinger C, Thiele H, Hohne W, Nurnberg G et al (2014) A missense mutation in the PISA domain of HsSAS-6 causes autosomal recessive primary microcephaly in a large consanguineous Pakistani family. Hum Mol Genet 23:5940–5949
153. Fu J, Glover DM (2012) Structured illumination of the interface between centriole and peri-centriolar material. Open Biol 2:120104
154. Conduit PT, Richens JH, Wainman A, Holder J, Vicente CC, Pratt MB, Dix CI, Novak ZA, Dobbie IM, Schermelleh L et al (2014b) A molecular mechanism of mitotic centrosome assembly in Drosophila. elife 3:e03399
155. Choi YK, Liu P, Sze SK, Dai C, Qi RZ (2010) CDK5RAP2 stimulates microtubule nucleation by the gamma-tubulin ring complex. J Cell Biol 191:1089–1095
156. Fong KW, Choi YK, Rattner JB, Qi RZ (2008) CDK5RAP2 is a pericentriolar protein that functions in centrosomal attachment of the gamma-tubulin ring complex. Mol Biol Cell 19:115–125
157. Buchman JJ, Tseng HC, Zhou Y, Frank CL, Xie Z, Tsai LH (2010) Cdk5rap2 interacts with pericentrin to maintain the neural progenitor pool in the developing neocortex. Neuron 66:386–402
158. Lizarraga SB, Margossian SP, Harris MH, Campagna DR, Han AP, Blevins S, Mudbhary R, Barker JE, Walsh CA, Fleming MD (2010) Cdk5rap2 regulates centrosome function and chromosome segregation in neuronal progenitors. Development 137:1907–1917
159. Lancaster MA, Renner M, Martin CA, Wenzel D, Bicknell LS, Hurles ME, Homfray T, Penninger JM, Jackson AP, Knoblich JA (2013) Cerebral organoids model human brain development and microcephaly. Nature 501:373–379
160. Barr AR, Kilmartin JV, Gergely F (2010) CDK5RAP2 functions in centrosome to spindle pole attachment and DNA damage response. J Cell Biol 189:23–39
161. Conduit PT, Feng Z, Richens JH, Baumbach J, Wainman A, Bakshi SD, Dobbelaere J, Johnson S, Lea SM, Raff JW (2014a) The centrosome-specific phosphorylation of Cnn by Polo/Plk1 drives Cnn scaffold assembly and centrosome maturation. Dev Cell 28:659–669
162. Conduit PT, Brunk K, Dobbelaere J, Dix CI, Lucas EP, Raff JW (2010) Centrioles regulate centrosome size by controlling the rate of Cnn incorporation into the PCM. Curr Biol 20:2178–2186
163. Conduit PT, Raff JW (2010) Cnn dynamics drive centrosome size asymmetry to ensure daughter centriole retention in Drosophila neuroblasts. Curr Biol 20:2187–2192
164. Yalgin C, Ebrahimi S, Delandre C, Yoong LF, Akimoto S, Tran H, Amikura R, Spokony R, Torben-Nielsen B, White KP et al (2015) Centrosomin represses dendrite branching by orienting microtubule nucleation. Nat Neurosci 18:1437–1445
165. Hussain MS, Baig SM, Neumann S, Peche VS, Szczepanski S, Nurnberg G, Tariq M, Jameel M, Khan TN, Fatima A et al (2013) CDK6 associates with the centrosome during mitosis and is mutated in a large Pakistani family with primary microcephaly. Hum Mol Genet 22:5199–5214
166. do Carmo Avides M, Glover DM (1999) Abnormal spindle protein, asp, and the integrity of mitotic centrosomal microtubule organizing centers. Science 283:1733–1735
167. Fish JL, Kosodo Y, Enard W, Paabo S, Huttner WB (2006) Aspm specifically maintains symmetric proliferative divisions of neuroepithelial cells. Proc Natl Acad Sci U S A 103:10438–10443
168. Gul A, Hassan MJ, Mahmood S, Chen W, Rahmani S, Naseer MI, Dellefave L, Muhammad N, Rafiq MA, Ansar M et al (2006) Genetic studies of autosomal recessive primary microcephaly in 33 Pakistani families: novel sequence variants in ASPM gene. Neurogenetics 7:105–110

169. Pulvers JN, Bryk J, Fish JL, Wilsch-Brauninger M, Arai Y, Schreier D, Naumann R, Helppi J, Habermann B, Vogt J et al (2010) Mutations in mouse Aspm (abnormal spindle-like microcephaly associated) cause not only microcephaly but also major defects in the germline. Proc Natl Acad Sci U S A 107:16595–16600
170. Rujano MA, Sanchez-Pulido L, Pennetier C, le Dez G, Basto R (2013) The microcephaly protein asp regulates neuroepithelium morphogenesis by controlling the spatial distribution of myosin II. Nat Cell Biol 15:1294–1306
171. Nicholas AK, Khurshid M, Desir J, Carvalho OP, Cox JJ, Thornton G, Kausar R, Ansar M, Ahmad W, Verloes A et al (2010) WDR62 is associated with the spindle pole and is mutated in human microcephaly. Nat Genet 42:1010–1014
172. Chen JF, Zhang Y, Wilde J, Hansen KC, Lai F, Niswander L (2014) Microcephaly disease gene Wdr62 regulates mitotic progression of embryonic neural stem cells and brain size. Nat Commun 5:3885
173. Ramdas Nair A, Singh P, Salvador Garcia D, Rodriguez-Crespo D, Egger B, Cabernard C (2016) The microcephaly-associated protein Wdr62/CG7337 is required to maintain centrosome asymmetry in Drosophila neuroblasts. Cell Rep 14:1100–1113
174. Hall JG, Flora C, Scott CI Jr, Pauli RM, Tanaka KI (2004) Majewski osteodysplastic primordial dwarfism type II (MOPD II): natural history and clinical findings. Am J Med Genet A 130A:55–72
175. Majewski F, Goecke T (1982) Studies of microcephalic primordial dwarfism I: approach to a delineation of the Seckel syndrome. Am J Med Genet 12:7–21
176. Majewski F, Goecke TO (1998) Microcephalic osteodysplastic primordial dwarfism type II: report of three cases and review. Am J Med Genet 80:25–31
177. Willems M, Genevieve D, Borck G, Baumann C, Baujat G, Bieth E, Edery P, Farra C, Gerard M, Heron D et al (2010) Molecular analysis of pericentrin gene (PCNT) in a series of 24 Seckel/microcephalic osteodysplastic primordial dwarfism type II (MOPD II) families. J Med Genet 47:797–802
178. Piane M, Della Monica M, Piatelli G, Lulli P, Lonardo F, Chessa L, Scarano G (2009) Majewski osteodysplastic primordial dwarfism type II (MOPD II) syndrome previously diagnosed as Seckel syndrome: report of a novel mutation of the PCNT gene. Am J Med Genet A 149A:2452–2456
179. Rauch A, Thiel CT, Schindler D, Wick U, Crow YJ, Ekici AB, van Essen AJ, Goecke TO, Al-Gazali L, Chrzanowska KH et al (2008) Mutations in the pericentrin (PCNT) gene cause primordial dwarfism. Science 319:816–819
180. Delaval B, Doxsey SJ (2010) Pericentrin in cellular function and disease. J Cell Biol 188:181–190
181. Griffith E, Walker S, Martin CA, Vagnarelli P, Stiff T, Vernay B, Al Sanna N, Saggar A, Hamel B, Earnshaw WC et al (2008) Mutations in pericentrin cause Seckel syndrome with defective ATR-dependent DNA damage signaling. Nat Genet 40:232–236
182. Tibelius A, Marhold J, Zentgraf H, Heilig CE, Neitzel H, Ducommun B, Rauch A, Ho AD, Bartek J, Kramer A (2009) Microcephalin and pericentrin regulate mitotic entry via centrosome-associated Chk1. J Cell Biol 185:1149–1157
183. Doxsey SJ, Stein P, Evans L, Calarco PD, Kirschner M (1994) Pericentrin, a highly conserved centrosome protein involved in microtubule organization. Cell 76:639–650
184. Purohit A, Tynan SH, Vallee R, Doxsey SJ (1999) Direct interaction of pericentrin with cytoplasmic dynein light intermediate chain contributes to mitotic spindle organization. J Cell Biol 147:481–492
185. Miyoshi K, Kasahara K, Miyazaki I, Shimizu S, Taniguchi M, Matsuzaki S, Tohyama M, Asanuma M (2009) Pericentrin, a centrosomal protein related to microcephalic primordial dwarfism, is required for olfactory cilia assembly in mice. FASEB J 23:3289–3297
186. Seckel HPG (1960) Bird-headed Dwarfs: studies in developmental anthropology including human proportions. Charles C Thomas, Springfield

187. Bond J, Roberts E, Mochida GH, Hampshire DJ, Scott S, Askham JM, Springell K, Mahadevan M, Crow YJ, Markham AF et al (2002) ASPM is a major determinant of cerebral cortical size. Nat Genet 32:316–320
188. Al-Dosari MS, Shaheen R, Colak D, Alkuraya FS (2010) Novel CENPJ mutation causes Seckel syndrome. J Med Genet 47:411–414
189. Kalay E, Yigit G, Aslan Y, Brown KE, Pohl E, Bicknell LS, Kayserili H, Li Y, Tuysuz B, Nurnberg G et al (2011) CEP152 is a genome maintenance protein disrupted in Seckel syndrome. Nat Genet 43:23–26
190. McIntyre RE, Lakshminarasimhan Chavali P, Ismail O, Carragher DM, Sanchez-Andrade G, Forment JV, Fu B, Del Castillo V-HM, Edwards A, van der Weyden L et al (2012) Disruption of mouse Cenpj, a regulator of centriole biogenesis, phenocopies Seckel syndrome. PLoS Genet 8:e1003022
191. Habedanck R, Stierhof YD, Wilkinson CJ, Nigg EA (2005) The polo kinase Plk4 functions in centriole duplication. Nat Cell Biol 7:1140–1146
192. Shaheen R, Al Tala S, Almoisheer A, Alkuraya FS (2014) Mutation in PLK4, encoding a master regulator of centriole formation, defines a novel locus for primordial dwarfism. J Med Genet 51:814–816
193. Scheidecker S, Etard C, Haren L, Stoetzel C, Hull S, Arno G, Plagnol V, Drunat S, Passemard S, Toutain A et al (2015) Mutations in TUBGCP4 alter microtubule organization via the gamma-tubulin ring complex in autosomal-recessive microcephaly with chorioretinopathy. Am J Hum Genet 96:666–674
194. Shaheen R, Faqeih E, Shamseldin HE, Noche RR, Sunker A, Alshammari MJ, Al-Sheddi T, Adly N, Al-Dosari MS, Megason SG et al (2012) POC1A truncation mutation causes a ciliopathy in humans characterized by primordial dwarfism. Am J Hum Genet 91:330–336
195. Koparir A, Karatas OF, Yuceturk B, Yuksel B, Bayrak AO, Gerdan OF, Sagiroglu MS, Gezdirici A, Kirimtay K, Selcuk E et al (2015) Novel POC1A mutation in primordial dwarfism reveals new insights for centriole biogenesis. Hum Mol Genet 24:5378–5387
196. Ferrante MI, Giorgio G, Feather SA, Bulfone A, Wright V, Ghiani M, Selicorni A, Gammaro L, Scolari F, Woolf AS et al (2001) Identification of the gene for oral-facial-digital type I syndrome. Am J Hum Genet 68:569–576
197. Romio L, Wright V, Price K, Winyard PJ, Donnai D, Porteous ME, Franco B, Giorgio G, Malcolm S, Woolf AS et al (2003) OFD1, the gene mutated in oral-facial-digital syndrome type 1, is expressed in the metanephros and in human embryonic renal mesenchymal cells. J Am Soc Nephrol: JASN 14:680–689
198. AlKattan WM, Al-Qattan MM, Bafaqeeh SA (2015) The pathogenesis of the clinical features of oral-facial-digital syndrome type I. Saudi Med J 36:1277–1284
199. Bicknell LS, Walker S, Klingseisen A, Stiff T, Leitch A, Kerzendorfer C, Martin CA, Yeyati P, Al Sanna N, Bober M et al (2011) Mutations in ORC1, encoding the largest subunit of the origin recognition complex, cause microcephalic primordial dwarfism resembling Meier-Gorlin syndrome. Nat Genet 43:350–355
200. Hemerly AS, Prasanth SG, Siddiqui K, Stillman B (2009) Orc1 controls centriole and centrosome copy number in human cells. Science 323:789–793
201. Hossain M, Stillman B (2012) Meier-Gorlin syndrome mutations disrupt an Orc1 CDK inhibitory domain and cause centrosome reduplication. Genes Dev 26:1797–1810
202. Barrera JA, Kao LR, Hammer RE, Seemann J, Fuchs JL, Megraw TL (2010) CDK5RAP2 regulates centriole engagement and cohesion in mice. Dev Cell 18:913–926
203. Niccols A (2007) Fetal alcohol syndrome and the developing socio-emotional brain. Brain Cogn 65:135–142
204. Pan American Health Organization (2015) Epidemiological alert. Neurological syndrome, congenital malformations, and Zika virus infection. Implications for public health in the Americas. Pan American Health Organization, Washington. Available from: http://www.paho.org/hq/index.phpoption=com_docman&task=doc_download&Itemid=&gid=32405 [cited 2016 Apr 13]

205. WHO (2015) Zika virus outbreaks in the Americas. Releve epidemiologique hebdomadaire/ Section d'hygiene du Secretariat de la Societe des Nations = Weekly epidemiological record/ Health Section of the Secretariat of the League of Nations 90:609–610
206. Schuler-Faccini L, Ribeiro EM, Feitosa IM, Horovitz DD, Cavalcanti DP, Pessoa A, Doriqui MJ, Neri JI, Neto JM, Wanderley HY et al (2016) Possible association between Zika virus infection and microcephaly – Brazil, 2015. MMWR Morb Mortal Wkly Rep 65:59–62
207. Bell TM, Field EJ, Narang HK (1971) Zika virus infection of the central nervous system of mice. Archiv fur die gesamte Virusforschung 35:183–193
208. Dick GW, Kitchen SF, Haddow AJ (1952) Zika virus. I Isolations and serological specificity. Trans R Soc Trop Med Hyg 46:509–520
209. Tang H, Hammack C, Ogden SC, Wen Z, Qian X, Li Y, Yao B, Shin J, Zhang F, Lee EM et al (2016) Zika virus infects human cortical neural progenitors and attenuates their growth. Cell Stem Cell. doi:10.1016/j.stem.2016.02.016
210. Baala L, Briault S, Etchevers HC, Laumonnier F, Natiq A, Amiel J, Boddaert N, Picard C, Sbiti A, Asermouh A et al (2007) Homozygous silencing of T-box transcription factor EOMES leads to microcephaly with polymicrogyria and corpus callosum agenesis. Nat Genet 39:454–456

Chapter 3
Dividing with Extra Centrosomes: A Double Edged Sword for Cancer Cells

Alexander D. Rhys and Susana A. Godinho

Abstract The presence of supernumerary centrosomes is a hallmark of human tumours. Recent work in animal models suggests that extra centrosomes are not just bystanders in cancer but can accelerate tumourigenesis in the absence of the tumour suppressor p53. Centrosome amplification could indeed actively participate in tumour progression through the induction of chromosome instability, disruption of tissue architecture and promoting cell invasion. Paradoxically, however, centrosome amplification is rather poorly tolerated in normal cells and there are several hurdles cells need to overcome in order to efficiently proliferate in the presence of extra centrosomes. Here, we review the adaptation mechanisms that allow cells to efficiently divide in the presence of extra centrosomes and how these could be exploited to develop selective cancer therapies.

Keywords Centrosome Amplification • Mitosis • Clustering • Cancer • Aneuploidy

3.1 Introduction

The centrosome is the main microtubule (MT) organising centre in animal cells, consisting of a pair of orthogonally positioned barrel-shaped centrioles, embedded in a proteinaceous matrix called the pericentriolar material (PCM), which provides the site for MT nucleation. The two centrioles are structurally different at their distal ends, with the older mother centriole containing distal and subdistal appendages required for MT anchorage and ciliogenesis [1]. In cycling cells, the centrosome is important for cell shape, motility and the formation of a bipolar spindle during mitosis [1]. In addition, in many differentiated cell types, the mother centriole acts as the basal body required for cilia formation [2].

A.D. Rhys • S.A. Godinho (✉)
Centre for Molecular Oncology, Barts Cancer Institute, Queen Mary University of London, EC1M 6BQ London, UK
e-mail: a.d.rhys@qmul.ac.uk; s.godinho@qmul.ac.uk

M. Gotta, P. Meraldi (eds.), *Cell Division Machinery and Disease*, Advances in Experimental Medicine and Biology 1002, DOI 10.1007/978-3-319-57127-0_3

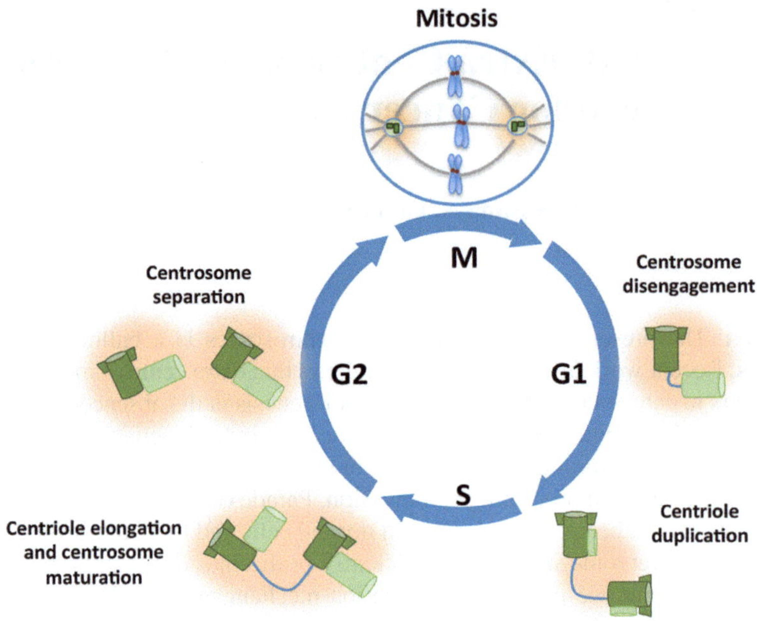

Fig. 3.1 Centrosome duplication. As cells enter G1 phase they have a single centrosome with two orthogonally positioned centrioles, the mother centriole (*dark green*) has distal appendages, whereas the daughter centriole (*light green*) does not. During G1 the centrioles disengage but remain connected by a fibrous structure. In late G1/S phase the centrioles duplicate in a semi-conservative manner by forming new centrioles perpendicularly to the existing ones. These daughter procentrioles then elongate and the centrosomes mature by recruiting PCM in G2. The fibrous tether no longer connects the two centrosomes and they separate to opposite poles of the cell for the formation of the mitotic spindle.

Centrosome duplication occurs much like DNA replication, in a semi-conservative manner during S-phase. The centrosome cycle was first elaborated using electron microscopy and can be sub-divided into five main steps: centrosome segregation in M-phase, centriole disengagement in late M-phase/G1-phase, centriole duplication in S-phase, centriole maturation and separation in G2-phase (Fig. 3.1) (reviewed in [3–5]). Centrosome duplication is limited to only once per cell cycle, so that upon cell division each daughter cell inherits a single centrosome. During early mitosis the two centrosomes move to opposite ends of the cell where they are then able to anchor and nucleate MTs required for the formation of the mitotic spindle, allowing faithful chromosome segregation [1].

Centrosomes were first described by Boveri in the 1880s as "*the organ for cell division*", being seen as a vital organelle required for cells to divide [6, 7]. Later studies started to challenge this view when seed plant cells were shown to not contain centrosomes [8]. Furthermore, using flattened primary spermatocytes from crane flies, which lack a centrosome at one or both poles, Dietz found that cells could still successfully divide their genetic material [9]. It is now established that acentrosomal cells can still form a bipolar spindle through microtubule nucleation from chromatin [10]. This mechanism, initially observed in *Xenopus* egg extracts [11, 12], is particularly important for the formation of bipolar spindles in female oocytes, which lack centrosomes prior to fertilisation [13].

Despite being dispensable for cell division, loss of centrosomes in somatic cells does not come without a cost. Acentrosomal *Drosophila* mutants, lacking DSas-4, a protein important for centriole duplication, are able to produce flies [14]. These flies go through development without any morphological abnormalities, however they lack cilia and flagella, and die prematurely. Further examination showed that neuroblasts undergoing mitoses have defects in spindle positioning and asymmetric cell division [14]. Previous work also showed that haploid cells derived from unfertilised *Drosophila* cells lacking centrosomes are inherently aneuploid, suggesting that centrosomes may be important for preserving genetic stability [15]. This concept was recently established in DT40 B chicken cells where centrosome loss led to slower mitoses with higher rates of chromosomal instability—suggesting that centrosomes are important for rapid segregation of genetic material whilst maintaining integrity [16]. This was further confirmed in cells treated with a selective inhibitor (centrinone) of the kinase PLK4, the master regulator of centrosome duplication [17, 18], which leads to centrosome loss and low levels of chromosome missegregation [19].

Although cancer cells can efficiently proliferate in the absence of centrosomes [19], centrosome loss hasn't been reported in cancer. In contrast, increased numbers of centrosomes are often observed in cancer cells and considered a feature of human tumours [20]. Intriguingly, and unlike loss of centrioles, amplified centrosome numbers can have catastrophic consequences for the cell division and survival of cancer cells [21]. In this review, we will focus on the mechanisms that promote efficient cell division and survival of cells with supernumerary centrosomes. We will discuss how, during mitosis, centrosome amplification can both be detrimental and promote tumourigenesis.

3.2 Centrosome Amplification and Cancer

A link between extra centrosomes and tumourigenesis was first postulated over a century ago by Theodor Boveri. Boveri suggested that extra centrosomes would lead to multipolar cell divisions, resulting in genetic instability and malignant transformation [6, 22]. Boveri based his theory on the observation that dispermic sea urchin eggs, which contain multiple centrosomes, formed multipolar spindles that led to asymmetric distribution of the genetic material. This was further supported by the observations of Gino Galeotti and David von Hansemann, who first investigated genetic instability in cancer and observed that abnormal mitotic divisions, including multipolar spindles, are common features of human tumours [22–24]. Based on this, Boveri theorised that extra centrosomes could drive aneuploidy from multipolar cell division and subsequently tumourigenesis [22, 23].

While Boveri's hypothesis remained untested for more than 100 years, an extensive body of work has established that the majority of solid and haematological malignancies display centrosome abnormalities and, in particular, excessive numbers of centrosomes [20, 25–31]. In most cases, centrosome amplification correlates with high-grade tumours and poor prognosis, however it has also been observed in some early low-grade lesions, giving weight to the argument that centrosome amplification could drive tumourigenesis [25, 28, 32–35]. Centrosome amplification is also associated with tumour recurrence and metastasis in some cancers, making it a potentially viable biomarker for advanced disease [20, 21, 35]. It is still unclear how centrosome abnormalities are acquired in cancer, but several mechanisms have been proposed to result in excessive numbers of centrosomes: centriole over-duplication, de novo centriole assembly, mitotic slippage, cell-cell fusion and cytokinesis failure (for review see [21]).

The recent development of transgenic mouse models to study the impact of centrosome amplification on tumourigenesis enabled Boveri's hypothesis to be tested. Transient overexpression of PLK4, which leads to centrosome amplification [17, 18, 36, 37], was shown to accelerate tumourigenesis in mice that lacked the tumour suppressor p53 [38, 39]. In the absence of p53, overexpression of PLK4 (PLK4OE) in the mouse epidermis leads to rapid skin tumour formation [38]. Importantly, after birth and prior to tumour formation, the percentage of epidermal cells containing extra centrosomes decreases dramatically, which coincides with loss of overexpressed PLK4 mRNA levels [38]. However, aneuploidy generated by this transient centrosome amplification is maintained, suggesting that it may play a role in the rapid development of these tumours. Similarly, inducible ubiquitous PLK4OE advances the onset of tumour formation, mainly lymphomas and sarcomas, in the absence of p53 [39]. Although other types of tumours were not detected, hyperplasia of the pancreas and skin was also observed in this model. It is possible that this could reflect a different time course for the development of pancreatic and skin neoplasias, since $p53^{-/-}$ mice succumb early to lymphomas [39]. In contrast, a different mouse model showed that ubiquitous centrosome amplification induced by PLK4OE did not cause any acceleration of tumour formation, even in the absence of p53 [40]. The nature of such differences is unclear at this point. However, one possibility could be that constitutive overexpression of PLK4 could may impair tumour acceleration [40], as in the model in which centrosome amplification results in rapid tumour formation PLK4OE is only transient [38, 39]. Notably, PLK4OE does not seem to induce the formation of new tumours, but rather accelerates the development of tumours that form upon p53 loss. Altogether, these results suggest that centrosome amplification alone can contribute to the tumour progression, either by promoting aneuploidy [41, 42], disrupting asymmetric cell division and polarity [37], altering cilia signalling [43], or by promoting the acquisition of invasive features characteristic of aggressive tumours [44].

3.3 Centrosome Amplification: A Double Edged Sword?

Tetraploid cells containing extra centrosomes undergo a negative selection overtime, suggesting that at least in vitro centrosome amplification is deleterious to cell survival [41, 44, 45]. This was further confirmed in a mouse model of centrosome amplification, where a strong selection pressure to lose extra centrosomes seems to exist, at least in the mouse epidermis [38]. This effect can be partially attributed to the fact that cells containing extra centrosomes are prone to catastrophic levels of aneuploidy resulting from a multipolar division and p53-mediated cell cycle arrest [21]. Thus, it is becoming increasingly clear that in order to efficiently proliferate cells need to adapt to centrosome amplification. There are two major barriers cells need to overcome in order to maintain high levels of centrosome amplification: firstly they need to bypass the cell cycle arrest induced by extra centrosomes and secondly they need to form a pseudo-bipolar spindle to survive.

3.3.1 p53-Dependent Cell Cycle Arrest

Centrosome amplification as a consequence of cytokinesis failure and tetraploidization [46, 47] or as a result of centriole over-duplication leads to the stabilisation of p53 and consequently p21 expression, causing a G1 cell cycle arrest and decreased cell proliferation [48]. Loss of p53 rescued the cell cycle defects, enabling cells to survive in the presence of, and maintain, supernumerary centrosomes. Activation of the p53 pathway does not seem to result from the aneuploidy generated by centrosome amplification, but from the centrosome amplification itself [48]. Indeed, a recent screen to identify modulators of p53 arrest mediated by tetraploid cells containing extra centrosomes identified the large tumour suppressor kinase 2 (LATS2) as an important factor for maintaining tetraploid cells arrested in G1. Moreover, the authors found that phosphorylation of LATS2, which leads to Hippo pathway activation, was observed in cells containing multiple centrosomes, suggesting that centrosome amplification could play a role on Hippo activation [49]. Activation of the Hippo pathway could partly be attributed to the decrease of RhoA activity observed in cells with extra centrosomes [44, 49], which was previously shown to activate this pathway (reviewed in [50]). Because Rac1 antagonises RhoA [51], hyperactivation of Rac1, as a consequence of increased microtubule nucleation, could explain the low levels of active RhoA observed in cells with extra centrosomes, [44]. Thus, it is possible that loss or attenuation of the Hippo pathway is an adaptation mechanism that facilitates the survival of cancer cells with extra centrosomes although this hypothesis needs to be tested in vivo.

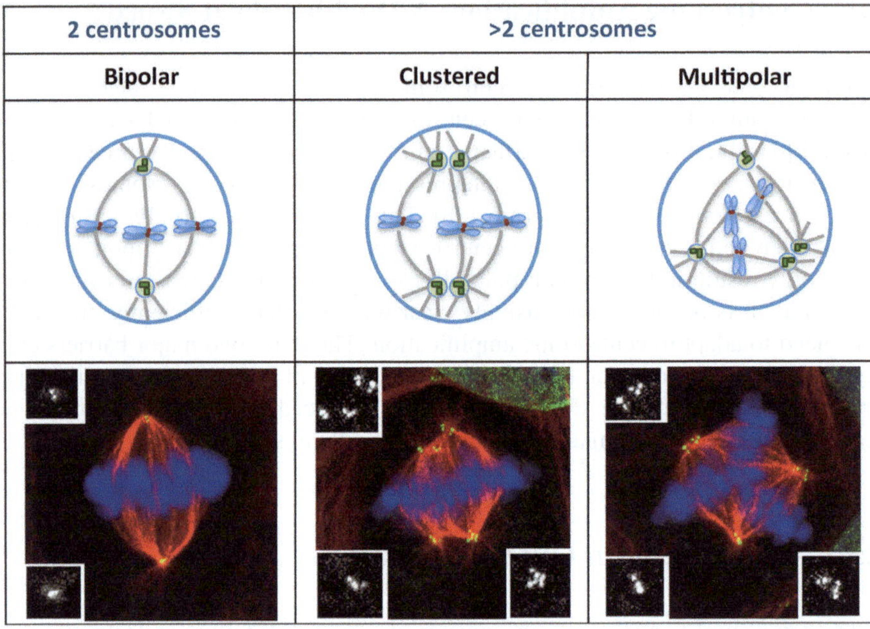

Fig. 3.2 Representation of mitotic cells with normal and supernumerary centrosomes. Cells with normal centrosome number (2) undergo bipolar cell division (*left*). Cells with supernumerary centrosomes (>2) in mammalian cells can either undergo multipolar cell divisions (*right*) or cluster their extra centrosomes into a pseudo-bipolar spindle (*middle*). Representative immunofluorescent images were taken in RPE-1 cells overexpressing PLK4 to induce centrosome amplification. Cells were stained for centrioles (centrin, *green*), microtubules (alpha-tubulin, *red*) and DNA (Hoescht, *blue*).

3.3.2 Dividing with Extra Centrosomes

Up until the early 1980s it was assumed that there would be a link between centrosome amplification and multipolar cell division. However, work performed using the N1E-115 mouse neuroblastoma cell line, in which almost all cells harbour supernumerary centrosomes, revealed a phenomenon now termed "centrosome clustering", where extra centrosomes remained closely associated through mitosis [52] (Fig. 3.2). The ability to cluster these extra centrosomes into two poles enabled the cells to form a pseudo-bipolar spindle, allowing for chromosome segregation with little to no aneuploidy [37, 41, 52, 53].

In addition to centrosome clustering, other distinct but non-exclusive mechanisms have been shown to allow cells to cope with extra centrosomes, including centrosome inactivation, centrosome loss and asymmetric segregation of centrosomes during division [54]. However, to date only centrosome clustering has been observed in tumours.

Table 3.1 Proteins involved in centrosome clustering.

Key proteins involved in centrosome clustering	Protein	References
Spindle assembly checkpoint and chromosomal passenger complex	Mad2	[37, 55]
	AuroraB	[57]
	Borealin	[57]
	INCENP	[57]
	Survivin	[57]
Kinetochore-microtubule tension	SPC24	[57]
	SPC25	[57]
	HEC1	[57]
	SGOL1	[57]
	CENPT	[57]
	Sororin	[57]
	HURP	[58]
Actin cytoskeleton	ILK	[59]
	TACC3	[59]
	ch-TOG	[59]
	Myo10	[55]
	MyoII	[55]
	Moesin	[60]
Microtubule associated proteins (MAPs)	NuMA	[56]
	Dynein	[56]
	HSET/KIFC1	[37, 55]

3.3.2.1 Centrosome Clustering

Centrosome clustering is the best-characterised mechanism of coping with supernumerary centrosomes. Since the experiments done in N1E-115 cells, many cancer cell lines which have a large proportion of cells containing extra centrosomes (>30%) have been shown to be able to cluster them efficiently [41, 52, 53, 55, 56].

To identify the key proteins involved in centrosome clustering, and to try and elucidate the mechanisms involved in this process, two genome-wide screens were carried out in *Drosophila* S2 cells and the UPCI:SCC114 human oral squamous carcinoma cell line [55, 57]. Multiple mechanisms were found to play important roles in centrosome clustering: spindle assembly checkpoint (SAC) and chromosomal passenger complex (CPC), kinetochore-microtubule tension, and actin cytoskeleton and microtubule associated proteins (MAPs; see Table 3.1).

Spindle Assembly Checkpoint and Chromosomal Passenger Complex

The SAC plays a role in delaying the onset of anaphase until all kinetochores are properly attached to the mitotic spindle [61–63]. In a normal mitotic cell with two centrosomes, the SAC is satisfied when a large number of kinetochore-microtubule interactions are made and stabilised, allowing the cells to progress through mitosis [64]. Cells with extra centrosomes take a longer time to form these stable interactions, leading to a prolonged SAC-dependent mitosis that facilitates centrosome clustering [37, 55, 65]. Moreover, loss of the SAC protein Mad2 was shown to prevent efficient centrosome clustering [37, 55, 65] leading to decreased viability in flies with extra centrosomes [37]. Importantly, treating Mad2 depleted cells with the proteasome inhibitor MG132, which delays anaphase onset, rescued the multipolarity observed, suggesting that it is the resultant delay in anaphase onset, and not the SAC itself, that is important to promote efficient centrosome clustering [55]. The role of the SAC in prolonging mitosis in cells with supernumerary centrosomes gives a potential hypothesis behind the increased mitotic index observed in tumours [66]. Supporting this idea, transformed SV40 human fibroblasts have increased levels of tetraploid cells containing supernumerary centrosomes and showed an increased mitotic index when compared to their non-transformed counterparts [65, 67]. However, not all cells with supernumerary centrosomes showed delays in mitosis. Binucleated rat kangaroo kidney epithelial cells (Ptk2) containing extra centrosomes divide without a mitotic delay. Interestingly, these cells do not cluster extra centrosomes and undergo multipolar division [68], further supporting the idea that delaying SAC inactivation might be an important mechanism to prevent multipolarity.

In addition to the SAC, a role in centrosome clustering has also been described for components of the CPC [57], which is involved in recognising and correcting incorrect chromosome-microtubule attachments that are not bi-orientated, such as merotely and syntely [62]. Knocking down any of the CPC components, such as AuroraB, borealin, INCENP and survivin, in the USCI:SCC114 cancer cell line resulted in de-clustering of supernumerary centrosomes [57]. It is possible that CPC components facilitate clustering through the destabilisation of syntelic and merotelic attachments that will elicit a SAC response resulting in an anaphase delay that promotes centrosome clustering.

Kinetochore-Microtubule Tension

Tension generated at the kinetochore-microtubule interface was shown to be important to maintain centrosomes clustered at the spindle poles. Knockdown of SPC24, SPC25 or HEC1, components of the Ndc80 complex important for stabilising microtubule and kinetochore attachments [69], resulted in clustering defects [57]. In addition, depletion of SGOL1, CENPT and sororin, involved in generating kinetochore tension resulted in a similar phenotype [57]. Furthermore, a role for hepatoma up-regulated protein (HURP), which has been shown to serve as a

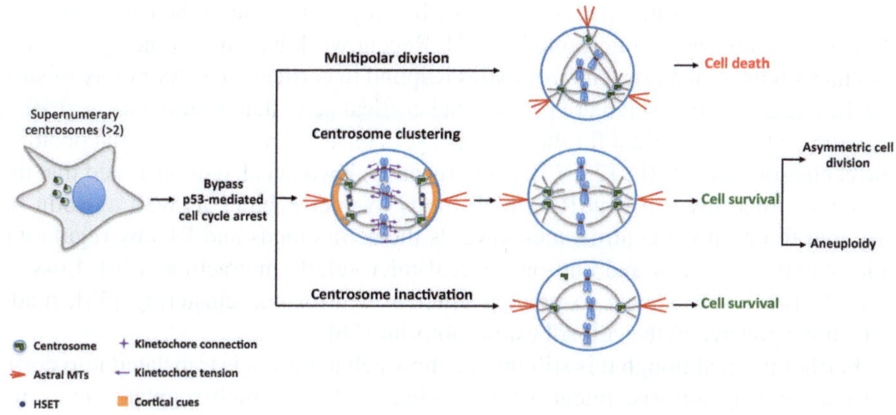

Fig. 3.3 Schematic representation of the mechanisms that allow cells to divide with extra centrosomes and how their consequences for cell division. Cells with supernumerary centrosomes (>2) can undergo different cell fates, once they overcome a p53-mediated cell cycle arrest. *Top*—cells can undergo multipolar cell division, leading to high levels of aneuploidy, resulting in cell death. *Middle*—cells can cluster their supernumerary centrosomes into two poles, this is determined by factors such as the interaction of astral microtubules with the cell cortex, minus-end directed motors such as HSET, cortical cues, SAC and kinetochore tension. Centrosome clustering results in viable daughter cells but can leads to aneuploidy and defective asymmetric cell division. *Bottom*—depletion of microtubule nucleating proteins results in inactivation of a centrosome, stopping the centrosome having an effect on spindle orientation, resulting in two daughter cells.

kinetochore-microtubule stabilizing factor during mitosis [70, 71], has also been identified in centrosome clustering [58]. Forces acting on the spindle poles are important to prevent multipolarity, even in the absence of centrosome amplification, by counteracting the traction forces generated during chromosome congression during prometaphase [72]. Thus, it is possible that clustered extra centrosomes at the spindle poles could affect the fine-tuning of spindle forces making cells with amplified centrosomes more reliable on some kinetochore proteins that regulate kinetochore-microtubule tension (Fig. 3.3).

Actin Cytoskeleton

The actin cytoskeleton has been shown to play a part in translating cortical cues into mitosis through the formation of retraction fibres (RFs) [73, 74]. Elegant work from Thery et al. using micro-contact printing showed that during mitotic cell rounding, the actin-rich RFs, which were linked to sites of adhesion, remain attached to the substrate, conferring a spatial footprint to the mitotic cell. Interaction of astral microtubules to the cortical cues specified by the RFs was shown to control spindle positioning [74]. Importantly, this footprint can determine whether cells with extra centrosomes cluster them or not. Cells that maintained a "bipolar" distribution of RFs after cell rounding were able to cluster supernumerary centrosomes more effectively than cells that had a more distributed pattern around the cell, a process that also requires astral microtubules [55]. Recent work has shed some light on the mechanism by which astral microtubules respond to cortical cues. Assembly of subcortical actin clouds, which are pools of subcortical actin that accumulate near RFs, was shown to be important for the pulling forces on astral microtubules that position the centrosomes near RFs [73, 75]. More recently, Kwon et al. demonstrated that the unconventional myosin Myo10, which binds to microtubules as well as actin, is required to orient the centrosomes towards the actin clouds and RFs by regulating microtubule dynamics and end-on cortical-microtubule interactions [76]. Loss of Myo10, previously shown to prevent efficient centrosome clustering [55], made cells unresponsive to the cell adhesion footprint [76].

Furthermore, although it is still unclear how cell adhesion is translated into cortical cues during mitosis, integrin-linked kinase (ILK), which regulates integrin-mediated cell adhesion and localises to both focal adhesions and centrosomes, facilitates clustering of supernumerary centrosomes through TACC3 and ch-TOG, two proteins involved in regulating the minus end of microtubules at the spindle poles [59]. Therefore, there is a potential role for cell adhesion proteins in controlling centrosome clustering, but further work is required to understand the interplay between the actin cytoskeleton, adhesion proteins and cortical cues in this process.

Inhibition of cortical contractility can prevent efficient centrosome clustering, suggesting that the actin cytoskeleton may also promote centrosome clustering by controlling cortical contractility [55]. However, the mechanism by which contractility facilitate clustering is unknown. Myosin II and actin are important for centrosome separation after centrosome duplication by providing key cortical forces [77]. Centrosome separation during mitosis is aided by inhibition of cortical myosin II contractility by the astral microtubules, which in turn would lead to asymmetric cellular contractility that drives the separation of the two centrosomes [77]. Whilst this model works well for a cell containing the normal number of centrosomes, where each is pushed to opposite sides of the cell, the model is more complicated when there are multiple centrosomes, where depending on the contractility, they could be pushed together, or apart. Further work is needed to fully understand the role of cortical contractility in centrosome clustering, and how those forces are distributed across the cell cortex.

Microtubule Associated Proteins (MAPs)

Cells with extra centrosomes depend on microtubule motors and associated proteins, which play important roles in the organisation of the mitotic spindle, to cluster supernumerary centrosomes [54, 78, 79]. The first microtubule proteins proposed to have a role in centrosome clustering were the minus-end directed motor dynein and the mitotic apparatus protein NuMA, a key component of the mitotic spindle [56]. NuMA regulates dynein localisation at the centrosomes, which in turn is important for centrosome clustering in human cells. Increasing NuMA levels in cancer cells containing supernumerary centrosomes results in dynein disassociation from the spindle poles and subsequent centrosome de-clustering and multipolar divisions [56]. Titrating the levels of NuMA restores dynein localisation and clustering, supporting the combined roles of the NuMA/dynein complex in this process [56].

The anaphase-promoter complex/cyclosome (APC/C) is an essential E3 ligase involved in proteasome-mediated protein degradation essential to regulate mitotic progression [80]. When bound to its co-activator CDH1, APC/C, essential to regulate metaphase to anaphase transition, was also shown to regulate centrosome clustering [81]. The microtubule motor Eg5, a substrate of APC/C-CDH1, is stabilised upon inhibition of APC/C-CDH1 leading to an imbalance of forces within the spindle that prevents efficient centrosome clustering [81]. This suggests that part of the centrosome clustering occurs at metaphase to anaphase transition, which is in agreement with our unpublished observations (Rhys and Godinho unpublished).

Another microtubule motor involved in centrosome clustering is KIFC1/HSET (also known as nonclaret disjunctional, Ncd, in flies) [55], a member of the minus-end directed kinesin-14 protein family. Depletion of KIFC1/HSET by siRNA results in increased multipolar mitoses in cells with extra centrosomes, but has no effect on cell division in cells with normal centrosome number, suggesting that HSET has a unique role in centrosome clustering [55]. Importantly, loss of Ncd does not affect viability of WT flies, but decreases survival of flies harbouring extra centrosomes [37, 82]. Thus, KIFC1/HSET is a promising drug target that could potentially target cancer cells.

3.3.2.2 Centrosome Inactivation

Alongside centrosome clustering, it has also been reported that centrosome inactivation can contribute to the formation of bipolar spindles in cells with extra centrosomes. *Drosophila* neuroblasts containing extra centrosomes, as a result of the overexpression of SAK/PLK4, efficiently undergo bipolar mitoses. Although the majority of extra centrosomes cluster into two poles, few un-clustered centrosomes showed reduced levels of PCM and low microtubule nucleation activity [37]. These observations suggest that extra centrosomes compete for PCM, and that centrosomes that do not contain enough PCM will be inactivated and will not contribute to the formation of the mitotic spindle. Centrosome inactivation has been previously described in polyspermic newt eggs, where a gradient of γ-tubulin, essential for centrosomal microtubule nucleation, ensures that only the centrosome associated with the principal sperm nucleus contributes to the assembly of the bipolar spindle [83]. Recent work has shed some light into the mechanisms involved in centrosome inactivation. In flies overexpressing SAK/PLK4, moesin, which is upregulated and localises to the centrosomes in epithelial wing discs but not in neuroblasts, was able to inhibit centrosome inactivation leading to increased multipolar divisions in epithelial cells [60]. Furthermore, a hypomorphic moesin mutant caused a decrease in the centrosomal accumulation of the PCM protein centrosomin, suggesting that the regulation of PCM by moesin is important for centrosome inactivation. This is the first evidence that the mechanisms to cope with centrosome amplification depend on the cellular context. However, it is still unclear whether the upregulation of moesin observed in the wing disks are a result of SAK/PLK4 or reflect intrinsic differences between epithelial and neuronal fly cells.

3.3.2.3 Centrosome Loss

Centrosome removal is a mechanism often utilised by cells during oogenesis, where the maternal centrosome is removed/destroyed to prevent the cell acquiring multiple centrosomes after fertilisation [13]. Centrosome loss is associated with decreased PCM and consequent loss of MT nucleation, probably leading to centrosome disintegration [84]. The exact mechanism for centrosome elimination has not been fully elucidated, however in *C. elegans*, it is reliant on the cyclin dependent kinase (cdk) inhibitor Cki-2. In Cki-2 deficient oocytes the maternal centriole remains, resulting in multipolar zygotic cell divisions [85]. More recently, loss of the helicase CHG-1 was shown to delay centrosome loss in *C. elegans*, probably by preventing the degradation of specific mRNA(s) involved in this process [84]. It has also been suggested that during the process of centrosome clustering, cells could asymmetrically divide their extra centrosomes [86]. This would potentially enable one daughter cell to inherit one centrosome, enabling this cell to successfully proliferate in future divisions [86]. In addition, centrosome extrusion from the cell has been also observed in *Dictyoistelium* [54, 87]. However, to date neither of these mechanisms has been shown to be important for the survival of cancer cells with extra centrosomes (Fig. 3.3).

There are therefore different mechanisms that allow cells to adapt and survive in the presence of extra centrosomes. However, whether these mechanisms function with similar efficiency in all cell types or if different cells have the same likelihood to tolerate extra centrosomes is still unknown.

3.4 Are there Benefits to Centrosome Amplification?

Cancer cells actively maintain their supernumerary centrosome population, suggesting that there is some advantage to maintaining them, despite the negative effects described above. It is thought that the role of extra centrosomes on chromosomal instability, asymmetric cell division and a role in increasing invasion could be beneficial effects of tolerating extra centrosomes.

3.4.1 Chromosome Instability (CIN)

It is well established that multipolar mitoses are poorly tolerated by cancer cells and it still remains to be fully determined the extent to which multipolar cell divisions could generate viable aneuploid cells [41, 45, 55]. It is therefore unlikely that

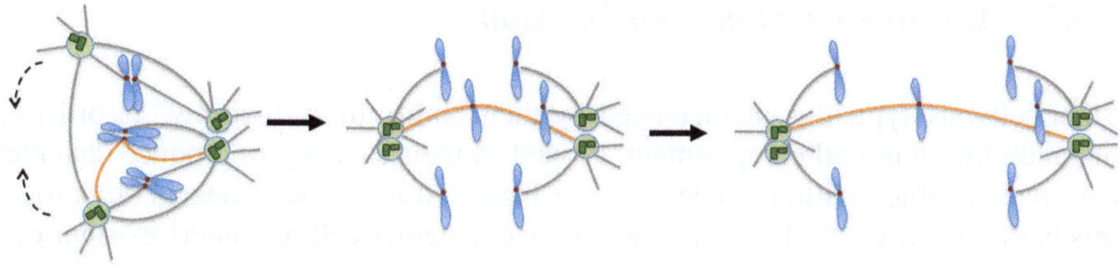

Fig. 3.4 Supernumerary centrosomes can give rise to merotelic attachments and lagging chromosomes. Supernumerary centrosomes can promote merotelic attachments (when one kinetochore attaches to microtubules that emanate from opposite spindle poles—*orange lines*) due to altered spindle geometry. This merotely can then result in lagging chromosomes and subsequently aneuploidy daughter cells.

the correlations observed between CIN, persistent rates of chromosomal alterations, and centrosome amplification in tumours arises from multipolar divisions. Indeed, centrosome amplification can drive CIN independently of multipolar divisions [20, 88]. Cells with supernumerary centrosomes form a transient multipolar spindle, which precedes centrosome clustering, that promote the formation of erroneous merotelic attachments [29, 41, 42]. These attachments, where a single kinetochore is attached to both spindle poles, can go unrepaired, as the SAC is satisfied once sister kinetochores are under tension and is not capable of determining whether that attachment is correct or not (Fig. 3.4) (recently reviewed in: [89]). Merotelic attachments can then result in lagging chromosomes, giving rise to aneuploid daughter cells [90]. Thus, even in the absence of multipolar divisions, centrosome amplification could lead to viable aneuploid cells, supporting Boveri's initial hypothesis. In addition, recent work suggests that lagging chromosomes, for example as a consequence of centrosome amplification, could have a broader impact in the generation of widespread CIN observed in tumours. DNA damage can occur as a result of a lagging chromosome being caught at the cytokinetic furrow [91] or when it is segregated into a micronucleus [92, 93]. Therefore, aneuploidy can be generated through DNA damage and subsequent mutations, or "chromothripsis", where a chromosome or chromosome arm is fragmented and reassembled in a random order [92, 94, 95].

The role of aneuploidy in cancer is not easily picked apart—aneuploidy and CIN can aid tumourigenesis, but at high levels they can inhibit it [96, 97]. In agreement, as a consequence of inefficient centrosome clustering and massive aneuploidy, mouse neuronal cells with extra centrosomes undergo apoptosis leading to brain developmental defects without any tumour formation [98]. In contrast, low levels of aneuploidy as a consequence of transient centrosome amplification results in skin carcinomas in mice lacking p53 [38]. In tumours, CIN can drive genetic heterogeneity, which could lead to the acquisition of advantageous features, allowing tumours to evolve [99]. Thus, the generation of CIN could be an important selective pressure to maintain extra centrosomes in tumours. Nevertheless, even in the context of centrosome amplification, cancer cells need to find the right balance to utilise aneuploidy to their advantage.

3.4.2 Asymmetric Stem Cell Division

Stem cells are typically characterised by their ability to "self-renew" in order to maintain the stem-cell compartment, as well as their capacity to produce daughter cells that are able to differentiate. One mechanism used by stem-cells in order to do this is asymmetric cell division, whereby one daughter cell maintains its stem-cell fate, whereas the other differentiates [100, 101]. Experiments in *Drosophila* showed that transplantation of neuroblasts with supernumerary centrosomes into the abdomen of adult flies resulted in fast growing tumours, with low levels of aneuploidy [37]. Indeed, induction of aneuploidy is not sufficient to drive tumourigenesis in this model [102]. Neuroblasts with supernumerary centrosomes, although able to cluster them efficiently, showed defects in asymmetric cell division [37, 102], which has previously been shown to lead to the expansion of the stem-cell compartment and tumourigenesis [103]. Although it is still unclear how extra centrosomes affect asymmetric cell division, several ideas have been proposed. The presence of supernumerary centrosomes could disrupt of the normal cues directed by the mother and daughter centrosome since the centrosomes are differentially segregated to the differentiating or stem cell [104]. In addition, disorganisation of the astral microtubules, which links the centrosome to the cell cortex could result in spindle orientation defects [105]. Finally, extra centrosomes could potentially change the localisation of polarity determinants in the cell [103]. Similar defects on spindle orientation were also observed in mouse epidermis cells containing extra centrosomes that undergo asymmetric cell division, but not in symmetric mitosis [106]. In contrast, centrosome amplification in mouse neuronal cells does not induce spindle mispositioning during cell division [98], suggesting that these defects can vary according to cell type.

3.4.3 Cell Invasion

In addition to the roles extra centrosomes play during mitosis (described above), recent work highlighted that the contribution of centrosome amplification to tumourigenesis might not be restricted to cell division. Indeed, multiple centrosomes were shown to lead to cell invasion in cells plated in 3D cultures [44]. Cells with extra centrosomes display higher levels of Rac1 activity as a consequence of increased microtubule nucleation capacity, which leads to of cell-cell contacts defects and cell invasion [44]. Similar results have also been observed in endothelial cells where extra centrosome caused Rac1 activation leading to polarisation defects during blood vessel formation [107]. As the microtubule cytoskeleton plays pleiotropic roles in cells, including cell migration, polarity, intracellular transport, it would not be surprising that microtubule alterations downstream of extra centrosomes could have a broader impact on tumourigensis than previously anticipated.

3.5 Therapeutic Advantage

Cells with extra centrosomes have unique survival requirements, such as centrosome clustering. De-clustering extra centrosomes, which leads to multipolar mitoses and massive aneuploidy, causes cell cycle arrest and cell death [41, 55, 108, 109]. Thus, centrosome clustering is an attractive therapeutic target for cancer treatment.

Several drugs have been proposed to de-cluster extra centrosomes. Griseofulvin, a nontoxic antifungal, has been shown to de-cluster centrosomes, as well as being anti-mitotic in a variety of human cancer cell lines [109], and selectively kills tumour cells at concentrations that are non-toxic to normal cells. A 2′-substituted derivative of griseofulvin, which has the highest potency, is also the most efficient at de-clustering supernumerary centrosomes, but further work needs to be done to better understand the correlation between the cell death observed and de-clustering [110]. Another set of compounds with potential for being used as de-clustering drugs are phenanthrene-derived poly-ADP-ribose polymerase (PARP) inhibitors [111]. PARP-1's normal function is to detect and initiate base-excision repair of DNA damage, as well as activation of DNA damage checkpoints and cell-death via apoptosis. In addition, different PARP related proteins have been identified as being required for centrosome clustering [55, 112]. In support of this, tumours with supernumerary centrosomes treated with the PARP-1 inhibitor PJ-34 have increased multipolar spindles, mitotic catastrophe and death [113]. In contrast, non-transformed cells treated under similar conditions showed no spindle morphology changes, and cell viability was maintained [114]. Interestingly, PJ-34 was recently shown to suppress the expression of KIFC1/HSET, essential for centrosome clustering, which could explain the de-clustering phenotype observed in breast cancer cell lines upon treatment with this inhibitor [115].

The identification of important proteins such as KIFC1/HSET that are involved in centrosome clustering, but do not appear to have an essential function in most normal cells, make them attractive drug targets. Newly developed KIFC1/HSET inhibitors induce multipolar divisions via centrosome de-clustering in cell lines harbouring supernumerary centrosomes, but not in cells with normal centrosome numbers [116]. Likewise, another allosteric KIFC1/HSET inhibitor, CW069, was shown to induce multipolar divisions in a panel of cell lines with supernumerary centrosomes, and significantly impaired the viability of these cells. However, whilst not affecting the spindle formation of MCF-7 cells which have a normal complement of centrosomes, CW069 did impair their growth, suggesting some toxicity associated with this drug [117]. Thus, although targeting KIFC1/HSET is an attractive therapeutic strategy, developing less toxic and more specific inhibitors is essential to assess the effectiveness of such a strategy in vivo. In general, the effectiveness of these de-clustering inhibitors in treating tumours containing extra centrosomes in vivo is still unclear and further work is required to assess the validity of such strategy as a cancer therapy.

3.6 Concluding Remarks

Centrosome amplification is a widespread characteristic of both solid and haematological malignancies. However, centrosome amplification does not come without a cost, as these cells tend to arrest or undergo a catastrophic multipolar mitosis. It is now clear that cells require specific mechanisms that allow their proliferation and survival in the presence of these abnormalities. This opens a window of opportunity to develop strategies that selectively kill these cells. However, it is still unclear how prevalent the different clustering mechanisms are and whether cancer cells require to adapt to centrosome amplification. The novel in vivo models that are now available would be instrumental in addressing these questions. Because centrosome amplification can accelerate tumourigenesis, induce CIN and promote cell invasion, defining strategies that could eradicate cells containing extra centrosomes within a tumour could positively impact cancer treatment. The discovery that centrosome amplification could redirect the microtubule cytoskeleton to induce cell invasion, emphasises how little we know about the impact of extra centrosomes and the cytoskeleton in cancer. To fully understanding the impact of centrosome amplification to tumourigenesis, it is imperative to dissect the cellular changes associated with extra centrosomes, in particular alterations of the microtubule cytoskeleton.

Acknowledgments We thank Hefin Rhys and members of the Godinho laboratory, Teresa Arnandis, Pedro Monteiro and Sophie Adams, for critically reading this manuscript. We apologise for the many authors whose work we were unable to cite due to space limitations. A.D.R. is supported by a CRUK PhD fellowship. S.A.G. is supported by the Higher Education Funding Council for England and a Medical Research Council Grant. S.A.G. is a Lister Institute Prize Fellow.

References

1. Bettencourt-Dias M, Glover DM (2007) Centrosome biogenesis and function: centrosomics brings new understanding. Nat Rev Mol Cell Biol 8:451–463
2. Kim S, Dynlacht BD (2013) Assembling a primary cilium. Curr Opin Cell Biol 25:506–511
3. Agircan FG, Schiebel E, Mardin BR (2014) Separate to operate: control of centrosome positioning and separation. Philos Trans R Soc Lond Ser B Biol Sci 369:20130461
4. Fu J, Hagan IM, Glover DM (2015) The centrosome and its duplication cycle. Cold Spring Harb Perspect Biol 7:a015800
5. Nigg EA, Stearns T (2011) The centrosome cycle: centriole biogenesis, duplication and inherent asymmetries. Nat Cell Biol 13:1154–1160. doi:10.1038/ncb2345
6. Boveri T (2008) Concerning the origin of malignant tumours by Theodor Boveri. Translated and annotated by Henry Harris. J Cell Sci 121(Supplement 1):1–84
7. Bignold LP, Coghlan BLD, Jersmann HPA (2006) Hansemann, Boveri, chromosomes and the gametogenesis-related theories of tumours. Cell Biol Int 30:640–644
8. Pickett-Heaps J (1971) The autonomy of the centriole: fact or fallacy? Cytobios 3:205–214
9. Dietz R (1966) The dispensability of the centrioles in the spermatocyte division of pales ferruginea (Nematocera). In: Darlington CD, Lewis KR (eds) Chromosomes today. Oliver and Boyd, London, pp 161–166

10. Cavazza T, Vernos I (2015) The RanGTP pathway: from nucleo-cytoplasmic transport to spindle assembly and beyond. Front Cell Dev Biol 3:82
11. Heald R, Tournebize R, Blank T, Sandaltzopoulos R, Becker P, Hyman A et al (1996) Self-organization of microtubules into bipolar spindles around artificial chromosomes in Xenopus egg extracts. Nature 382:420–425
12. Karsenti E, Newport J, Kirschner M (1984) Respective roles of centrosomes and chromatin in the conversion of microtubule arrays from interphase to metaphase. J Cell Biol 99:47s–54s
13. Manandhar G, Schatten H, Sutovsky P (2005) Centrosome reduction during gametogenesis and its significance. Biol Reprod 72:2–13
14. Basto R, Lau J, Vinogradova T, Gardiol A, Woods CG, Khodjakov A et al (2006) Flies without centrioles. Cell 125:1375–1386
15. Debec A (1978) Haploid cell cultures of *Drosophila melanogaster*. Nature 274:255–256
16. Sir J-H, Pütz M, Daly O, Morrison CG, Dunning M, Kilmartin JV et al (2013) Loss of centrioles causes chromosomal instability in vertebrate somatic cells. J Cell Biol 203:747–756
17. Bettencourt-Dias M, Rodrigues-Martins A, Carpenter L, Riparbelli M, Lehmann L, Gatt MK et al (2005) SAK/PLK4 is required for centriole duplication and flagella development. Curr Biol 15:2199–2207
18. Habedanck R, Stierhof Y-D, Wilkinson CJ, Nigg EA (2005) The Polo kinase Plk4 functions in centriole duplication. Nat Cell Biol 7:1140–1146
19. Wong YL, Anzola JV, Davis RL, Yoon M, Motamedi A, Kroll A et al (2015) Reversible centriole depletion with an inhibitor of Polo-like kinase 4. Science 348:1155–1160
20. Chan JY (2011) A clinical overview of centrosome amplification in human cancers. Int J Biol Sci 7:1122–1144
21. Godinho SA, Pellman D (2014) Causes and consequences of centrosome abnormalities in cancer. Philos Trans R Soc Lond Ser B Biol Sci 369:20130467
22. Boveri T (1888) Zellen-studien 2: die befrunchtung und telung des eies von Ascaris megalocephala. Jenaische Zeitschr Med Naturw 22:685–882
23. Boveri T. Uber die befunchtung der eier von Ascaris megalocephala. Sitz Ber Ges Morph Phys Munchen. 1887;3:71–80.
24. Hansemann D (1890) Ueber asymmetrische zelltheilung in epithelkrebsen und deren biologische bedeutung. Arch Für Pathol Anat Und Physiol Und Für Klin Med 119:299–326
25. Giehl M, Fabarius A, Frank O, Hochhaus A, Hafner M, Hehlmann R et al (2005) Centrosome aberrations in chronic myeloid leukemia correlate with stage of disease and chromosomal instability. Leukemia 19:1192–1197
26. Hsu L-C, Kapali M, DeLoia JA, Gallion HH (2005) Centrosome abnormalities in ovarian cancer. Int J Cancer 113:746–751
27. Krämer A, Neben K, Ho AD (2005) Centrosome aberrations in hematological malignancies. Cell Biol Int 29:375–383
28. Lingle WL, Lutz WH, Ingle JN, Maihle NJ, Salisbury JL (1998) Centrosome hypertrophy in human breast tumors: implications for genomic stability and cell polarity. Proc Natl Acad Sci U S A 95:2950–2955
29. Nigg EA (2002) Centrosome aberrations: cause or consequence of cancer progression? Nat Rev Cancer 2:815–825
30. Pihan GA, Purohit A, Wallace J, Knecht H, Woda B, Quesenberry P et al (1998) Centrosome defects and genetic instability in malignant tumors. Cancer Res 58:3974–3985
31. Sato N, Mizumoto K, Nakamura M, Nakamura K, Kusumoto M, Niiyama H et al (1999) Centrosome abnormalities in pancreatic ductal carcinoma. Clin Cancer Res 5:963–970
32. D'Assoro AB, Barrett SL, Folk C, Negron VC, Boeneman K, Busby R et al (2002) Amplified centrosomes in breast cancer: a potential indicator of tumor aggressiveness. Breast Cancer Res Treat 75:25–34
33. Pihan GA, Wallace J, Zhou Y, Doxsey SJ (2003) Centrosome abnormalities and chromosome instability occur together in pre-invasive carcinomas. Cancer Res 63:1398–1404

34. Segat D, Cassaro M, Dazzo E, Cavallini L, Romualdi C, Salvador R et al (2010) Pericentriolar material analyses in normal esophageal mucosa, Barrett's metaplasia and adenocarcinoma. Histol Histopathol 25:551–560
35. Yamamoto Y, Matsuyama H, Furuya T, Oga A, Yoshihiro S, Okuda M et al (2004) Centrosome hyperamplification predicts progression and tumor recurrence in bladder cancer. Clin Cancer Res 10:6449–6455
36. Kleylein-Sohn J, Westendorf J, Le Clech M, Habedanck R, Stierhof Y-D, Nigg EA (2007) Plk4-induced centriole biogenesis in human cells. Dev Cell 13:190–202
37. Basto R, Brunk K, Vinadogrova T, Peel N, Franz A, Khodjakov A et al (2008) Centrosome amplification can initiate tumorigenesis in flies. Cell 133:1032–1042
38. Serçin Ö, Larsimont J-C, Karambelas AE, Marthiens V, Moers V, Boeckx B et al (2015) Transient PLK4 overexpression accelerates tumorigenesis in p53-deficient epidermis. Nat Cell Biol 18:100
39. Coelho PA, Bury L, Shahbazi MN, Liakath-Ali K, Tate PH, Wormald S et al (2015) Overexpression of Plk4 induces centrosome amplification, loss of primary cilia and associated tissue hyperplasia in the mouse. Open Biol 5:150209
40. Vitre B, Holland AJ, Kulukian A, Shoshani O, Hirai M, Wang Y et al (2015) Chronic centrosome amplification without tumorigenesis. Proc Natl Acad Sci U S A 112:E6321–E6330
41. Ganem NJ, Godinho SA, Pellman D (2009) A mechanism linking extra centrosomes to chromosomal instability. Nature 460:278–282
42. Silkworth WT, Nardi IK, Scholl LM, Cimini D (2009) Multipolar spindle pole coalescence is a major source of kinetochore mis-attachment and chromosome mis-segregation in cancer cells. PLoS One 4:e6564
43. Mahjoub MR, Stearns T (2012) Supernumerary centrosomes nucleate extra cilia and compromise primary cilium signaling. Curr Biol 22:1628–1634
44. Godinho SA, Picone R, Burute M, Dagher R, Su Y, Leung CT et al (2014) Oncogene-like induction of cellular invasion from centrosome amplification. Nature 510:167–171
45. Krzywicka-Racka A, Sluder G (2011) Repeated cleavage failure does not establish centrosome amplification in untransformed human cells. J Cell Biol 194:199–207
46. Andreassen PR, Lohez OD, Lacroix FB, Margolis RL (2001) Tetraploid state induces p53-dependent arrest of nontransformed mammalian cells in G1. Mol Biol Cell 12:1315–1328
47. Fujiwara T, Bandi M, Nitta M, Ivanova EV, Bronson RT, Pellman D (2005) Cytokinesis failure generating tetraploids promotes tumorigenesis in p53-null cells. Nature 437:1043–1047
48. Holland AJ, Fachinetti D, Zhu Q, Bauer M, Verma IM, Nigg EA et al (2012) The autoregulated instability of Polo-like kinase 4 limits centrosome duplication to once per cell cycle. Genes Dev 26:2684–2689
49. Ganem NJ, Cornils H, Chiu S-Y, O'Rourke KP, Arnaud J, Yimlamai D et al (2014) Cytokinesis failure triggers hippo tumor suppressor pathway activation. Cell 158:833–848
50. Bolgioni AF, Ganem NJ (2015) The interplay between centrosomes and the hippo tumor suppressor pathway. Chromosom Res 24:93–104
51. Sander EE, ten Klooster JP, van Delft S, van der Kammen RA, Collard JG (1999) Rac downregulates rho activity: reciprocal balance between both GTPases determines cellular morphology and migratory behavior. J Cell Biol 147:1009–1022
52. Ring D, Hubble R, Kirschner M (1982) Mitosis in a cell with multiple centrioles. J Cell Biol 94:549–556
53. Brinkley BR (2001) Managing the centrosome numbers game: from chaos to stability in cancer cell division. Trends Cell Biol 11:18–21
54. Godinho SA, Kwon M, Pellman D (2009) Centrosomes and cancer: how cancer cells divide with too many centrosomes. Cancer Metastasis Rev 28:85–98
55. Kwon M, Godinho SA, Chandhok NS, Ganem NJ, Azioune A, Thery M et al (2008) Mechanisms to suppress multipolar divisions in cancer cells with extra centrosomes. Genes Dev 22:2189–2203

56. Quintyne NJ, Reing JE, Hoffelder DR, Gollin SM, Saunders WS (2005) Spindle multipolarity is prevented by centrosomal clustering. Science 307:127–129
57. Leber B, Maier B, Fuchs F, Chi J, Riffel P, Anderhub S et al (2010) Proteins required for centrosome clustering in cancer cells. Sci Transl Med 2:33ra38
58. Breuer M, Kolano A, Kwon M, Li C-C, Tsai T-F, Pellman D et al (2010) HURP permits MTOC sorting for robust meiotic spindle bipolarity, similar to extra centrosome clustering in cancer cells. J Cell Biol 191:1251–1260
59. Fielding AB, Dobreva I, McDonald PC, Foster LJ, Dedhar S (2008) Integrin-linked kinase localizes to the centrosome and regulates mitotic spindle organization. J Cell Biol 180:681–689
60. Sabino D, Gogendeau D, Gambarotto D, Nano M, Pennetier C, Dingli F et al (2015) Moesin is a major regulator of centrosome behavior in epithelial cells with extra centrosomes. Curr Biol 25:879–889
61. Maresca TJ, Salmon ED (2010) Welcome to a new kind of tension: translating kinetochore mechanics into a wait-anaphase signal. J Cell Sci 123:825–835
62. Nezi L, Musacchio A (2009) Sister chromatid tension and the spindle assembly checkpoint. Curr Opin Cell Biol 21:785–795
63. Weaver BAA, Cleveland DW (2005) Decoding the links between mitosis, cancer, and chemotherapy: the mitotic checkpoint, adaptation, and cell death. Cancer Cell 8:7–12
64. Lara-Gonzalez P, Westhorpe FG, Taylor SS (2012) The spindle assembly checkpoint. Curr Biol 22:R966–R980
65. Yang Z, Loncarek J, Khodjakov A, Rieder CL (2008) Extra centrosomes and/or chromosomes prolong mitosis in human cells. Nat Cell Biol 10:748–751
66. Lambert RA (1913) Comparative studies upon cancer cells and normal cells: II. The character of growth in vitro with special reference to cell division. J Exp Med 17:499–510
67. Levine DS, Sanchez CA, Rabinovitch PS, Reid BJ (1991) Formation of the tetraploid intermediate is associated with the development of cells with more than four centrioles in the elastase-simian virus 40 tumor antigen transgenic mouse model of pancreatic cancer. Proc Natl Acad Sci U S A 88:6427–6431
68. Sluder G, Thompson EA, Miller FJ, Hayes J, Rieder CL (1997) The checkpoint control for anaphase onset does not monitor excess numbers of spindle poles or bipolar spindle symmetry. J Cell Sci 110(Pt 4):421–429
69. Ciferri C, Musacchio A, Petrovic A (2007) The Ndc80 complex: hub of kinetochore activity. FEBS Lett 581:2862–2869
70. Silljé HHW, Nagel S, Körner R, Nigg EA (2006) HURP is a ran-importin beta-regulated protein that stabilizes kinetochore microtubules in the vicinity of chromosomes. Curr Biol 16:731–742
71. Wong J, Fang G (2006) HURP controls spindle dynamics to promote proper interkinetochore tension and efficient kinetochore capture. J Cell Biol 173:879–891
72. Logarinho E, Maffini S, Barisic M, Marques A, Toso A, Meraldi P et al (2012) CLASPs prevent irreversible multipolarity by ensuring spindle-pole resistance to traction forces during chromosome alignment. Nat Cell Biol 14:295–303
73. Mitchison TJ (1992) Actin based motility on retraction fibers in mitotic PtK2 cells. Cell Motil Cytoskeleton 22:135–151
74. Théry M, Racine V, Pépin A, Piel M, Chen Y, Sibarita J-B et al (2005) The extracellular matrix guides the orientation of the cell division axis. Nat Cell Biol 7:947–953
75. Fink J, Carpi N, Betz T, Bétard A, Chebah M, Azioune A et al (2011) External forces control mitotic spindle positioning. Nat Cell Biol 13:771–778
76. Kwon M, Bagonis M, Danuser G, Pellman D (2015) Direct microtubule-binding by myosin-10 orients centrosomes toward retraction fibers and subcortical actin clouds. Dev Cell 34:323–337

77. Rosenblatt J, Cramer LP, Baum B, McGee KM (2004) Myosin II-dependent cortical movement is required for centrosome separation and positioning during mitotic spindle assembly. Cell 117:361–372
78. Krämer A, Maier B, Bartek J (2011) Centrosome clustering and chromosomal (in)stability: a matter of life and death. Mol Oncol 5:324–335
79. Marthiens V, Piel M, Basto R (2012) Never tear us apart--the importance of centrosome clustering. J Cell Sci 125:3281–3292
80. Pines J (2011) Cubism and the cell cycle: the many faces of the APC/C. Nat Rev Mol Cell Biol 12:427–438
81. Drosopoulos K, Tang C, Chao WCH, Linardopoulos S (2014) APC/C is an essential regulator of centrosome clustering. Nat Commun 5:3686
82. Endow SA, Komma DJ (1998) Assembly and dynamics of an anastral:astral spindle: the meiosis II spindle of Drosophila oocytes. J Cell Sci 111(Pt 17):2487–2495
83. Iwao Y, Murakawa T, Yamaguchi J, Yamashita M (2002) Localization of gamma-tubulin and cyclin B during early cleavage in physiologically polyspermic newt eggs. Develop Growth Differ 44:489–499
84. Mikeladze-Dvali T, von Tobel L, Strnad P, Knott G, Leonhardt H, Schermelleh L et al (2012) Analysis of centriole elimination during *C. elegans* oogenesis. Development 139:1670–1679
85. Kim DY, Roy R (2006) Cell cycle regulators control centrosome elimination during oogenesis in *Caenorhabditis elegans*. J Cell Biol 174:751–757
86. Chiba S, Okuda M, Mussman JG, Fukasawa K (2000) Genomic convergence and suppression of centrosome hyperamplification in primary p53−/− cells in prolonged culture. Exp Cell Res 258:310–321
87. Gräf R, Euteneuer U, Ho T-H, Rehberg M (2003) Regulated expression of the centrosomal protein DdCP224 affects microtubule dynamics and reveals mechanisms for the control of supernumerary centrosome number. Mol Biol Cell 14:4067–4074
88. Zyss D, Gergely F (2009) Centrosome function in cancer: guilty or innocent? Trends Cell Biol 19:334–346
89. Wang Y, Jin F, Higgins R, McKnight K (2014) The current view for the silencing of the spindle assembly checkpoint. Cell Cycle 13:1694–1701
90. Cimini D (1786) Merotelic kinetochore orientation, aneuploidy, and cancer. Biochim Biophys Acta 2008:32–40
91. Janssen A, van der Burg M, Szuhai K, Kops GJPL, Medema RH (2011) Chromosome segregation errors as a cause of DNA damage and structural chromosome aberrations. Science 333:1895–1898
92. Crasta K, Ganem NJ, Dagher R, Lantermann AB, Ivanova EV, Pan Y et al (2012) DNA breaks and chromosome pulverization from errors in mitosis. Nature 482:53–58
93. Hatch EM, Fischer AH, Deerinck TJ, Hetzer MW (2013) Catastrophic nuclear envelope collapse in cancer cell micronuclei. Cell 154:47–60
94. Stephens PJ, Greenman CD, Fu B, Yang F, Bignell GR, Mudie LJ et al (2011) Massive genomic rearrangement acquired in a single catastrophic event during cancer development. Cell 144:27–40
95. Zhang C-Z, Spektor A, Cornils H, Francis JM, Jackson EK, Liu S et al (2015) Chromothripsis from DNA damage in micronuclei. Nature 522:179–184
96. Weaver BAA, Silk AD, Montagna C, Verdier-Pinard P, Cleveland DW (2007) Aneuploidy acts both oncogenically and as a tumor suppressor. Cancer Cell 11:25–36
97. Holland AJ, Cleveland DW (2012) Losing balance: the origin and impact of aneuploidy in cancer. EMBO Rep 13:501–514
98. Marthiens V, Rujano MA, Pennetier C, Tessier S, Paul-Gilloteaux P, Basto R (2013) Centrosome amplification causes microcephaly. Nat Cell Biol 15:731–740
99. Thompson SL, Compton DA (2011) Chromosomes and cancer cells. Chromosom Res 19:433–444

100. Betschinger J, Knoblich JA (2004) Dare to be different: asymmetric cell division in Drosophila, *C. elegans* and vertebrates. Curr Biol 14:R674–R685
101. Morrison SJ, Kimble J (2006) Asymmetric and symmetric stem-cell divisions in development and cancer. Nature 441:1068–1074
102. Castellanos E, Dominguez P, Gonzalez C (2008) Centrosome dysfunction in Drosophila neural stem cells causes tumors that are not due to genome instability. Curr Biol 18:1209–1214
103. Caussinus E, Gonzalez C (2005) Induction of tumor growth by altered stem-cell asymmetric division in *Drosophila melanogaster*. Nat Genet 37:1125–1129
104. Reina J, Gonzalez C (2014) When fate follows age: unequal centrosomes in asymmetric cell division. Philos Trans R Soc Lond Ser B Biol Sci 369:20130466
105. Siller KH, Doe CQ (2009) Spindle orientation during asymmetric cell division. Nat Cell Biol 11:365–374
106. Kulukian A, Holland AJ, Vitre B, Naik S, Cleveland DW, Fuchs E (2015) Epidermal development, growth control, and homeostasis in the face of centrosome amplification. Proc Natl Acad Sci U S A 112:E6311–E6320
107. Kusner EJ, Ferro LS, Yu Z, Bautch VL (2016) Excess centrosomes perturb dynamic endothelial cell repolarization during blood vessel formation. Mol Biol Cell 27(12):1911–1920
108. Karna P, Rida PCG, Pannu V, Gupta KK, Dalton WB, Joshi H et al (2011) A novel microtubule-modulating noscapinoid triggers apoptosis by inducing spindle multipolarity via centrosome amplification and declustering. Cell Death Differ 18:632–644
109. Rebacz B, Larsen TO, Clausen MH, Rønnest MH, Löffler H, Ho AD et al (2007) Identification of griseofulvin as an inhibitor of centrosomal clustering in a phenotype-based screen. Cancer Res 67:6342–6350
110. Raab MS, Breitkreutz I, Anderhub S, Rønnest MH, Leber B, Larsen TO et al (2012) GF-15, a novel inhibitor of centrosomal clustering, suppresses tumor cell growth in vitro and in vivo. Cancer Res 72:5374–5385
111. Castiel A, Visochek L, Mittelman L, Dantzer F, Izraeli S, Cohen-Armon M (2011) A phenanthrene derived PARP inhibitor is an extra-centrosomes de-clustering agent exclusively eradicating human cancer cells. BMC Cancer 11:412
112. Tong WM, Yang YG, Cao WH, Galendo D, Frappart L, Shen Y et al (2007) Poly(ADP-ribose) polymerase-1 plays a role in suppressing mammary tumourigenesis in mice. Oncogene 26:3857–3867
113. Castiel A, Visochek L, Mittelman L, Zilberstein Y, Dantzer F, Izraeli S et al (2013) Cell death associated with abnormal mitosis observed by confocal imaging in live cancer cells. J Vis Exp (78):e50568
114. Pannu V, Rida PCG, Celik B, Turaga RC, Ogden A, Cantuaria G et al (2014) Centrosome-declustering drugs mediate a two-pronged attack on interphase and mitosis in supercentrosomal cancer cells. Cell Death Dis 5:e1538
115. Li Y, Lu W, Chen D, Boohaker RJ, Zhai L, Padmalayam I et al (2015) KIFC1 is a novel potential therapeutic target for breast cancer. Cancer Biol Ther 16:1316–1322
116. Wu J, Mikule K, Wang W, Su N, Petteruti P, Gharahdaghi F et al (2013) Discovery and mechanistic study of a small molecule inhibitor for motor protein KIFC1. ACS Chem Biol 8:2201–2208
117. Watts CA, Richards FM, Bender A, Bond PJ, Korb O, Kern O et al (2013) Design, synthesis, and biological evaluation of an allosteric inhibitor of HSET that targets cancer cells with supernumerary centrosomes. Chem Biol 20:1399–1410

Chapter 4
Kinetochore Malfunction in Human Pathologies

Bas de Wolf and Geert J.P.L. Kops

Abstract The cell cycle culminates in mitosis with the purpose of dividing the cell's DNA content equally over two daughter cells. Error-free segregation relies on correct connections between chromosomes and spindle microtubules. Kinetochores are complex multi-protein assemblies that mediate these connections and are the platforms for attachment-error-correction and spindle assembly checkpoint signaling. Proper kinetochore function is therefore key in preventing aneuploidization. Mutations in genes encoding kinetochore proteins are associated with several severe developmental disorders associated with microcephaly, and kinetochore defects contribute to chromosomal instability in certain cancers. This chapter gives an overview of the processes necessary for faithful chromosome segregation and how kinetochore malfunction causes various human pathologies.

Keywords Aneuploidy • Development • Kinetochore • Cell division • Chromosomal instability • Microcephaly • Cancer

B. de Wolf
Hubrecht Institute – KNAW (Royal Netherlands Academy of Arts and Sciences), Uppsalalaan 8, 3584 CT Utrecht, The Netherlands
e-mail: b.dewolf@hubrecht.eu

G.J.P.L. Kops (✉)
Hubrecht Institute – KNAW (Royal Netherlands Academy of Arts and Sciences), Uppsalalaan 8, 3584 CT Utrecht, The Netherlands

Cancer Genomics Netherlands, University Medical Center Utrecht, 3584 CG Utrecht, The Netherlands

Center for Molecular Medicine, University Medical Center Utrecht, 3584 CG Utrecht, The Netherlands
e-mail: g.kops@hubrecht.eu

M. Gotta, P. Meraldi (eds.), *Cell Division Machinery and Disease*, Advances in Experimental Medicine and Biology 1002, DOI 10.1007/978-3-319-57127-0_4

4.1 Mitosis and Mitotic Surveillance Mechanisms

During cell division, accurate chromosome segregation ensures that each daughter cell receives a complete and identical copy of the genetic code, a process that is essential for cell and organismal viability. Chromosome segregation errors result in apoptosis or an abnormal DNA content referred to as aneuploidy. Aneuploidy, in turn, leads to abnormal gene dosage, causing human conditions such as Down syndrome and Turner syndrome, exposes detrimental recessive mutations, and is associated with cancer [1, 2].

Chromosome segregation is mediated by microtubules that emanate from the spindle poles and attach to chromosomes through large protein complexes, known as kinetochores, that are assembled on centromeric DNA [3]. Vital for correct segregation is the stable attachment of each of the sister chromatids to the opposing spindle poles (*amphitelic* attachments), a state called bi-orientation. The initial capture of microtubules by kinetochores is stochastic and individual kinetochores typically bind multiple microtubules (a fully attached HeLa cell kinetochore is bound by approximately 20 microtubules, also known as k-fibers) [4]. This frequently results in erroneous attachments configurations such as *merotely* (one kinetochore bound to microtubules from both spindle poles) or *syntely* (both kinetochores bound to microtubules from the same spindle pole). Such attachments are corrected by the error-correction machinery, orchestrated by the Aurora B kinase [5].

Until proper bi-orientation has been achieved, the lack of stable attachments needs to be communicated to the cell cycle machinery, which cannot be permitted to prematurely initiate chromosome segregation (anaphase) and exit from mitosis. Anaphase and mitotic exit are initiated by the E3 ubiquitin ligase anaphase-promoting complex (APC/C) [6]. The surveillance mechanism responsible for this *wait-anaphase* signal is the spindle assembly checkpoint (SAC, also known as the mitotic checkpoint). The SAC is a feedback-control system that monitors kinetochore-microtubule attachments and if needed delays anaphase onset by generating an inhibitor of the APC/C, known as the mitotic checkpoint complex (MCC) [7, 8]. Once the last kinetochores achieve stable attachments, inhibitor production is shut down, the APC/C is liberated, and anaphase can ensue.

4.1.1 Kinetochores

Kinetochores are complex multi-subunit structures that consist of more than a 100 different proteins in human cells, each at up to hundreds of copies per kinetochore. Metazoan and fungal kinetochores are built on specialized centromeric nucleosomes that contain the histone H3 variant CENP-A/CenH3 [9]. These nucleosomes direct assembly of an inner-kinetochore network known as the CCAN (constitutive centromere-associated network) that is composed of 16 'CENP' subunits (CENP-C, CENP-S-X-T-W, CENP-H-I-K-M, CENP-L-N, and CENP-O-P-Q-R-U) [3, 9].

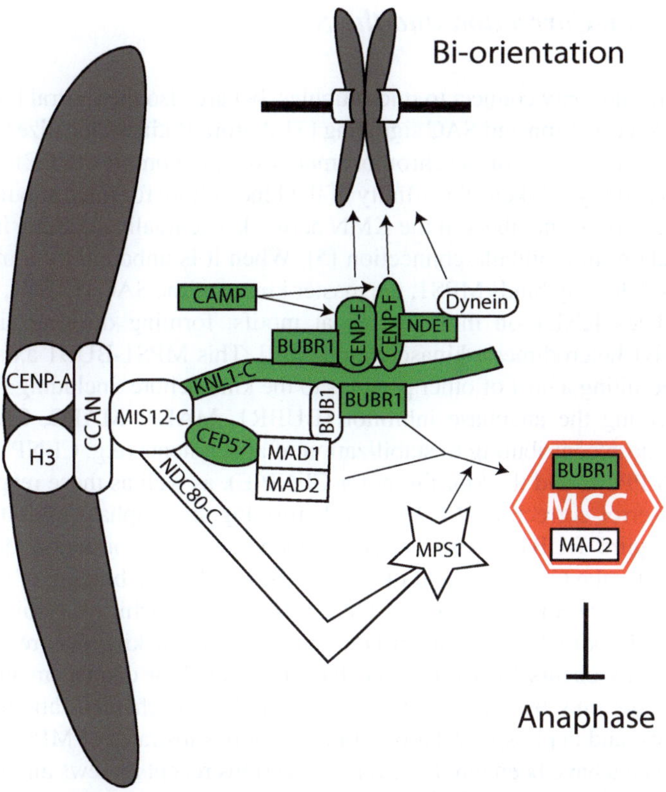

Fig. 4.1 SAC signaling and contributions to chromosome bi-orientation at the kinetochore. At an unattached kinetochore, MPS1 phosphorylation recruits BUB1, BUBR1, MAD1 and MAD2 to the kinetochore and triggers MCC formation. In parallel CAMP, CENP-E, CENP-F and NDE1-Dynein are also recruited to the kinetochore, where they (and BUBR1) contribute to chromosome bi-orientation (*Note*: the interactions that contribute to localization of CAMP, CENP-E and CEN-F are not known). The kinetochore proteins that are associated with disease (as discussed in Sect. 4.3) are depicted in green

CCAN complexes in turn recruit outer kinetochore protein subcomplexes including the KMN network that contains direct microtubule-binding proteins (Fig. 4.1). The KMN network consist of the subcomplexes KNL1-C (KNL1-Zwint1), Mis12-C (Mis12, Dsn1, Nsl1, Pmf1), and Ndc80-C (Ndc80/Hec1, Nuf2, Spc24, Spc25). The Ndc80-C is the main interface of kinetochore-microtubule interactions responsible for load-bearing attachments, with some contributions from the KNL1 protein [10]. The MIS12-C is required to properly target Ndc80-C and KNL1-C to the kinetochore by connecting them to the CCAN, and enhances their microtubule-binding affinities [11].

4.1.2 Error-Correction and the SAC

Kinetochores not only connect to microtubules, but are also the central hubs for integrating error-correction and SAC signaling [7]. Aurora B kinase localizes to the inner centromere as a member of the chromosomal passenger complex (CPC) during early mitosis. Its activity weakens the affinity of the kinetochore for microtubules by phosphorylating various members of the KMN network, eventually resulting in release of the kinetochore-microtubule connection [5]. When it is unbound by a microtubule, the Ndc80-C directly binds MPS1, the master kinase of the SAC [12, 13], which then phosphorylates KNL1 on multiple repeat motifs, forming docking sites for the BUB3-BUB1 heterodimeric kinase complex [8]. This MPS1-BUB1 axis is responsible for recruiting a host of other proteins to the kinetochore, including those essential for forming the anaphase inhibitor (BUBR1, MAD1, MAD2, CDC20, RZZ complex), those contributing to stabilization of attachments (e.g. CENP-F, BUBR1-bound B56-PP2A, and the kinesin motor CENP-E), as well as those responsible for attachment-error-correction [14–19]. While forming the anaphase inhibitor does not seem to require BUB1 kinase activity, attachment-error-correction does: BUB1 phosphorylates the C-terminal tail of Histone H2B, allowing binding of adaptor proteins of the Aurora B kinase [20, 21]. Once a kinetochore achieves stable attachment to spindle microtubules, the surveillance mechanisms at kinetochores have to be quenched. This occurs by various mechanisms including displacement of MPS1, removal of proteins by the Dynein motor complex, mechanical alteration within kinetochores, and dephosphorylation of kinetochore substrates of MPS1 and Aurora B. These events have been amply covered in various recent reviews and we refer the reader to those for more details [22–25].

4.1.3 Microtubule Motors at Kinetochores

Chromosome movements and stability of kinetochore-microtubule attachments involve kinetochore-localized motor proteins (Fig. 4.1). The plus-end directed kinesin CENP-E, for example, promotes chromosome congression to the spindle equator and contributes to full stability of attachment [26]. It is unclear how it localizes to kinetochores, but one of its binding partners there is BUBR1 [27]. Kinetochores also bind the minus-end directed motor complex Dynein, where it promotes microtubule attachment, chromosome movement during anaphase, and spindle checkpoint silencing [26, 28]. Dynein binds to kinetochores via several receptors, most notably the RZZ-Spindly and the CENPF-NdeL1-Nde1 pathways [28]. How these pathways cooperate in ensuring proper Dynein localization and function is currently unclear.

4.2 Aneuploidy and Chromosomal Instability

Defects in any of the kinetochore-related processes can cause whole-chromosome segregation errors in mitosis or meiosis and lead to aneuploid progeny [29]. Aneuploidy (a chromosome number that is not an exact multiple of the haploid set) is a hallmark of human cancers, and is a common cause for spontaneous abortions and developmental disorders [2]. Specific aneuploidies (for instance trisomy or monosomy of a certain chromosome) present throughout the body are referred to as *constitutional* aneuploidies. Such aneuploidies are caused by chromosome segregation errors during germ cell formation in one of the parents [30]. *A somatic* aneuploidy, on the other hand, is the result of a mitotic error and is present in only a subset of an individual's cells. Most constitutional aneuploidies cause embryonic lethality, with trisomy 21 (also known as Down Syndrome, MIM 190685) being the most notable exception. The consequences of somatic aneuploidies can also be severe. Losing chromosome X during early embryonic development (monosomy X), for instance, can cause Mosaic Turner Syndrome in women, a milder version of Turner Syndrome which is characterized by short stature and an early loss of ovarian function [31]. Another example is Mosaic Variegated Aneuploidy (MVA), which will be more extensively discussed in Sect. 4.3.5. Other disorders characterized by increased aneuploidy include those with defects in DNA repair mechanisms, such as Ataxia Telangiectasia (MIM 208900) and Nijmegen Breakage Syndrome (MIM 251260) [32], and cohesinopathies, such as Cornelia de Lange Syndrome (MIM 122470), Roberts Syndrome (MIM 268300) and Warsaw Breakage Syndrome (MIM 613398) [33, 34].

Aneuploidy is a result of chromosome segregation errors, and cells prone to missegregate chromosomes (either as a whole or in parts) are referred to as displaying a chromosomal instability (CIN) phenotype. It is important to emphasize that aneuploidy represents a karyotype state, while CIN refers to the elevated chance of making segregation errors. Down syndrome patient cells are aneuploid but not CIN. While this seems intuitive, CIN is often erroneously inferred from aneuploid karyotypes, most notably in cancer studies.

On a cellular level, CIN and aneuploidy can have numerous effects. Errors in chromosome segregation lead to p53 activation and G1 arrest [35], likely through DNA damage [36] or elevated ROS levels [37]. Gains or losses of chromosomes in most cases result in a corresponding change in the expression of both genes located on that chromosome [38], as well as genes on chromosomes other than the aneuploid ones, the latter likely through alteration of transcriptional networks [38, 39]. In addition to these chromosome-specific changes, studies of aneuploid yeast, plant, mice and human cells have revealed conserved gene expression responses to the aneuploid state, most notably those related to cell growth, proliferation, nucleic-acid metabolism, and stress responses [40]. Strikingly, complex aneuploidies elicit the same transcriptional changes as a trisomy [41]. This all ties in with an aneuploidy-induced proteotoxic stress response that is caused by limitations in protein folding and protein turnover, leading to an accumulation of unfolded and misfolded proteins [42]. All in all, aneuploidy usually results in reduced proliferation.

On top of whole-chromosome aneuploidy, chromosome segregation errors can cause additional genomic insults, depending on the type of error made. For example, chromosomes that lag behind in anaphase or that end up in micronuclei can acquire extensive DNA damage by various mechanisms, including cytokinesis-induced double-strand breaks and incomplete replication [36, 43]. The results can be devastating: unbalanced translocations and even chromosome shattering, also known as chromothripsis [36, 44, 45]. Errors in cell division can therefore lead to whole-chromosomal aneuploidy as well as structural aneuploidy, and may thus be at the root of complexes genomic alterations in some cancers and congenital diseases [46, 47].

Given these various far-reaching adverse effects, it is not surprising that aneuploid cells are rare in healthy tissues. Recent single-cell sequencing efforts have shown that the prevalence of aneuploid cells in healthy skin, liver and brain tissue is less than 5% [48, 49]. It is unknown if this is due to very high fidelity of the chromosome segregation process throughout all phases of development and in our adult stem cell populations, or whether aneuploid cells are cleared efficiently, as recently shown in a CIN mouse model [50].

4.3 Congenital Kinetochore Syndromes

Mutations in the kinetochore genes *KNL1*, *CENP-E*, *CENP-F*, *NDE1*, *CHAMP1*, *BUB1B*, and *CEP57* (Fig. 4.1, Table 4.1) are associated with various congenital diseases. Each of these syndromes is described in more detail below. Strikingly, a common characteristic of patients carrying these mutations is microcephaly: a condition in which the brain does not develop properly resulting in a smaller than normal head circumference. This seems to be a common outcome of mutations in components of mitotic structures (centrosomes, spindle, kinetochores). The molecular details behind this causal relationship are unknown, but a leading hypothesis is related to the neural stem cells, the cells that generate the various differentiated lineages in the brain, including neurons. Asymmetric divisions of these stem cells by spindle mispositioning, or aneuploidy by chromosome segregation errors can cause neural stem cell death and depletion of the stem cell pool [51, 52]. In support of this, Drosophila brains developed to a smaller size when aneuploidy was introduced in the neural stem cells [53]. A plausible scenario therefore is that kinetochore gene mutations cause microcephaly by aneuploidy-induced clearance of a fraction of the brain stem cells.

Of note: mutations in *SGO1* (a centromere and kinetochore protein important for sister-chromatid cohesion) are known to cause CAID syndrome (MIM 616201), a disorder characterized by chronic atrial and intestinal dysrhythmia [54]. However, since the patient mutations do not seem to affect kinetochore function, we have not included this disorder in the following discussion.

Table 4.1 Kinetochore gene mutations and associated disorders

Gene	Mutations	Disorder	Refs
CASC5/KNL1	c.6125G>A; p.M2041I c.6673-19T>A; p.M2225Ifs*7	MCPH4	[58]–[60, 132]
CENP-E	c.2797G>A:p.D933N c.4063A>G:p.K1355E	MCPH13	[61]
CENP-F	c.165_193del; p.N57Kfs*11 c.574-2A>C c.1744G>T; p.E582* c.2734G>T; p.E912X c.8692C>T; p.R2898* c.9280C>T; p.R3094*	Strømme syndrome	[65, 66, 133]
NDE1	c.-43-3548_83+622del	Microhydranencephaly	[134]
	c.83+1G>T; p.A29Qfs*114 c.684_685delAC; p.P229Wfs*85 c.733dup; p.L245Pfs*70	Microlissencephaly	[71, 73]
CHAMP1	c.635delC; p.P212Lfs*7 c.1192C>T; R398* c.1768C>T; p.Q590* c.1866_1867delCA; p.D622Efs*8	Mental retardation-40	[77, 135]
BUB1B	c.107G>A; p.R36Q c.580C>T; p.R194* c.670C>T; p.R224X c.IVS10−1G>T; p.Q467fs*483 c.IVS10-5A>G; W468fs*480 c.1649G>A; p.R550Q c.1833delT; p.F611fs*625 c.2211-2insGTTA; p.S738fs*753 c.2441G>A; p.R814H c.2530C>T; p.L844F c.2726T>C; p.I909T c.2763G>C; Q921H c.3035T>C; p.L1012P 6G3 haplotype	MVA	[83, 136]
CEP57	c.241C>T; p.R81* c.520_521delGA; p.E174Tfs*4 c.915_925dup11; p.L309Pfs*9	MVA	[85, 137]

c coding DNA sequence, *p* protein sequence

4.3.1 *Primary Microcephaly (MCPH):* CASC5/KNL1 *and* CENP-E

Autosomal recessive primary microcephaly (MCPH, MIM 251200) is a disorder characterized by an unusually small head circumference, intellectual disability, and in some cases mild facial dysmorphism and/or short stature. 16 loci are associated with MCPH; these include many genes that are important for proper cell division, such as those implicated in centrosome biogenesis and mitotic spindle assembly, as well as proteins that affect neuronal migration, DNA replication and the DNA damage response. Two of the 16 loci are genes encoding the kinetochore proteins CENP-E and CASC5/KNL1 (see below).

Mutations in *CASC5/KNL1* (hereafter named *KNL1*) are associated with the microcephaly variant syndrome MCPH4 (MIM 604321) (Table 4.1). MCPH4 patients exhibit mild to severe cognitive impairment, normal to short stature, and in some cases some dysmorphic features. KNL1 (kinetochore null 1) is a large, multifunctional kinetochore scaffold protein [11] that, together with Zwint1, forms the KNL1-C, a part of the KMN network. KNL1 is involved in the formation of kinetochore-microtubule attachments and SAC signaling among others by providing docking sites for the BUB1 and BUBR1 proteins [11]. In experimental systems, penetrant depletion of KNL1 induces chromosome misalignment and premature mitotic exit, causing chromosome segregation errors and thus aneuploidy [55–57].

All MCPH4 patients carry homozygous *KNL1* mutations that lead to incidental skipping of exon 18 or 25, resulting in a frameshift and subsequent truncation of the protein (Table 4.1). The mutation that causes incidental skipping of exon 18 also causes a M2041I substitution (in both full length and truncated protein), located in the Zwint1 interaction motif. Nevertheless, the mutation had no significant effect on total KNL1 levels, and, though conclusive live cell imaging data is lacking, quantification of micronuclei suggested that chromosome segregation fidelity was not substantially affected, at least in the lymphoblastoid cell lines analyzed [58]. Possibly therefore, the effect of this mutation reveals itself only in brain tissue. On the other hand, skipping of exon 25 resulted in a reduction of total KNL1 levels to about 50% of the controls. In fibroblast and lymphoblastoid cells derived from these patients, mitotic delays, increased micronuclei and an abnormal DNA damage response were observed [59, 60]. The latter could be a novel function of KNL1 or could be related to damage caused by mitotic segregation errors [36, 43].

Mutations in *CENPE* cause MCPH13 (MIM 616051) in two siblings. Both exhibited profound short stature and microcephaly (also known as microcephalic primordial dwarfism) associated with developmental delay, simplified gyri and other isolated abnormalities. In both patients compound heterozygous point mutations were identified (D993N and K1355E, in the central coiled-coil region; Table 4.1) which did not adversely affect CENP-E (centromere protein E) protein expression or stability [61].

CENP-E is a microtubule plus-end-directed motor protein required for proper kinetochore-microtubule attachments and the congression of chromosomes to the spindle equator [62, 63]. Experimental depletion of CENP-E impairs kinetochore-microtubule attachment, slows chromosome congression, and consequently prolongs mitosis. CENP-E also affects centrosome stability as their fragmentation increases upon CENP-E knockdown [27]. Patient mitotic lymphocytes displayed reduced kinetochore levels of mutant CENP-E, which correlated with substantial spindle multipolarity, chromosome segregation errors and micronuclei [61].

4.3.2 *Strømme Syndrome:* CENP-F

Strømme syndrome (MIM 243605) is characterized by microcephaly, and affects multiple systems with features of a ciliopathy (a disorder characterized by defects in the microtubule/membrane-based protrusions known as cilia that mediate cell motility and transduce sensory information [64]). Affected individuals suffer from microcephaly, intestinal atresia (absence of part of the intestine), variable ocular abnormalities, and occasionally defects in other systems, such as the kidneys and heart. The causative mutations were found to reside in the *CENP-F* gene (Table 4.1), which severely affected protein expression [65, 66].

CENP-F (centromere protein F) is a large coiled-coil protein with two unusual microtubule-binding domains that it uses for multiple processes in mitosis. CENP-F couples mitochondria to polymerizing microtubules to allow for their segregation during mitosis [67], and it enables kinetochores to remain bound to shortening microtubules during anaphase [68]. Besides binding microtubules, CENP-F plays a role in the recruitment of the two kinetochore motors Dynein (via NDE1) and CENP-E [69] that mediate correct kinetochore-microtubule attachment and chromosome alignment. Depletion of CENP-F also results in premature chromatid separation, suggesting a role in the protection of sister-chromatid cohesion [17]. Finally, CENP-F is required for ciliogenesis, possibly by targeting IFT88 to the ciliary axoneme [65]. No cell biological analyses were performed on cells derived from the patients, so it is unknown which processes are affected in the patients. Nevertheless, they may suffer from a combination of cellular defects that together cause the wide array of symptoms observed in these patients.

4.3.3 *Microhydranencephaly and Microlissencephaly:* NDE1

Eleven patients in five families were diagnosed with microhydranencephaly or microlissencephaly. Microhydranencephaly (MIM 605013) is a disorder in which microcephaly is associated with severe hydranencephaly (where missing parts of brain hemispheres are filled with fluid, causing enlarged head size, also known as hydrocephalus). In microlissencephaly (MIM 614019), microcephaly is associated with lissencephaly (where brain folds and grooves are underdeveloped). All patients have homozygous truncating mutations in the *NDE1* gene that result in the expression of an unstable protein.

NDE1 (NudE neurodevelopment protein 1) is an adaptor protein of the Dynein microtubule motor complex. NDE1 is found on centrosomes and the nuclear envelope (NE) in interphase, where it ensures NE-centrosome anchoring, which is necessary for proper spindle formation. In mitosis, it is found on kinetochores, where it interacts with CENP-F, the protein mutated in Strømme syndrome (Sect. 4.3.2), to enable Dynein recruitment. Depletion of NDE1 leads to an increase in chromosomes that are incorrectly attached to the mitotic spindle, causing a significant increase in chromosome segregation errors [69]. Interestingly, NDE1 was reported to be a negative regulator of ciliogenesis [70]. The unstable, truncated NDE1 proteins expressed in the 11 patients lack the domains responsible for the interactions with both CENP-F and Dynein (Table 4.1) [71–73]. NDE1 malfunction in patient cells may thus lead to mitotic defects as well as impairment of (interphasic) Dynein function. The latter inhibits neuronal migration, a likely cause for lissencephaly [74].

4.3.4 Mental Retardation-40: CHAMP1

Autosomal dominant mental retardation-40 (MRD40, MIM 616579) disorder is characterized by intellectual disability, global developmental delay and dysmorphic facial features, as well as microcephaly [75–77]. All patients carried heterozygous truncating mutations in the *CHAMP1* gene, which encodes the CAMP protein (Table 4.1).

Relatively little is known about the function of CAMP (chromosome alignment maintaining phosphoprotein). It localizes to chromosomes, the mitotic spindle and kinetochores. CAMP binds MAD2L2 (MAD2B/hRev7), a protein involved in regulating mitotic entry and DNA repair at telomeres [78]. CAMP is also required for the kinetochore localization of CENP-E and CENP-F, though direct interactions have not been observed [78]. Knockdown of CAMP results in abnormal spindle formation and chromosome alignment defects, possibly by misregulation of CENP-E and CENP-F [78]. This role of CAMP may further involve Aurora B activation, as CAMP interacts with HP1 (Heterochromatin protein 1) and POGZ (POGO transposable element with ZNF domain), both of which impact Aurora B localization and sister chromatid cohesion [79, 80]. The *CHAMP1* mutations in MRD40 patients affected transcript level, and the resulting protein lacks the domains necessary for its localization to chromosomes and the mitotic spindle as well as its interaction with POGZ and HP1 [75]. Intriguingly, loss-of-function mutations in *POGZ* have been identified in individuals affected by neurodevelopmental disorders with a similar phenotype (intellectual disability, microcephaly) [81]. Loss of the POGZ-CAMP interaction and misregulation of CENP-E and CENP-F function (both associated with microcephaly syndromes, see Sect. 4.3.1 and Sect. 4.3.2) may thus be key molecular causes of MRD40.

4.3.5 Mosaic Variegated Aneuploidy: BUB1B *and* CEP57

Mosaic Variegated Aneuploidy syndrome (MVA, MIM 257300) is a rare autosomal recessive disorder characterized by aneuploidies of mostly random chromosomes in different tissues. Although the proportion of aneuploid cells usually is more than

25%, there does not seem to be a correlation between the clinical phenotype and the observed karyotypes [82]. MVA patients usually suffer from intrauterine growth retardation and microcephaly. Eye anomalies, mild dysmorphism, variable developmental delay, and a broad spectrum of additional congenital abnormalities and medical conditions are also frequently part of the MVA phenotype. Furthermore, a subset of MVA patients has developed childhood cancers such as Wilms tumor, rhabdomyosarcoma, and leukemia [83]. To date, pathogenic mutations in two genes have been identified in MVA patients: *BUB1B* and *CEP57* (Table 4.1), both of which are associated with the kinetochore [84, 85]. The phenotypes associated with *BUB1B* and *CEP57* mutations are broadly similar, although only *BUB1B* mutations are strongly associated with cancer. Aneuploidy seems the most likely culprit causing the symptoms, but MVA has also been suggested to be a ciliopathy [86] or a progeria (premature ageing) syndrome [87].

BUB1B (BUB1 mitotic checkpoint serine/threonine kinase B) encodes BUBR1, a multi-domain pseudokinase implicated in various, mostly mitotic processes. BUBR1 localizes to mitotic kinetochores via KNL1 and binds CENP-E, the genes mutated in MCPH4/13 (Sect. 4.3.1). BUBR1 is required for SAC signaling by directly inhibiting APC/C. It also promotes stable kinetochore-microtubule attachments through direct binding of the B56-PP2A phosphatase [88]. In interphase, BUBR1 is thought to prevent inappropriate centrosome amplification [89] and to positively regulate ciliogenesis through the suppression of disheveled (DVL), canonical Wnt-activation and cell cycle progression [86]. *BUB1B* mutations in MVA result in low protein expression, which is the primary molecular cause of BUBR1 malfunction [90]. The low expression is due to either biallelic combinations of a truncating mutation that leads to RNA decay with a missense mutation that destabilizes the protein, to a combination of two missense mutations that destabilize the protein, or to one truncating mutation together with a haplotype that is associated with low BUBR1 expression. All mutations analyzed impair SAC function and chromosome alignment [90]. Furthermore, cells derived from patients with *BUB1B* mutations displayed centrosome amplification and multipolar mitoses as well as impaired ciliogenesis [86, 89].

CEP57 (centrosomal protein 57) was initially identified as an intracellular transporter of fibroblast growth factor 2 [91], but has recently acquired more attention for its roles at centrosomes and kinetochores. CEP57 binds and stabilizes microtubules and thereby promotes spindle assembly, spindle pole integrity and central spindle organization [92]. At kinetochores, CEP57 promotes SAC signaling, although it is not absolutely required for it. It does so by binding the KMN network member MIS12 and functioning as a scaffold for MAD1-MAD2 [93]. The MVA-associated mutations in *CEP57* are homozygous truncating mutations [94]. It is unknown whether they affect protein expression, though it is very likely that its microtubule-binding capacity is impaired. Unfortunately, functional assays on cells derived from these patients have not been performed so it is currently unknown if the cellular defects of *CEP57* mutant cells are similar to those of *BUB1B* patient cells.

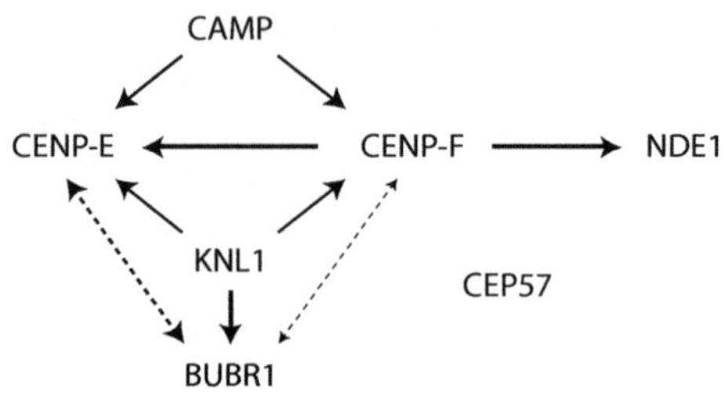

Fig. 4.2 Model showing dependencies of kinetochore localization of disease-associated kinetochore proteins. *Dotted arrows* connecting CENP-E and BUBR1 signify a direct interaction (exact contributions to localization are unknown) [27]. Thin *dotted arrows* connecting CENP-F and BUBR1 signify interaction found in a yeast two-hybrid screen (direct interaction has not been observed) [138]

4.3.6 Kinetochore Mutations in Microcephaly Syndromes: One Common Pathway?

Strikingly, all proteins mutated in the congenital syndromes discussed have been functionally linked in various studies (Fig. 4.2). CENP-E depends on KNL1, CAMP and CENP-F for its kinetochore recruitment and is known to interact with BUBR1. CENP-F likewise depends on KNL1 and CAMP for its localization and, in turn, recruits NDE1 to the kinetochore. A common molecular defect in almost all kinetochore syndromes (with the exception of CEP57) may therefore be malfunction of the kinetochore motor proteins CENP-E and Dynein (via CENP-F and NDE1), both of which promote kinetochore-microtubule attachments and chromosome bi-orientation. With the caveat of relatively low number of patients, this begs the question if there is a particular reason why this pathway as opposed to others involved in similar kinetochore functions is found more frequently mutated. Mutation in components of this pathway may simply result in a 'just right' level of aneuploidy that causes disease without early embryonic death, while mutations in components of other pathways do not. Alternatively, there is a functional difference between this pathway and others, for example impacting spindle assembly or cilia function, in addition to aneuploidization. Organismal studies of these and other mutations will be required to answer this question.

4.4 Kinetochore Proteins and Cancer

One of the hallmarks of cancer is the presence of highly abnormal karyotypes that are characterized by structural and numerical aneuploidies. Over 90% of solid tumors and 50% of blood tumors is aneuploid, with a quarter of the genome of a typical cancer cell affected by either whole-arm or whole-chromosome copy number alterations [95, 96]. Aneuploidy may signal ongoing CIN, which is thought to contribute to tumorigenesis, tumor progression, and the development of therapy resistance [97]. Several mechanisms contribute to CIN in laboratory cancer cell lines, including replication stress [98], centrosome amplification [99], chromatid cohesion loss [80], increased spindle microtubule dynamics [100], impaired SAC signaling [101], and erroneous kinetochore-microtubule attachments [102–104]. Those related to kinetochore functions are discussed below.

4.4.1 Premature Sister Chromatid Separation (PSCS)

Impaired SAC signaling and cohesion loss have in common that both defects lead to premature separation of the sister chromatids. Experimental reduction of SAC protein expression in mice has not lead to conclusive insight into a potential role of SAC defects in cancer. Whereas some models develop benign tumors late in life in specific tissues, others suffer from earlier and more widespread tumor formation, and yet others do not develop tumors at all or only with carcinogen challenges [105]. The actual levels of CIN and the sensitivity of particular tissues to those levels may explain the differences between animal models. Although SAC defects can cause cancer, they may not be a common cause of tumor formation in humans. SAC defects do not appear to be a common characteristic of laboratory CIN cancer cells [106] and with the possible exception of MAD2 in gastric tumors [107], mutations in kinetochore genes are a rare occurrence [29, 108]. Moreover, though numerous studies have reported the deregulation of SAC gene expression in different types of cancer, there is no consistency between reports, and some reported overexpressions may reflect a general activation of the cell division machinery [109].

Sister chromatids are held together by ring-shaped cohesin protein complexes. The majority of cohesion complexes are removed from the DNA, but centromeric cohesin is protected from removal by members of the inner centromere–shugoshin (ICS) network until anaphase is initiated [110]. Premature cohesin removal in cancer cells has been reported to occur at two levels. First, STAG2, a component of the cohesin ring, is mutated frequently in cancers [111]. Targeted correction of the mutant alleles of *STAG2* led to reduced cohesion defects and enhanced chromosomal stability [112]. Second, a recent study found that ICS network impairment is a cause for CIN in various laboratory cancer cell lines derived from lung, colon, skin and bone tissues [80]. Mutations in components of the ICS network (*BUB1*, *SGO1*, CPC) are rare in human cancers [113], but misregulation of the network at other levels may occur.

4.4.2 Persistence of Erroneous Kinetochore-Microtubule Attachments

The type of segregation error most frequently seen in laboratory cancer cell lines is lagging of chromosomes, most likely the result of a merotelic attachment state [114]. Merotely can be caused by various means, including altered microtubule dynamics. Measurements of kinetochore-microtubule attachment turnover rates revealed that attachments were generally more stable in CIN cancer cell lines than in those that are non-cancerous and genetically stable. Reducing k-fiber stability in the CIN cells was sufficient to restore chromosomal stability [100]. K-fiber stability is regulated by various processes, including Aurora B-dependent regulation of the microtubule-binding protein Ndc80/Hec1 and the microtubule depolymerase MCAK [5]. Genes such as *AURKB* and *MCAK* are rarely mutated or inactivated in human cancer [113], but deregulated genes may indirectly promote attachment stability. Hec1, for instance, is frequently overexpressed human cancer, as is *MAD2*, which, instead of mediating hyper-activity of the SAC, was found to hyperstabilize kinetochore-microtubule attachments [115]. However, as mentioned in Sect. 4.4.1 it is worth to keep in mind that some reported overexpressions may reflect a general activation of the cell division machinery rather than a specific selection for overexpression of the gene in question [109]. Finally, the deregulation of the ICS network deregulates Aurora B localization to inner centromeres, possibly affecting k-fiber stability. Another means to achieve merotely is the presence of supernumerary centrosomes that cause transient multipolar spindles which are enriched for merotelic attachments [116, 117]. Late centrosome clustering then bipolarizes the spindle but does not resolve some of the merotelic attachments. Centrosome amplification has been shown to occur frequently in tumors and has a strong correlation with CIN [103].

In colorectal cancer cell lines, attachment errors and lagging chromosomes appear to be caused by an increase in microtubule plus end assembly rates [118]. This mechanism has similarities to merotely caused by centrosome amplification as it also triggers transient spindle geometry abnormalities that facilitate the generation of erroneous kinetochore attachments. It seems likely that increased assembly rates involve altered kinetochore function, but it is unclear if this is the case and what the mechanism is.

4.4.3 Tumor-Suppressor Loss and Kinetochore Function

Altered expression of kinetochore proteins rather than mutations may be a more common mechanism for CIN in tumor cells. Recent studies have implied various transcriptional and post-transcriptional alterations due to loss of tumor suppressor genes as a potential cause for this. Mutations in *BRCA2* predispose to breast cancer [119] as well as cancers in the ovary and pancreas [120]. One of its many functions is the recruitment of P/CAF acetyltransferase to BUBR1 at the kinetochore, which facilitates BUBR1 acetylation and protects it from proteasomal degradation. *BRCA2*-deficient cells exhibit lower BUBR1 levels and impaired SAC signaling [121]. Inhibition of the BRCA2–BUBR1 interaction in mice led to spontaneous tumorigenesis [121, 122]. In another example, mutations in *TP53* or *RB1* increase *MAD2* expression, leading to hyperstabilization of kinetochore-microtubule attachments [123]. This is mediated by the transcription factor E2F, which is inhibited by Rb, as well as indirectly by p53. *MAD2* upregulation is required for CIN in mutant

TP53 and *RB1* cancer mouse models [124, 125]. Conversely, *MAD2* expression is also downregulated in certain types of cancer, for instance in neuro- and medulloblastomas. In these tumors, the *repressor-element-1-silencing transcription factor (REST)* is frequently overexpressed, which negatively regulates *MAD2* levels [126].

Finally, the tumor suppressor APC, inactivated in most colorectal tumors, localizes to kinetochores in mouse embryonic stem cells, where it ensures proper spindle assembly and k-fiber stability [127]. Whether there is a role for APC mutations in CIN in human cancers is debatable, as APC loss in human intestinal organoid cultures did not cause substantial CIN [128].

4.5 Conclusions and Future Perspectives

Recent next generation sequencing approaches have significantly contributed to the rapid discovery of mutations in many new genes associated with a wide variety of diseases. This has resulted in an increasingly longer list of mutated kinetochores genes, especially in congenital neurodevelopmental disorders. Although the mechanistic relation between the mutations and the disorders is unknown in all cases, most affected genes control chromosome bi-orientation, spindle assembly, or both. Consistent with this idea, many of the MCPH-associated genes encode centrosome-associated proteins [129], raising the possibility that spindle assembly defects are the common denominator of the microcephaly-causing gene mutations. Mosaic Variegated Aneuploidy and Strømme Syndrome are characterized by a much wider variety of symptoms than the other mentioned disorders. Both have been suggested to be ciliopathies, and it will be of interest to further investigate whether ciliary defects are common in these patients and whether this contributes to their symptoms. Aneuploidy-induced senescence may also contribute to the different disease phenotypes, and may explain some of the progeroid-like symptoms of MVA patients [130]. We expect that the discovery of disease-associated mutations will continue to expand rapidly over the coming years. It will be interesting to see whether mutations in kinetochore genes continue to converge in certain functional pathways and whether they contribute to diseases other than neurodevelopmental disorders.

A major unanswered question is how kinetochore defects contribute to cancer. CIN seems the most parsimonious explanation, but there is currently a discrepancy between levels of CIN/aneuploidy in models of kinetochore malfunction and the extent of tumor development/progression. An animal model in which CIN can be conditionally induced to various levels and monitoring CIN in the relevant tissues will be crucial in addressing this conundrum. Finally, it is not understood why, given the widespread occurrence of aneuploidy in human cancers, mutations in kinetochore components are rarely found in cancer genomes. We speculate that such mutations do occur but are either neutral or purged as a result of causing too much aneuploidy. Loss of a single allele of kinetochore proteins in mice is generally well tolerated while loss of both causes early embryonic death [131]. A level of aneuploidy that eventually is beneficial to tumor cells may thus be more easily attainable by other mechanisms, including altered expression levels, for example by mutating transcriptional regulators. Whether this is true awaits identification of the molecular causes of CIN in human cancers.

References

1. Siegel JJ, Amon A (2012) New insights into the troubles of aneuploidy. Annu Rev Cell Dev Biol 28:189–214
2. Ricke RM, van Deursen JM (2013) Aneuploidy in health, disease, and aging. J Cell Biol 201(1):11–21
3. Cheeseman IM (2014) The kinetochore. Cold Spring Harb Perspect Biol 6(7):a015826
4. Wendell KL, Wilson L, Jordan MA (1993) Mitotic block in HeLa cells by vinblastine: ultrastructural changes in kinetochore-microtubule attachment and in centrosomes. J Cell Sci 104(2):261–274
5. Carmena M, Wheelock M, Funabiki H, Earnshaw WC (2012) The chromosomal passenger complex (CPC): from easy rider to the godfather of mitosis. Nat Rev Mol Cell Biol 13(12):789–803
6. Primorac I, Musacchio A (2013) Panta rhei: the APC/C at steady state. J Cell Biol 201(2):177–189
7. Sacristan C, Kops GJPL (2015) Joined at the hip: kinetochores, microtubules, and spindle assembly checkpoint signaling. Trends Cell Biol 25(1):21–28
8. London N, Biggins S (2014) Signalling dynamics in the spindle checkpoint response. Nat Rev Mol Cell Biol 15(11):736–747
9. McKinley KL, Cheeseman IM (2016) The molecular basis for centromere identity and function. Nat Rev Mol Cell Biol 17(1):16–29
10. Varma D, Salmon ED (2012) The KMN protein network–chief conductors of the kinetochore orchestra. J Cell Sci 125(Pt 24):5927–5936
11. Caldas GV, DeLuca JG (2014) KNL1: bringing order to the kinetochore. Chromosoma 123(3):169–181
12. Hiruma Y, Sacristan C, Pachis ST, Adamopoulos A, Kuijt T, Ubbink M, von Castelmur E, Perrakis A, Kops GJPL (2015) CELL DIVISION CYCLE. Competition between MPS1 and microtubules at kinetochores regulates spindle checkpoint signaling. Science 348(6240):1264–1267
13. Ji Z, Gao H, Yu H, Cayirlioglu P, Kadow IG, Vosshall LB, Stornetta RL, West GH, Guyenet PG, Isacson O, Nishiyama A, Mulkey DK, Koschnitzky JE (2015) CELL DIVISION CYCLE. Kinetochore attachment sensed by competitive Mps1 and microtubule binding to Ndc80C. Science 348(6240):1260–1264
14. Vleugel M, Hoek T, Tromer E, Sliedrecht T, Groenewold V, Omerzu M, Kops GJPL (2015) Dissecting the roles of human BUB1 in the spindle assembly checkpoint. J Cell Sci 128:2975–2982
15. Zhang G, Lischetti T, Hayward DG, Nilsson J (2015) Distinct domains in Bub1 localize RZZ and BubR1 to kinetochores to regulate the checkpoint. Nat Commun 6:7162
16. Johnson VL, Scott MIF, Holt SV, Hussein D, Taylor SS (2004) Bub1 is required for kinetochore localization of BubR1, Cenp-E, Cenp-F and Mad2, and chromosome congression. J Cell Sci 117(Pt 8):1577–1589
17. Holt SV, Vergnolle MA, Hussein D, Wozniak MJ, Allan VJ, Taylor SS (2005) Silencing Cenp-F weakens centromeric cohesion, prevents chromosome alignment and activates the spindle checkpoint. J Cell Sci 118(Pt 20):4889–4900
18. Suijkerbuijk SJE, Vleugel M, Teixeira A, Kops GJPL (2012) Integration of kinase and phosphatase activities by BUBR1 ensures formation of stable kinetochore-microtubule attachments. Dev Cell 23(4):745–755

19. Akera T, Goto Y, Sato M, Yamamoto M, Watanabe Y (2015) Mad1 promotes chromosome congression by anchoring a kinesin motor to the kinetochore. Nat Cell Biol 17(9):1124–1133
20. Kawashima SA, Yamagishi Y, Honda T, Ishiguro K, Watanabe Y (2010) Phosphorylation of H2A by Bub1 prevents chromosomal instability through localizing shugoshin. Science 327(5962):172–177
21. Yamagishi Y, Honda T, Tanno Y, Watanabe Y (2010) Two histone marks establish the inner centromere and chromosome bi-orientation. Science 330(6001):239–243
22. Kops GJPL, Shah JV (2012) Connecting up and clearing out: how kinetochore attachment silences the spindle assembly checkpoint. Chromosoma 121(5):509–525
23. Funabiki H, Wynne DJ (2013) Making an effective switch at the kinetochore by phosphorylation and dephosphorylation. Chromosoma 122(3):135–158
24. Foley EA, Kapoor TM (2013) Microtubule attachment and spindle assembly checkpoint signalling at the kinetochore. Nat Rev Mol Cell Biol 14(1):25–37
25. Etemad B, Kops GJPL (2016) Attachment issues: kinetochore transformations and spindle checkpoint silencing. Curr Opin Cell Biol 39:101–108
26. Kops GJPL, Saurin AT, Meraldi P (2010) Finding the middle ground: how kinetochores power chromosome congression. Cell Mol Life Sci 67(13):2145–2161
27. Yao X, Abrieu A, Zheng Y, Sullivan KF, Cleveland DW (2000) CENP-E forms a link between attachment of spindle microtubules to kinetochores and the mitotic checkpoint. Nat Cell Biol 2:484–491
28. Bader JR, Vaughan KT (2010) Dynein at the kinetochore: timing, interactions and functions. Semin Cell Dev Biol 21(3):269–275
29. Thompson SL, Bakhoum SF, Compton DA (2010) Mechanisms of chromosomal instability. Curr Biol 20(6):R285
30. Al Alwan I, Khadora M, Amir I, Nasrat G, Omair A, Brown L, Al Dubayee M, Badri M (2014) Turner syndrome genotype and phenotype and their effect on presenting features and timing of diagnosis. Int J Health Sci (Qassim) 8(2):195–202
31. Knoch J, Kamenisch Y, Kubisch C, Berneburg M (2012) Rare hereditary diseases with defects in DNA-repair. Eur J Dermatol 22(4):443–455
32. Cucco F, Musio A (2016) Genome stability: what we have learned from cohesinopathies. Am J Med Genet C Semin Med Genet 172:171
33. Zakari M, Yuen K, Gerton JL (2015) Etiology and pathogenesis of the cohesinopathies. Wiley Interdiscip Rev Dev Biol 4:489
34. Thompson SL, Compton DA (2010) Proliferation of aneuploid human cells is limited by a p53-dependent mechanism. J Cell Biol 188(3):369–381
35. Janssen A, van der Burg M, Szuhai K, Kops GJPL, Medema RH (2011) Chromosome segregation errors as a cause of DNA damage and structural chromosome aberrations. Science 333(6051):1895–1898
36. Li M, Fang X, Baker DJ, Guo L, Gao X, Wei Z, Han S, van Deursen JM, Zhang P (2010) The ATM-p53 pathway suppresses aneuploidy-induced tumorigenesis. Proc Natl Acad Sci U S A 107(32):14188–14193
37. Stingele S, Stoehr G, Peplowska K, Cox J, Mann M, Storchova Z (2012) Global analysis of genome, transcriptome and proteome reveals the response to aneuploidy in human cells. Mol Syst Biol 8:608
38. Upender MB, Habermann JK, McShane LM, Korn EL, Barrett JC, Difilippantonio MJ, Ried T (2004) Chromosome transfer induced aneuploidy results in complex dysregulation of the cellular transcriptome in immortalized and cancer cells. Cancer Res 64(19):6941–6949
39. Dürrbaum M, Storchová Z (2015) Effects of aneuploidy on gene expression: implications for cancer. FEBS J 283:791–802
40. Dürrbaum M, Kuznetsova AY, Passerini V, Stingele S, Stoehr G, Storchová Z (2014) Unique features of the transcriptional response to model aneuploidy in human cells. BMC Genomics 15(1):139

41. Santaguida S, Amon A (2015) Short- and long-term effects of chromosome mis-segregation and aneuploidy. Nat Rev Mol Cell Biol 16(8):473–485
42. Crasta K, Ganem NJ, Dagher R, Lantermann AB, Ivanova EV, Pan Y, Nezi L, Protopopov A, Chowdhury D, Pellman D (2012) DNA breaks and chromosome pulverization from errors in mitosis. Nature 482(7383):53–58
43. Storchová Z, Kloosterman WP (2016) The genomic characteristics and cellular origin of chromothripsis. Curr Opin Cell Biol 40:106–113
44. Zhang C-Z, Spektor A, Cornils H, Francis JM, Jackson EK, Liu S, Meyerson M, Pellman D (2015) Chromothripsis from DNA damage in micronuclei. Nature 522(7555):179–184
45. De Pagter MS, Van Roosmalen MJ, Baas AF, Renkens I, Duran KJ, Van Binsbergen E, Tavakoli-Yaraki M, Hochstenbach R, Van Der Veken LT, Cuppen E, Kloosterman WP (2015) Chromothripsis in healthy individuals affects multiple protein-coding genes and can result in severe congenital abnormalities in offspring. Am J Hum Genet 96(4):651–656
46. Kloosterman WP, Guryev V, van Roosmalen M, Duran KJ, de Bruijn E, Bakker SCM, Letteboer T, van Nesselrooij B, Hochstenbach R, Poot M, Cuppen E (2011) Chromothripsis as a mechanism driving complex de novo structural rearrangements in the germline. Hum Mol Genet 20(10):1916–1924
47. Knouse KA, Wu J, Whittaker CA, Amon A (2014) Single cell sequencing reveals low levels of aneuploidy across mammalian tissues. Proc Natl Acad Sci U S A 111(37):13409–13414
48. McConnell MJ, Lindberg M, Brennand KJ, Piper JC, Voet T, Cowing-Zitron C, Shumilina S, Lasken RS, Vermeesch JR, Hall IM, Gage FH (2013) Mosaic copy number variation in human neurons. Science 342:632–638
49. Pfau SJ, Silberman RE, Knouse KA, Amon A (2016) Aneuploidy impairs hematopoietic stem cell fitness and is selected against in regenerating tissues in vivo. Genes Dev 30:1395–1408
50. Marthiens V, Rujano MA, Pennetier C, Tessier S, Paul-Gilloteaux P, Basto R (2013) Centrosome amplification causes microcephaly. Nat Cell Biol 15(7):731–740
51. Woods CG, Basto R (2014) Microcephaly. Curr Biol 24(23):R1109–R1111
52. Gogendeau D, Siudeja K, Gambarotto D, Pennetier C, Bardin AJ, Basto R (2015) Aneuploidy causes premature differentiation of neural and intestinal stem cells. Nat Commun 6:8894
53. Chetaille P, Preuss C, Burkhard S, Côté J-M, Houde C, Castilloux J, Piché J, Gosset N, Leclerc S, Wünnemann F, Thibeault M, Gagnon C, Galli A, Tuck E, Hickson GR, El Amine N, Boufaied I, Lemyre E, de Santa Barbara P, Faure S, Jonzon A, Cameron M, Dietz HC, Gallo-McFarlane E, Benson DW, Moreau C, Labuda D, Zhan SH, Shen Y, Jomphe M, Jones SJM, Bakkers J, Andelfinger G (2014) Mutations in SGOL1 cause a novel cohesinopathy affecting heart and gut rhythm. Nat Genet 46(11):1245–1249
54. Silió V, McAinsh AD, Millar JB (2015) KNL1-Bubs and RZZ provide two separable pathways for checkpoint activation at human kinetochores. Dev Cell 35(5):600–613
55. Kiyomitsu T, Obuse C, Yanagida M (2007) Human Blinkin/AF15q14 is required for chromosome alignment and the mitotic checkpoint through direct interaction with Bub1 and BubR1. Dev Cell 13(5):663–676
56. Vleugel M, Tromer E, Omerzu M, Groenewold V, Nijenhuis W, Snel B, Kops GJPL (2013) Arrayed BUB recruitment modules in the kinetochore scaffold KNL1 promote accurate chromosome segregation. J Cell Biol 203(6):943–955
57. Genin A, Desir J, Lambert N, Biervliet M, Van der Aa N, Pierquin G, Killian A, Tosi M, Urbina M, Lefort A, Libert F, Pirson I, Abramowicz M (2012) Kinetochore KMN network gene CASC5 mutated in primary microcephaly. Hum Mol Genet 21(24):5306–5317
58. Szczepanski S, Hussain MS, Sur I, Altmüller J, Thiele H, Abdullah U, Waseem SS, Moawia A, Nürnberg G, Noegel AA, Baig SM, Nürnberg P (2016) A novel homozygous splicing mutation of CASC5 causes primary microcephaly in a large Pakistani family. Hum Genet 135(2):157–170
59. Saadi A, Verny F, Siquier-Pernet K, Bole-Feysot C, Nitschke P, Munnich A, Abada-Dendib M, Chaouch M, Abramowicz M, Colleaux L (2016) Refining the phenotype associated with CASC5 mutation. Neurogenetics 17(1):71–78

60. Mirzaa GM, Vitre B, Carpenter G, Abramowicz I, Gleeson JG, Paciorkowski AR, Cleveland DW, Dobyns WB, O'Driscoll M (2014) Mutations in CENPE define a novel kinetochore-centromeric mechanism for microcephalic primordial dwarfism. Hum Genet 133(8):1023–1039
61. Kapoor TM, Lampson MA, Hergert P, Cameron L, Cimini D, Salmon ED, McEwen BF, Khodjakov A (2006) Chromosomes can congress to the metaphase plate before biorientation. Science 311(5759):388–391
62. Yen TJ, Li G, Schaar BT, Szilak I, Cleveland DW (1992) CENP-E is a putative kinetochore motor that accumulates just before mitosis. Nature 359:536–539
63. Powles-Glover N (2014) Cilia and ciliopathies: classic examples linking phenotype and genotype-an overview. Reprod Toxicol 48:98–105
64. Waters AM, Asfahani R, Carroll P, Bicknell L, Lescai F, Bright A, Chanudet E, Brooks A, Christou-Savina S, Osman G, Walsh P, Bacchelli C, Chapgier A, Vernay B, Bader DM, Deshpande C, Sullivan MO', Ocaka L, Stanescu H, Stewart HS, Hildebrandt F, Otto E, Johnson CA, Szymanska K, Katsanis N, Davis E, Kleta R, Hubank M, Doxsey S, Jackson A, Stupka E, Winey M, Beales PL (2015) The kinetochore protein, CENPF, is mutated in human ciliopathy and microcephaly phenotypes. J Med Genet 52(3):147–156
65. Filges I, Bruder E, Brandal K, Meier S, Undlien DE, Waage TR, Hoesli I, Schubach M, de Beer T, Sheng Y, Hoeller S, Schulzke S, Røsby O, Miny P, Tercanli S, Oppedal T, Meyer P, Selmer KK, Strømme P (2016) Strømme syndrome is a ciliary disorder caused by mutations in CENPF. Hum Mutat 37(4):359–363
66. Kanfer G, Courthéoux T, Peterka M, Meier S, Soste M, Melnik A, Reis K, Aspenström P, Peter M, Picotti P, Kornmann B (2015) Mitotic redistribution of the mitochondrial network by Miro and Cenp-F. Nat Commun 6:8015
67. Volkov VA, Grissom PM, Arzhanik VK, Zaytsev AV, Renganathan K, McClure-Begley T, Old WM, Ahn N, Richard McIntosh J (2015) Centromere protein F includes two sites that couple efficiently to depolymerizing microtubules. J Cell Biol 209(6):813–828
68. Vergnolle MAS, Taylor SS (2007) Cenp-F links kinetochores to Ndel1/Nde1/Lis1/dynein microtubule motor complexes. Curr Biol 17(13):1173–1179
69. Kim S, Zaghloul NA, Bubenshchikova E, Oh EC, Rankin S, Katsanis N, Obara T, Tsiokas L (2011) Nde1-mediated inhibition of ciliogenesis affects cell cycle re-entry. Nat Cell Biol 13(4):351–360
70. Bakircioglu M, Carvalho OP, Khurshid M, Cox JJ, Tuysuz B, Barak T, Yilmaz S, Caglayan O, Dincer A, Nicholas AK, Quarrell O, Springell K, Karbani G, Malik S, Gannon C, Sheridan E, Crosier M, Lisgo SN, Lindsay S, Bilguvar K, Gergely F, Gunel M, Woods CG (2011) The essential role of centrosomal NDE1 in human cerebral cortex neurogenesis. Am J Hum Genet 88(5):523–535
71. Guven A, Gunduz A, Bozoglu TM, Yalcinkaya C, Tolun A (2012) Novel NDE1 homozygous mutation resulting in microhydranencephaly and not microlyssencephaly. Neurogenetics 13(3):189–194
72. Alkuraya FS, Cai X, Emery C, Mochida GH, Al-Dosari MS, Felie JM, Hill RS, Barry BJ, Partlow JN, Gascon GG, Kentab A, Jan M, Shaheen R, Feng Y, Walsh CA (2011) Human mutations in NDE1 cause extreme microcephaly with lissencephaly. Am J Hum Genet 88(5):536–547
73. Dantas TJ, Carabalona A, Jun-Kit Hu D, Vallee RB (2016) Emerging roles for motor proteins in progenitor cell behavior and neuronal migration during brain development. Cytoskeleton 73:566–576
74. Isidor B, Küry S, Rosenfeld JA, Besnard T, Schmitt S, Joss S, Davies SJ, Roger Lebel R, Henderson A, Schaaf CP, Streff HE, Yang Y, Jain V, Chida N, Latypova X, Le Caignec C, Cogné B, Mercier S, Vincent M, Colin E, Bonneau D, Denommé AS, Parent P, Gilbert-Dussardier B, Odent S, Toutain A, Piton A, Dina C, Donnart A, Lindenbaum P, Charpentier E, Redon R, Iemura K, Ikeda M, Tanaka K, Bézieau S (2016) De novo truncating mutations in the kinetochore-microtubules attachment gene CHAMP1 cause syndromic intellectual disability. Hum Mutat 37(4):354–358

75. Tanaka AJ, Cho MT, Retterer K, Jones JR, Nowak C, Douglas J, Jiang Y-H, McConkie-Rosell A, Schaefer GB, Kaylor J, Rahman OA, Telegrafi A, Friedman B, Douglas G, Monaghan KG, Chung WK (2016) De novo pathogenic variants in CHAMP1 are associated with global developmental delay, intellectual disability, and dysmorphic facial features. Mol Case Stud 2(1):a000661
76. Hempel M, Cremer K, Ockeloen CW, Lichtenbelt KD, Herkert JC, Denecke J, Haack TB, Zink AM, Becker J, Wohlleber E, Johannsen J, Alhaddad B, Pfundt R, Fuchs S, Wieczorek D, Strom TM, Van Gassen KLI, Kleefstra T, Kubisch C, Engels H, Lessel D (2015) De Novo mutations in CHAMP1 cause intellectual disability with severe speech impairment. Am J Hum Genet 97(3):493–500
77. Itoh G, Kanno S, Uchida KSK, Chiba S, Sugino S, Watanabe K, Mizuno K, Yasui A, Hirota T, Tanaka K (2011) CAMP (C13orf8, ZNF828) is a novel regulator of kinetochore-microtubule attachment. EMBO J 30(1):130–144
78. Nozawa R-S, Nagao K, Masuda H-T, Iwasaki O, Hirota T, Nozaki N, Kimura H, Obuse C (2010) Human POGZ modulates dissociation of HP1alpha from mitotic chromosome arms through Aurora B activation. Nat Cell Biol 12(7):719–727
79. Tanno Y, Susumu H, Kawamura M, Sugimura H, Honda T, Watanabe Y (2015) The inner centromere-shugoshin network prevents chromosomal instability. Science 349(6253):1237–1240
80. Ye Y, Cho MT, Retterer K, Alexander N, Ben-Omran T, Al-Mureikhi M, Cristian I, Wheeler PG, Crain C, Zand D, Weinstein V, Vernon HJ, McClellan R, Krishnamurthy V, Vitazka P, Millan F, Chung WK (2015) De novo POGZ mutations are associated with neurodevelopmental disorders and microcephaly. Mol Case Stud 1(1):a000455
81. Jacquemont S, Bocéno M, Rival JM, Méchinaud F, David A (2002) High risk of malignancy in mosaic variegated aneuploidy syndrome. Am J Med Genet 109(1):17–21
82. García-Castillo H, Vásquez-Velásquez AI, Rivera H, Barros-Núñez P (2008) Clinical and genetic heterogeneity in patients with mosaic variegated aneuploidy: delineation of clinical subtypes. Am J Med Genet A 146A(13):1687–1695
83. Hanks S, Coleman K, Reid S, Plaja A, Firth H, Fitzpatrick D, Kidd A, Méhes K, Nash R, Robin N, Shannon N, Tolmie J, Swansbury J, Irrthum A, Douglas J, Rahman N (2004) Constitutional aneuploidy and cancer predisposition caused by biallelic mutations in BUB1B. Nat Genet 36(11):1159–1161
84. Snape K, Hanks S, Ruark E, Barros-Núñez P, Elliott A, Murray A, Lane AH, Shannon N, Callier P, Chitayat D, Clayton-Smith J, Fitzpatrick DR, Gisselsson D, Jacquemont S, Asakura-Hay K, Micale MA, Tolmie J, Turnpenny PD, Wright M, Douglas J, Rahman N (2011) Mutations in CEP57 cause mosaic variegated aneuploidy syndrome. Nat Genet 43(6):527–529
85. Miyamoto T, Porazinski S, Wang H, Borovina A, Ciruna B, Shimizu A, Kajii T, Kikuchi A, Furutani-Seiki M, Matsuura S (2011) Insufficiency of BUBR1, a mitotic spindle checkpoint regulator, causes impaired ciliogenesis in vertebrates. Hum Mol Genet 20(10):2058–2070
86. Baker DJ, Jeganathan KB, Cameron JD, Thompson M, Juneja S, Kopecka A, Kumar R, Jenkins RB, de Groen PC, Roche P, van Deursen JM (2004) BubR1 insufficiency causes early onset of aging-associated phenotypes and infertility in mice. Nat Genet 36(7):744–749
87. Suijkerbuijk SJE, Vleugel M, Teixeira A, Kops GJPL (2012) Integration of kinase and phosphatase activities by BUBR1 ensures formation of stable kinetochore-microtubule attachments. Dev Cell 23(4):745–755
88. Izumi H, Matsumoto Y, Ikeuchi T, Saya H, Kajii T, Matsuura S (2009) BubR1 localizes to centrosomes and suppresses centrosome amplification via regulating Plk1 activity in interphase cells. Oncogene 28(31):2806–2820
89. Suijkerbuijk SJE, van Osch MHJ, Bos FL, Hanks S, Rahman N, Kops GJPL (2010) Molecular causes for BUBR1 dysfunction in the human cancer predisposition syndrome mosaic variegated aneuploidy. Cancer Res 70(12):4891–4900
90. Bossard C, Laurell H, Van den Berghe L, Meunier S, Zanibellato C, Prats H (2003) Translokin is an intracellular mediator of FGF-2 trafficking. Nat Cell Biol 5:433–439

91. Momotani K, Khromov AS, Miyake T, Stukenberg PT, Somlyo AV (2008) Cep57, a multidomain protein with unique microtubule and centrosomal localization domains. Biochem J 412(2):265–273
92. Zhou H, Wang T, Zheng T, Teng J, Chen J (2016) Cep57 is a Mis12-interacting kinetochore protein involved in kinetochore targeting of Mad1–Mad2. Nat Commun 7:10151
93. Snape K, Hanks S, Ruark E, Barros-Núñez P, Elliott A, Murray A, Lane AH, Shannon N, Callier P, Chitayat D, Clayton-Smith J, Fitzpatrick DR, Gisselsson D, Jacquemont S, Asakura-Hay K, Micale MA, Tolmie J, Turnpenny PD, Wright M, Douglas J, Rahman N (2011) Mutations in CEP57 cause mosaic variegated aneuploidy syndrome. Nat Genet 43(6):527–529
94. Beroukhim R, Mermel C, Porter D (2010) The landscape of somatic copy-number alteration across human cancers. Nature 463:899–905
95. Mitelman F, Johansson B, Mertens F (2016) Mitelman Database of Chromosome Aberrations and Gene Fusions in Cancer. [Online]. Available: http://cgap.nci.nih.gov/Chromosomes/Mitelman
96. Duijf PHG, Benezra R (2013) The cancer biology of whole-chromosome instability. Oncogene 32(40):4727–4736
97. Burrell RA, McClelland SE, Endesfelder D, Groth P, Weller M-C, Shaikh N, Domingo E, Kanu N, Dewhurst SM, Gronroos E, Chew SK, Rowan AJ, Schenk A, Sheffer M, Howell M, Kschischo M, Behrens A, Helleday T, Bartek J, Tomlinson IP, Swanton C (2013) Replication stress links structural and numerical cancer chromosomal instability. Nature 494(7438):492–496
98. Fukasawa K (2005) Centrosome amplification, chromosome instability and cancer development. Cancer Lett 230(1):6–19
99. Bakhoum SF, Thompson SL, Manning AL, Compton DA (2009) Genome stability is ensured by temporal control of kinetochore-microtubule dynamics. Nat Cell Biol 11(1):27–35
100. Kops GJPL, Weaver BAA, Cleveland DW (2005) On the road to cancer: aneuploidy and the mitotic checkpoint. Nat Rev Cancer 5(10):773–785
101. Bakhoum SF, Genovese G, Compton DA (2009) Deviant kinetochore microtubule dynamics underlie chromosomal instability. Curr Biol 19(22):1937–1942
102. Gordon DJ, Resio B, Pellman D (2012) Causes and consequences of aneuploidy in cancer. Nat Rev Genet 13(3):189–203
103. Bastians H (2015) Causes of chromosomal instability. In: Ghadimi B, Ried T (eds) Chromosomal instability in cancer cells, vol 200. Springer, Cham, pp 95–113
104. Simon JE, Bakker B, Foijer F (2015) CINcere modelling: what have mouse models for chromosome instability taught us? In: Chromosomal instability in cancer cells. Springer, Cham, pp 39–60
105. Tighe A, Johnson VL, Albertella M, Taylor SS (2001) Aneuploid colon cancer cells have a robust spindle checkpoint. EMBO Rep 2(7):609–614
106. Kim HS, Kyung HP, Kim SA, Wen J, Seung WP, Park B, Gham CW, Woo JH, Sung HN, Ho KK, Song SY (2005) Frequent mutations of human Mad2, but not Bub1, in gastric cancers cause defective mitotic spindle checkpoint. Mutat Res Fundam Mol Mech Mutagen 578(1–2):187–201
107. Suijkerbuijk SJE, Kops GJPL (2008) Preventing aneuploidy: the contribution of mitotic checkpoint proteins. Biochim Biophys Acta Rev Cancer 1786(1):24–31
108. Thiru P, Kern DM, McKinley KL, Monda JK, Rago F, Su K-C, Tsinman T, Yarar D, Bell GW, Cheeseman IM (2014) Kinetochore genes are coordinately up-regulated in human tumors as part of a FoxM1-related cell division program. Mol Biol Cell 25(13):1983–1994
109. Trivedi P, Stukenberg PT (2016) A centromere-signaling network underlies the coordination among mitotic events. Trends Biochem Sci 41(2):160–174
110. Hill VK, Kim J-S, Waldman T (2016) Cohesin mutations in human cancer. Biochim Biophys Acta Rev Cancer 1866(1):1–11
111. Solomon DA, Kim T, Diaz-Martinez LA, Fair J, Elkahloun AG, Harris BT, Toretsky JA, Rosenberg SA, Shukla N, Ladanyi M, Samuels Y, James CD, Yu H, Kim J-S, Waldman T

(2011) Mutational inactivation of STAG2 causes aneuploidy in human cancer. Science 333(6045):1039–1043
112. Lawrence MS, Stojanov P, Mermel CH, Robinson JT, Garraway LA, Golub TR, Meyerson M, Gabriel SB, Lander ES, Getz G (2014) Discovery and saturation analysis of cancer genes across 21 tumour types. Nature 505(7484):495–501
113. Bakhoum SF, Silkworth WT, Nardi IK, Nicholson JM, Compton DA, Cimini D (2014) The mitotic origin of chromosomal instability. Curr Biol 24(4):R148
114. Kabeche L, Compton DA (2012) Checkpoint-independent stabilization of kinetochore-microtubule attachments by Mad2 in human cells. Curr Biol 22(7):638–644
115. Silkworth WT, Nardi IK, Scholl LM, Cimini D (2009) Multipolar spindle pole coalescence is a major source of kinetochore mis-attachment and chromosome mis-segregation in cancer cells. PLoS One 4(8):e6564
116. Ganem NJ, Godinho SA, Pellman D (2009) A mechanism linking extra centrosomes to chromosomal instability. Nature 460(7252):278–282
117. Ertych N, Stolz A, Stenzinger A, Weichert W, Kaulfuß S, Burfeind P, Aigner A, Wordeman L, Bastians H (2014) Increased microtubule assembly rates influence chromosomal instability in colorectal cancer cells. Nat Cell Biol 16(8):779–791
118. Wooster R, Bignell G, Lancaster J, Swift S, Seal S, Mangion J, Collins N, Gregory S, Gumbs C, Micklem G (1995) Identification of the breast cancer susceptibility gene BRCA2. Nature 378(6559):789–792
119. T. Breast Cancer Linkage Consortium (1999) Cancer risks in BRCA2 mutation carriers. JNCI J Natl Cancer Inst 91(15):1310–1316
120. Choi E, Park PG, Lee HO, Lee YK, Kang GH, Lee JW, Han W, Lee HC, Noh DY, Lekomtsev S, Lee H (2012) BRCA2 fine-tunes the spindle assembly checkpoint through reinforcement of BubR1 acetylation. Dev Cell 22(2):295–308
121. Park I, Lee HO, Choi E, Lee YK, Kwon MS, Min J, Park PG, Lee S, Kong YY, Gong G, Lee H (2013) Loss of BubR1 acetylation causes defects in spindle assembly checkpoint signaling and promotes tumor formation. J Cell Biol 202(2):295–309
122. Hernando E, Nahlé Z, Juan G, Diaz-Rodriguez E, Alaminos M, Hemann M, Michel L, Mittal V, Gerald W, Benezra R, Lowe SW, Cordon-Cardo C (2004) Rb inactivation promotes genomic instability by uncoupling cell cycle progression from mitotic control. Nature 430(7001):797–802
123. Sotillo R, Schvartzman J-M, Socci ND, Benezra R (2010) Mad2-induced chromosome instability leads to lung tumour relapse after oncogene withdrawal. Nature 464(7287):436–440
124. Schvartzman JM, Duijf PHG, Sotillo R, Coker C, Benezra R (2011) Mad2 is a critical mediator of the chromosome instability observed upon Rb and p53 pathway inhibition. Cancer Cell 19(6):701–714
125. Guardavaccaro D, Frescas D, Dorrello NV, Peschiaroli A, Multani AS, Cardozo T, Lasorella A, Iavarone A, Chang S, Hernando E, Pagano M (2008) Control of chromosome stability by the beta-TrCP-REST-Mad2 axis. Nature 452(7185):365–369
126. Caldwell CM, Kaplan KB (2009) The role of APC in mitosis and in chromosome instability. Adv Exp Med Biol 656:51–64
127. Drost J, van Jaarsveld RH, Ponsioen B, Zimberlin C, van Boxtel R, Buijs A, Sachs N, Overmeer RM, Offerhaus GJ, Begthel H, Korving J, van de Wetering M, Schwank G, Logtenberg M, Cuppen E, Snippert HJ, Medema JP, Kops GJPL, Clevers H (2015) Sequential cancer mutations in cultured human intestinal stem cells. Nature 521(7550):43–47
128. Morris-Rosendahl DJ, Kaindl AM (2015) What next-generation sequencing (NGS) technology has enabled us to learn about primary autosomal recessive microcephaly (MCPH). Mol Cell Probes 29(5):271–281
129. Childs BG, Durik M, Baker DJ, van Deursen JM (2015) Cellular senescence in aging and age-related disease: from mechanisms to therapy. Nat Med 21(12):1424–1435
130. Foijer F, Draviam VM, Sorger PK (2008) Studying chromosome instability in the mouse. Biochim Biophys Acta Rev Cancer 1786(1):73–82

131. Jamieson CR, Govaerts C, Abramowicz MJ (1999) Primary autosomal recessive microcephaly: homozygosity mapping of MCPH4 to chromosome 15. Am J Hum Genet 65(5):1465–1469
132. Stromme P, Dahl E, Flage T, Stene-Johansen H (1993) Apple peel intestinal atresia in siblings with ocular anomalies and microcephaly. Clin Genet 44(4):208–210
133. Kavaslar GN, Onengut S, Derman O, Kaya A, Tolun A (2000) The novel genetic disorder microhydranencephaly maps to chromosome 16p13.3-12.1. Am J Hum Genet 66(5):1705–1709
134. Rauch A, Wieczorek D, Graf E, Wieland T, Endele S, Schwarzmayr T, Albrecht B, Bartholdi D, Beygo J, Di Donato N, Dufke A, Cremer K, Hempel M, Horn D, Hoyer J, Joset P, Röpke A, Moog U, Riess A, Thiel CT, Tzschach A, Wiesener A, Wohlleber E, Zweier C, Ekici AB, Zink AM, Rump A, Meisinger C, Grallert H, Sticht H, Schenck A, Engels H, Rappold G, Schröck E, Wieacker P, Riess O, Meitinger T, Reis A, Strom TM (2012) Range of genetic mutations associated with severe non-syndromic sporadic intellectual disability: an exome sequencing study. Lancet 380(9854):1674–1682
135. Ochiai H, Miyamoto T, Kanai A, Hosoba K, Sakuma T, Kudo Y, Asami K, Ogawa A, Watanabe A, Kajii T, Yamamoto T, Matsuura S (2013) TALEN-mediated single-base-pair editing identification of an intergenic mutation upstream of BUB1B as causative of PCS (MVA) syndrome. Proc Natl Acad Sci U S A 111:1461
136. Pinson L, Mannini L, Willems M, Cucco F, Sirvent N, Frebourg T, Quarantotti V, Collet C, Schneider A, Sarda P, Geneviève D, Puechberty J, Lefort G, Musio A (2014) CEP57 mutation in a girl with mosaic variegated aneuploidy syndrome. Am J Med Genet A 164(1):177–181
137. Chan GKT, Schaar BT, Yen TJ (1998) Characterization of the kinetochore binding domain of CENP-E reveals interactions with the kinetochore proteins CENP-F and hBUBR1. J Cell Biol 143(1):49–63

Chapter 5
The Elephant in the Room: The Role of Microtubules in Cancer

Luca Cirillo, Monica Gotta, and Patrick Meraldi

Abstract Microtubules are the backbone of all eukaryotic cells cytoskeleton. Their dynamic behaviour constitutes the basis for many biological processes such as cellular motility, cytoplasmic transport and cell division. Some the most effective chemotherapeutics, such as the taxanes, are microtubule interfering drugs. Moreover, many studies suggest that microtubule dynamics are altered in cancer cell divisions and linked to chromosomal instability, aneuploidy and development of drug resistances. The elephant in the room, however, is that despite all these evidences, the exact role of microtubules in malignancies remains elusive, partially due to the lack of clear genetic alterations linking microtubules to cancer. This review will discuss the molecular mechanisms that might alter microtubule dynamics in cancer cells, the pro and cons of the different theories linking these alterations to cancer progression, and the possible directions to address future key questions.

Keywords Microtubules • Aneuploidy • Chromosome instability • Dynamic instability • Cancer • Microtubule-associated proteins • Tubulin

5.1 Introduction

Microtubules are the main element of the mitotic spindle, which govern all aspects of cell division and chromosome segregation. Microtubule polymers linearly assemble from tubulin dimers composed of α- and β-tubulin; typically each polymer contains 13 proto-filaments that assemble into a hollow cylinder with a diameter of 22 nanometres. Microtubules are polarized with β-tubulin subunits exposed to the solvent at the plus end, while the minus-end is capped by α-tubulin subunits (see Fig. 5.1). Microtubules are nucleated at their minus ends via the γ-tubulin complex, which accumulates on the centrosome, the major microtubule organizing

L. Cirillo • M. Gotta (✉) • P. Meraldi (✉)
Department of Cell Physiology and Metabolism, Faculty of Medicine, Geneva University, 1211 Geneva, Switzerland
e-mail: monica.gotta@unige.ch; patrick.meraldi@unige.ch

M. Gotta, P. Meraldi (eds.), *Cell Division Machinery and Disease*, Advances in Experimental Medicine and Biology 1002, DOI 10.1007/978-3-319-57127-0_5

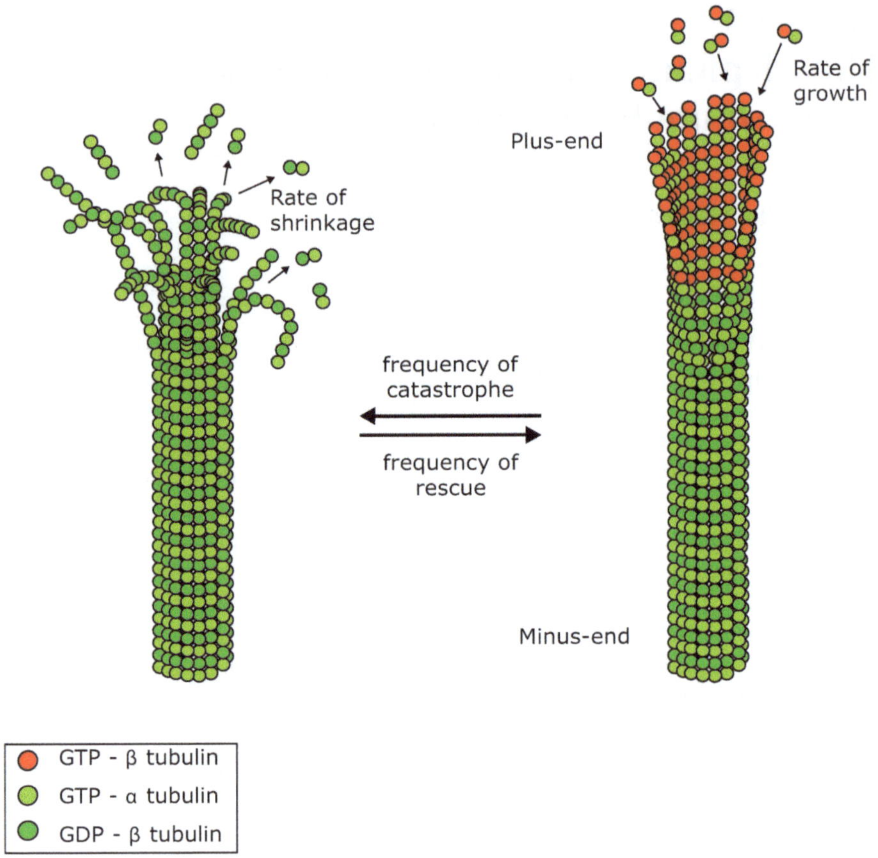

Fig. 5.1 Microtubules and their dynamic instability. Schematic representation of the dynamic instability of microtubules. Dynamic instability is governed by four parameters: growth and shrinkage rates, and catastrophe and rescue frequencies. Microtubules elongate adding tubulin dimers to their plus end, in an end-to-end manner, as the α tubulin moiety binds to the β tubulin in the preceding dimer. The resulting lattice is a cylindrical structure made of 13 tubulin protofilaments. The switch between growth and shrinkage depends on the hydrolysis of GTP bound to β tubulin. Microtubules grow as long as a cap of GTP-β tubulin protects their plus end. If β tubulin hydrolyzes GTP at the plus-end tip, depolymerisation ensues

centre in animal cells. Microtubules mostly grow and shrink from the plus end through addition and removal of tubulin-dimers. Both α- and β-tubulin bind to GTP, but only β-tubulin stochastically hydrolyses GTP to GDP once incorporated into microtubules, inducing a conformational switch that favours microtubule depolymerization. As long as the microtubule plus-end is capped with GTP-loaded dimers, the microtubule will continue to grow; in case GTP hydrolysis in the cap is faster than α/β-tubulin dimer addition, the plus-end loses its GTP-cap resulting in microtubule depolymerization. This behaviour forms the basis for the dynamic instability of microtubules, as microtubules stochastically grow, shrink and switch between these two states autonomously [1]. Microtubule dynamics are described by four parameters: the growth and shrinkage rates, the catastrophe frequency (switching from growth to shrinkage) and the rescue frequency (switching from shrinkage to growth) (Fig. 5.1). These dynamics allow microtubules to explore the three-dimensional space and led Mitchison and Kirschner to postulate the "search and capture" model, by which microtubules rapidly grow and shrink during mitosis until they bind a chromosome [2]. This binding stabilizes microtubules by suppressing microtubule dynamics, allowing a robust attachment of chromosomes to the mitotic spindle. This model was refined and complemented by additional pathways over the years, but the control of microtubule dynamics remains a key part of our understanding of chromosome segregation. Altered microtubule dynamics can lead to chromosome segregation errors, and have been linked to genetic instability, a key hallmark of cancer cells. The idea that errors in the chromosome segregation machinery might be linked to cancer is an old concept, since descriptions of cancer cells in the late nineteenth century already reported centrosome abnormalities (see accompanying review by Rhys and Godinho [3–5]). Studies in the last 20 years rediscovered this theory, based on the fact that cancer cells often show aneuploidy [6], which may arise from defects in microtubule dynamics during mitosis [7]. Here, we review how control of microtubule dynamics governs mitosis, and we highlight the potential mechanisms by which this regulation is altered in cancer cells, and how it might contribute to cancer formation. We discuss the enzymes regulating microtubule dynamics, the microtubule-associated proteins, the tubulin isoforms and the post-translational modifications in the normal and pathological context. Those interested in a broader view on microtubules and cancer should refer to the review of Parker et al. [8].

5.2 Mitosis in a Nutshell

Cell division depends on the faithful segregation of chromosomes during mitosis, a phase in which microtubules undergo a spectacular transformation (Fig. 5.2). As cells are about to enter mitosis, the two centrosomes enhance their microtubule nucleation capacity and migrate to opposite sides of the nucleus, while the chromosomes condense their DNA. After nuclear envelope breakdown, cells assemble a bipolar spindle and spindle microtubules bind chromosomes via kinetochores, large multi-protein complexes that assemble on centromeres and bind to the plus end of spindle microtubules (see accompanying review by De Wolf and Kops). The aim is to bind the chromosomes in a bipolar manner and to align them on the metaphase plate, at equal distance between the two spindle poles. During this process a surveillance mechanism known as the spindle assembly checkpoint (SAC) ensures that the cohesion between the sister-chromatids remains intact until the last kinetochore is bound by spindle microtubules [9, 10]. Once the SAC is satisfied the sister chromatids are pulled apart to separate poles, allowing the re-formation of the nuclear envelope before the two daughter cells are separated through cytokinesis.

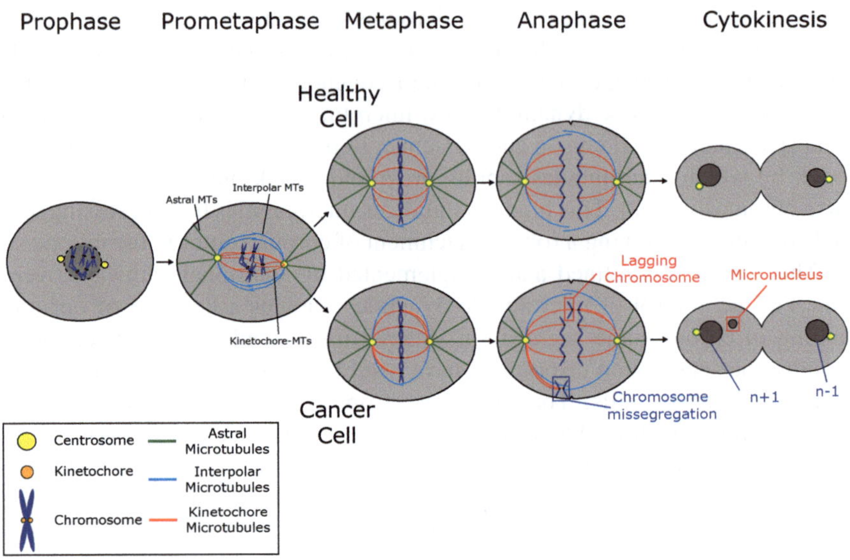

Fig. 5.2 Mitotic origin of chromosomal instability in cancer cells. Depicted is a model illustrating the link between chromosome segregation defects and altered K-MT stability. Mitosis is a dynamic process in which the genetic material must be equally segregated onto two daughter cells. To achieve this goal, the microtubule-based spindle undergoes a series of dynamic changes during mitotic progression. During prophase, the DNA condenses, the kinetochores assemble and the centrosomes migrate to opposite poles of the nucleus. In prometaphase, the nuclear envelope breaks down and the chromosome are bound by microtubules emanating from the centrosomes to form the mitotic spindle. The mitotic spindle consists of three populations of microtubules: astral-, interpolar- and kinetochore-microtubules. These three populations of microtubules exert forces that allow the alignment of chromosome onto the metaphase plate. Once all the chromosomes are aligned in the center of the cell, the segregation of genetic material can start. During anaphase, the sister chromatids of each chromosome are pulled towards opposite poles of the dividing cell and the two centrosomes are pushed apart. In telophase, the nuclear envelope reforms around the chromosomes, before the cytoplasm is divided during cytokinesis. The correct attachment of microtubules to kinetochores is crucial for the faithful segregation of the genetic material. Erroneous attachments may arise and, if not corrected may result in lagging and missegregation of entire chromosomes, and/or the formation of micronuclei, as is often the case in cancer cells. The former phenomenon is the root of whole chromosomal instability (w-CIN) while the latter may lead to structural chromosomal instability (s-CIN) in cancer cells

During each of these steps microtubule dynamics change. As cells transition from interphase to mitosis, microtubules become more dynamic, increasing their turnover rate, due to increased catastrophe and rescue frequencies [11]. Mitosis also sees the emergence of different types of microtubules: the astral, inter-polar, and kinetochore-microtubules (Fig. 5.2). Kinetochore microtubules are characterized by a higher stability, as their half-life is in the order of 2–3 min in prometaphase, in contrast to the other microtubule populations that have a half-life of about 10–15 s [12]. As cells progress through mitosis the kinetochore-microtubules further stabilize, increasing their half-life to 3–4 min in metaphase, before losing most of their dynamicity in anaphase, with half-lives in the order of 10–15 min [12–14]. Although kinetochore fibres do not undergo rapid depolymerizations at this stage, they nevertheless get shorter to pull the chromosomes towards the spindle poles, in a combination of minus-end depolymerization at centrosomes and a kinetochore-driven slow plus-end depolymerisation [15].

5.3 Deregulation of Microtubule Dynamics and Chromosome Instability in Cancer

Any deregulation of microtubule dynamics can result in chromosome segregation errors that may lead to chromosome instability (CIN). CIN comes in two flavours: structural-CIN (s-CIN) and whole chromosome-CIN (w-CIN). The former refers to changes in the structure of a chromosome, such as duplications, deletions, inversions or translocations. The latter, also known as numerical CIN, is the result from a gain or a loss of one or more entire chromosomes. CIN may cause karyotypic variations (aneuploidies) that may increase the probability for a cancer cell to acquire advantageous features, such as a higher proliferation rate or the resistance to a drug [16–21]. Aneuploidy is very frequent in solid human cancers [6, 22, 23] and cancers displaying CIN are associated with aggressiveness and poor prognosis [24–29], raising the possibility that frequent CIN may contribute to cancer progression. The most common cause of CIN in human cancer cells are kinetochore-microtubule attachment defects that may arise through deregulation of microtubule dynamics [7, 16, 20, 30–33].

Normally, the two kinetochores of each chromosome must achieve bi-orientation, with each sister-kinetochore bound by microtubules emanating from opposites spindle poles (amphitely; Fig. 5.3). However, due to the stochastic nature of the attachment process, erroneous kinetochore-microtubule attachments may arise: syntelic attachments, in which both sister-kinetochores are bound my microtubules emanating from one spindle pole, and merotelic attachments, in which one kinetochore is bound by microtubules originating from both spindle poles (Fig. 5.3; [34–36]). Both types of errors must be corrected to prevent chromosome missegregation. Syntelic attachments are destabilized by the centromeric kinase Aurora B via phosphorylation of the outer kinetochore proteins of the Knl1-Ndc80-Mis12 (KMN) network, resulting in unattached kinetochores that activate the SAC [37, 38]. It is generally assumed that the lack of inter-kinetochore tension in syntelic chromosomes brings Aurora B in proximity to its substrates in the outer kinetochore proteins, however other mechanisms might also contribute (Fig. 5.3; for more details see [39–42]). Merotelic attachments, in contrast, are most likely not specifically destabilized by Aurora B and are not detected by the SAC. Instead, error correction relies on the normal detachment rate of kinetochore-microtubules, which is controlled by microtubule dynamics. Given that the geometrical constraints of a sister-kinetochore pair favour amphitelic attachments [43], a sufficiently rapid turnover of kinetochore-microtubules will replace merotelic kinetochore-microtubule attachments over time. Therefore, even though syntelic and merotelic attachments are corrected via

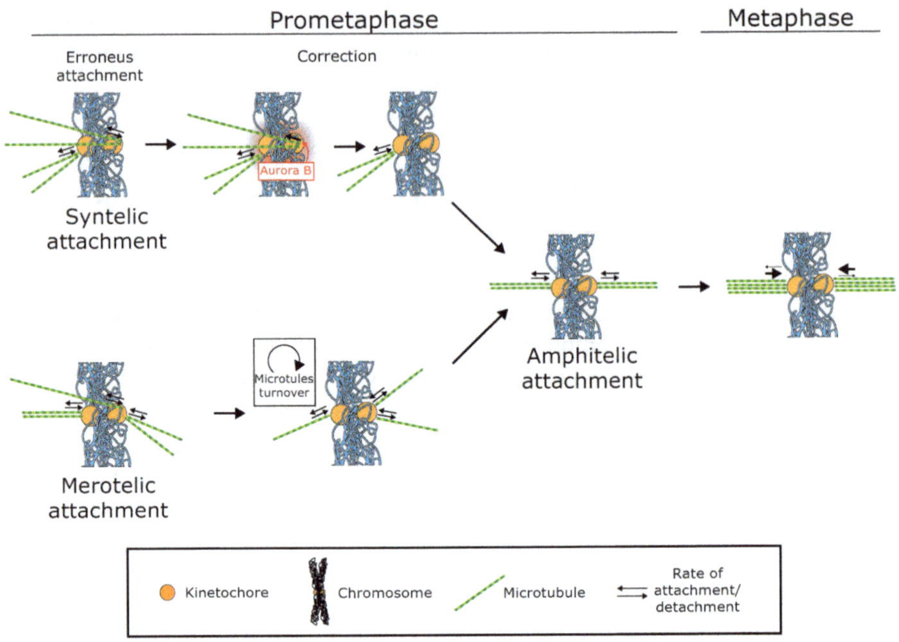

Fig. 5.3 Erroneous kinetochore-microtubule attachments and their correction. Shown are the most common erroneous kinetochore-microtubule attachments and their correction mechanisms. In the case of a syntelic attachment, both sister kinetochores are bound by microtubules emanating from the same centrosome. Syntelic attachment are corrected by the activity of the protein kinase Aurora B, possibly due to the lack of tension between the two sister-kinetochores, which brings the kinetochore-proteins under the influence of the Aurora-B activity gradient in between the two kinetochores. Aurora B phosphorylates kinetochore proteins favoring microtubule detachment. This gives the unattached kinetochore the possibility to bind new microtubules, eventually emanating from opposite poles. This greatly increases the chances to form a correct, amphitelic attachment in which the sister kinetochores are bound to microtubules from opposite spindle poles. In the case of a merotelic attachment, a single kinetochore is bound to microtubules originating from both spindle poles. Error correction of merotelic attachment depends on microtubule turnover, as kinetochore-microtubule attachments will detach in a stochastic manner. Since kinetochore geometry favours amphitelic attachments, merotelic attachments are corrected over time if the turnover rate is high enough

different mechanisms, they both rely on the dynamicity of microtubules. As a consequence, any stabilization of kinetochore microtubules, as seen after the depletion of the kinesins-13 MCAK (mitotic centromere-associated kinesin) or Kif2b (kinesin family member 2B), two mitotic microtubule depolymerases, will impair error correction, leading to chromosome segregation defects [44].

Recent studies suggest that cells fine-tune this general correction mechanism by differentially controlling kinetochore-microtubules dynamics in space and time. At the spatial level kinetochore-microtubule attachments of mis-aligned chromosomes close to centrosomes are destabilized by the centrosomal kinase Aurora A, a paralog of Aurora B [45–47]. At the temporal level the half-live of kinetochore-microtubules increases as cells progress from prometaphase to metaphase, destabilizing the erroneous attachments in early mitosis and to stabilize correct attachments in metaphase. This increase is under the negative control of Cyclin A, a co-activator of the Cylin-dependent kinase Cdk1 that is gradually degraded after mitotic entry [48]. The critical target of Cdk1/Cyclin A is, however, not known. Another pathway that regulates kinetochore stability in time is the molecular switch made by CLASP1, a microtubule plus-end binding protein [49]. In early mitosis CLASP1 is associated to the microtubule depolymerase Kif2b, which destabilizes kinetochore-microtubule attachments; as mitosis progress to metaphase CLASP1 binds to Astrin and promotes microtubule plus-end growth, thus stabilizing kinetochore-microtubule attachments [49, 50]. Nevertheless, metaphase kinetochore-microtubules are still under the control of the microtubule depolymerase MCAK [7, 51, 52], providing the required balance in microtubule stability. MCAK is a direct substrate of Aurora-B, which acts as a master regulator of kinetochore-microtubule stability [52–54], reviewed in [42], as reflected by the fact that Aurora-B inhibition over-stabilises kinetochore-microtubules, resulting in massive chromosome segregation errors [55–60].

The identification of the link between microtubule dynamics and error correction have led to the hypothesis that cancer cells accumulate chromosome segregation errors and become aneuploid because they over-stabilize their kinetochore-microtubules, preventing efficient correction of merotelic and syntelic kinetochore-microtubule attachments. This hypothesis was tested by the Compton group which found that kinetochore-microtubules in cancer cells tend to have a higher half-life and that such an increase in kinetochore-microtubule stability is sufficient to induce CIN [7]. Moreover overexpression of the microtubule-depolymerases MCAK or Kif2b suppressed CIN in cancer cell lines as it destabilized kinetochore-microtubule attachments [13]. Microtubule stabilization has been also linked to the DNA damage response, which is frequently active in cancer cells, as DNA damage in mitosis increases kinetochore-microtubule stability and induces CIN via the mitotic kinases Plk1 (Polo-Like Kinase 1) and Aurora A [61]. Finally, overexpression of the spindle assembly checkpoint protein Mad2 (Mitotic Arrest Deficient 2), a condition that can arise after loss of the tumour suppressor Rb [62], inhibits Aurora B function, also leading to microtubule stabilization and CIN [48].

An alternative deregulation of microtubule dynamics that may also contribute to CIN in cancer cells is the microtubule assembly rate. Holger Bastians and collaborators found that this parameter is higher in colorectal cancer cells displaying CIN than in non-CIN colorectal cancer cells [31]. Conversely, addition of taxol, which freezes the microtubule assembly rate, suppressed the CIN phenotype. Deregulation of microtubule dynamics occurs in these cells via a pathway in which the checkpoint kinase-2 (Chk2) phosphorylates the DNA repair factor breast cancer 1 (BRCA1), resulting in a reduction of Aurora A activity [63–65]. Loss of Chk2 (frequent in colorectal cancer), BRCA1 (frequent in colorectal and breast cancer) or overexpression of Aurora A (frequent in breast and colorectal cancer) [66–70], all increase the microtubule assembly rate leading to CIN [31, 63]. In this case chromosome segregation errors do not arise as a consequence of an impaired error correction, but result from a defective or delayed assembly of the bipolar spindle [31]. Indeed, cells that fail to rapidly assemble a bipolar spindle, with fully separated spindle poles at the opposite ends of spindle axis, form numerous merotelic kinetochore-microtubule attachments that saturate the error-correction machinery [32, 71, 72].

Overall this suggests that chromosome instability due to altered kinetochore-microtubule or spindle microtubule dynamics is a key element of the chromosomal instability seen in cancer cells. By acting as a "mutator" phenotype, such instability could accelerate tumorigenesis or contribute to drug resistances frequently seen in late stage cancers.

5.4 Microtubule Dynamics, CIN, Aneuploidy and Cancer: Evidence for and Against

Although the model linking changes in microtubule dynamics, chromosome segregation errors, CIN, aneuploidy and cancer is attractive, there are controversial points and many unresolved questions. At the phenomenological level although there is a correlation between the presence of merotelic chromosomes and CIN [7, 13, 33], it is striking that merotelic chromosomes rarely missegregate in anaphase, raising the question to which extent they give rise to whole chromosome CIN [73–75]. Nevertheless, lagging merotelic chromosomes in anaphase often end up forming micronuclei in the ensuing interphase. In micronuclei chromosomes experience severe DNA damage with major chromosomal rearrangements, a phenomenon called chymotrypsis [76, 77]. This could suggest that lagging chromosomes are more a cause of structural CIN, whereas whole chromosome CIN may have an independent origin. Therefore, it will be important in the future to test the exact relationship between altered kinetochore microtubule stability and whole chromosome CIN in cancer tissues.

A second question is whether CIN is sufficient to induce aneuploidy and ultimately cancer formation. Indeed, chromosome missegregation, alone, is not sufficient to drive aneuploidy in stable diploid cell lines [33]. It is nevertheless possible that these conditions do not resemble those found in vivo. Many more generations of CIN may be needed to result in an advantageous karyotype that will emerge, and/or another event may be required to establish aneuploidy. A prime suspect is loss of the tumour suppressor p53, which blocks the cell cycle after chromosome missegregation [78, 79]. A second suspect is tetraploidy, which in conjunction with p53 loss, favours the appearance of aneuploidy [80]. Consistent with this idea genetic backgrounds in mice leading to a pure CIN phenotype have only led to cancer formation at a late age [81–85]. In conjunction with other tumorigenic conditions, in some cases it accelerated, but in other cases suppressed cancer formation, suggesting that CIN is a double-edged sword that must be evaluated within a defined genetic background (reviewed in [86]).

Finally, whether aneuploidy on its own gives rise and contributes to cancer is matter of debate (for a review refer to [87, 88]). Yeast strains carrying extra chromosomes are defective in cell cycle progression and show poor viability, as well as aneuploid mammalian cells [89, 90]. Moreover, single chromosome aneuploidy acts as a tumour suppressor [91] arguing against an advantage for cancer cells of an aneuploid karyotype. Nevertheless, almost 70% of all solid tumours are aneuploid [6] and many observations suggest that aneuploidy can trigger cell transformation [17, 21, 92–94]. Moreover, human diseases that lead to aneuploidy are often associated with increased cancer risk. For instance, people affected by mosaic variegated aneuploidy syndrome (MVA) are predisposed to childhood cancer [95–98]. Therefore, the link between aneuploidy and cancer remains to be fully understood.

At the molecular and cellular level, there are a number of unresolved points in the link between kinetochore-microtubule stability, CIN, aneuploidy and cancer. First, although the two models linking changes in microtubule dynamics and CIN (change in microtubule stability and change in microtubule polymerization rate) may appear very similar, they are not the same. In the microtubule stability model, taxol leads to a higher microtubule stability and a higher rate of chromosome segregation errors [13]; in contrast in the microtubule polymerization rate model, taxol reduces this polymerization rate, leading to a lower CIN rate [31]. This suggests that both models cannot be true at the same time in the same cell. However, both models were tested in different cell lines (osteosarcoma and glioblastoma for the stability model vs. colorectal cancer cell lines in the case of microtubule polymerization rate), which raises the possibility that both types of deregulation exist, depending on the cancer type. Moreover, excessive destabilization of kinetochore-microtubules, has been also linked to the appearance of lagging chromosomes [99], implying that multiple types of microtubule deregulation could lead to the same phenotype. A second case of disagreement focuses on the exact function of the factors that regulate kinetochore-microtubule stability. Plk1 has been proposed to destabilize microtubules via Kif2b [100], or conversely to stabilize kinetochore-microtubule attachment [61, 101]. A similar controversy concerns Aurora A, as it contributes to an increased microtubule stability and microtubule growth rate in some situations [31, 61, 102, 103] but has also been shown to activate the microtubule destabilizer MCAK [104] and to destabilize kinetochore-microtubule attachments near the centrosomes [45–47]. This could suggest that Plk1 and Aurora A fine-tune kinetochore-microtubule stability depending on the mitotic phase and the location of the respective kinetochore. Another difficulty is that measurements of microtubule dynamics by photo-activation experiments can be quite variable, leading to large confidence intervals: for example it is generally accepted that kinetochore-microtubules have a slower turnover in metaphase than in prometaphase based on measurements in U2OS and RPE1 cells [48, 105]; however other studies from the same laboratory found no difference for DAOY cells or even the opposite for U251 cells [13]. Conversely the same studies reported either large, twofold differences in kinetochore-microtubule dynamics between cancerous U2OS and non-cancerous RPE1 cells [13] or barely any differences at all [48]. This suggests that better tools must be developed to measure microtubule dynamics in the future.

Finally, in terms of cancer genetics, one would expect that regulators that decrease kinetochore-microtubule stability should have been identified as tumour suppressors, if hyper-stable kinetochore-microtubule attachments were to unequivocally lead to CIN and cancer. Although Kif2b and MCAK overexpression restores chromosome stability in cancer cell lines with CIN [13] MCAK is not a tumour suppressor gene. Indeed, MCAK has been found upregulated in gastric and breast cancer [106, 107], while in colon cancer its expression correlates with metastatization and poor prognosis [108]. Moreover, MCAK promotes resistance to the microtubule stabilizing drug taxol in cancer cell lines [109, 110]. Among the genes known to destabilize kinetochore-microtubule attachments Aurora A, Aurora B and Cyclin A have been found upregulated in cancer while Plk1 is considered a proto-oncogene, and Kif2b has never been associated to cancer [111–116]. In fact, the only two *bona fide* tumour suppressor genes involved in microtubule stability are APC and VHL, which both stabilize kinetochore-microtubules (see below).

On the same track, if hyper-stable kinetochore-microtubule attachments were to favour cancer progression, one would expect to find regulators that stabilize them upregulated in cancer. This is the case for a number of these proteins, such as Astrin, Bub3 (Mitotic checkpoint protein BUB3), Hec1 (Highly Expressed in Cancer 1), HURP (Hepatoma Up-Regulate Protein), Mad2, SKAP (Small Kinetochore-Associated Protein) [117–123]. However, for all of these proteins there is no indication that they are more overexpressed than other mitotic genes, a class of proteins that tend to all be overexpressed in cancer cells, due to their higher proliferation rate [124]. At present stage, there is no clear genetic fingerprint linking regulators of kinetochore-microtubule stability to malignancies. This contradiction may be due to the fact that kinetochore-microtubule dynamics can only be changed within a given range that is not compatible with classical loss of function of tumour suppressors or the strong overexpression of oncogenes. One might thus need to look for more subtle changes in protein expression, particularly in microtubule binding proteins, which might provide an easier way to fine-tune kinetochore-microtubules (see below). Alternatively, changes in kinetochore-microtubule dynamics may play different roles at different steps of cancer progression. One may speculate that initially hyper-stable kinetochore-microtubule attachments may favour the tumour. As the cancer progresses, or upon drug selection, cells with less stable kinetochore-microtubule attachments may gain an advantage and replace the original population. It is also worth remembering that most of the regulators involved in kinetochore-microtubule stability have other roles both in interphase and mitosis: Mad2 is the main player of the SAC, Cyclin A an important cell cycle regulator, Plk1 and the Aurora kinases play several roles in spindle assembly and in mitotic timing. The other roles may contribute to cancer more than their function in kinetochore-microtubule dynamics. Thus, it will be crucial to test in animal models if changes in regulators of kinetochore-microtubule dynamics, such as MCAK or Kif2b, contribute to cancer formation.

5.5 Microtubule-Associated Proteins

The second major class of proteins potentially involved in deregulation of microtubule dynamics in cancer cells is the family of microtubule-associated proteins (MAPs). The functions of MAPs are broad: some MAPs have a structural role, linking microtubules with other cell components or forming microtubule bundles; others can affect the polymerization rate of tubulin or transport cargoes along microtubules. MAPs with a microtubule-stabilizing function are often found upregulated in cancer, supporting the idea of increased microtubule stability in malignances. One caveat to this is, however, that mitotic genes are generally up-regulated in cancers as a result of an increased proliferation rate [124].

MAP2, Mdp3 (microtubule-associated protein 7 (MAP7) domain-containing 3 (Mdp3)), CLIP-170 (Cytoplasmic Linker Protein of 170 kDa), EB1 (End Binding 1), Survivin [125–131], and Ch-TOG are some of the MAPs which stabilize microtubules and are up-regulated in cancer. Tau is a neuronal MAP that stabilizes microtubules binding at the interface between tubulin dimers [132, 133]. Its role is best characterized in neurodegenerative diseases: it is mutated in fronto-temporal dementia and found hyperphosphorylated in Alzheimer disease [134]. Tau can be found expressed in breast cancer where it correlates with poor prognosis [135]. Similarly, in melanoma and neuroblastoma, the increased expression of another neuronal MAP, MAP2, which induces the formation of microtubules bundles, has been associated with reduced sensitivity to the microtubule stabilizing drugs taxanes [136, 137]. Mdp3 stabilizes microtubules increasing their assembly rate [130] and its expression in breast cancer promotes metastasis and cancer growth [138]. CLIP-170, which promotes microtubules growth [128, 139], has been related to Hodgkin lymphomas [140, 141]. More recently CLIP-170 has been found to induce tumor angiogenesis and increase the sensitivity to taxanes in breast cancer cells [142, 143]. EB1 stimulates microtubules nucleation and growth [144] and is upregulated in many kinds of cancers [145–151]. Survivin stimulates nucleation of microtubules [129] and is overexpressed in nearly all human malignances [129, 152]. The MAP Ch-TOG (colonic and hepatic Tumor Overexpressed Gene), has been isolated as a gene overexpressed in tumors [153] and was later found to be the homologue of XMAP215. XMAP-215-Ch-TOG use specific regions of the TOG domain to bind the microtubule lattice on one side and free tubulin on the other to promote microtubule polymerization at the plus end [154–156]. Ch-TOG proteins bind to TACC protein family members (Transforming acidic coiled coil), which are important for microtubule stability. TACC proteins have been found to be both upregulated and downregulated in cancers where they are linked to genomic instability. On the mitotic spindle, TACC3 and Clathrin (which normally coats transport vesicles for protein trafficking) form together, in an Aurora-A regulated manner, a microtubule-binding site that interacts with the microtubules of the spindle, targets Ch-TOG to the complex and stabilizes K-fibers [157–160] reviewed in [161]. However, all TACCs have also potential roles in transcription, which could contribute to their tumorigenesis effect. Considering the diverse functions of TACC proteins and the fact that both higher and reduced expression are observed in cancers, it is not possible to conclude whether TACC proteins behave as oncogene, as tumor suppressors or both [162]. However, since depleting clathrin results in mitotic delays, targeting this complex with drugs may overcome some of the toxicity and resistance arising from using microtubule drugs [163].

The work summarized above may suggest that MAPs that stabilizes MTs behave as oncogenes. There are however microtubule stabilizing MAPs with tumor suppressor functions and, *vice versa*, microtubule destabilizers that are up-regulated in cancer. For instance, VHL (Von Hippel–Lindau tumor suppressor) stabilizes microtubules [164] but its loss is associated with a dominant cancer syndrome [165, 166]. Germline mutations of the tumor suppressor Adenomatous polyposis coli (APC) result in familial *adenomatous polyposis*, a syndrome linked with the development of colorectal cancer and other malignances. APC loss results in CIN in human cells [167, 168] and microtubules destabilization in many systems [169–171] although one study also reported that it stabilizes microtubules [13]. Instead, Stathmin, which has been shown to destabilize microtubules [172], is commonly up-regulated in cancer [173–178].

Therefore, it is hard to find a simple correlation between the expression of a MAP set and cancer development. First, to our knowledge, there is no MAP linked to MT stability that is up-regulated in cancer more than other cell division genes, suggesting that the up-regulation is a consequence of the high proliferation rate rather than the cause of it. Second, nearly all MAPs play different roles. The tumor suppressor APC can block DNA repair (reviewed in [179, 180]) and regulates Wnt signaling, the main function of VHL is to trigger the degradation of the hypoxia inducible factor HIF-1α, preventing its oncogenic effects (reviewed in [181]), TACC proteins are involved in regulating transcription (reviewed in [162]) and Ch-TOG is important in vesicle trafficking. These roles may contribute to cancer independently from the stabilization-destabilization effect on microtubules.

5.6 Mutations of Tubulin Genes

Since microtubules are crucial in many physiological processes and their dynamics are affected in cancer cells, one would expect to find alterations in tubulin genes during cancer progression as it happens for other genes, such as p53. Although some tubulin mutations have been correlated with drug resistance in cancer cell lines [182–187] their importance in cancer progression remains controversial. β-tubulin mutations do not occur or are very rare in lung cancers [188, 189], ovarian cancers [189–191], breast cancers [192, 193] and gastric cancers [194]. The data suggest that tubulin mutations can confer resistance to drugs in vitro but this does not represent a common mechanism by which cancer acquires resistance in patients, possibly because mutations in tubulin genes might not be compatible with life. Therefore, if altered MT stability can spark cancer initiation, it is unlikely to depend on mutations of tubulin genes.

5.7 Altered Expression of Tubulin Isotypes in Malignancies

In humans, the tubulin dimer subunits are encoded by at least seven α-tubulin and eight β-tubulin genes [195]. Each gene is translated into a different protein, giving rise to several tubulin isotypes. Tubulin isotypes expression and levels change between different organs and tissues. For instance, tubulin beta class I (TUBB) is broadly expressed [196, 197] while tubulin beta 2A class IIa (TUBB2A) is limited to the nervous tissue and muscles [198] and tubulin beta 3 class III (TUBB3) is almost exclusively expressed in neurons [199]. Some tubulin isotypes have been proposed to play specific roles in specific tissues or subcellular localization. For example, tubulin beta 4A class IVa and IVb (TUBB4A and TUBB4B) constitute the axoneme of cilia and flagella. It must be noted that this does not mean that microtubules in those cells or structures are exclusively composed of a specific tubulin isotype, but rather that this isotype is prevalent. Here we focus on the tubulin isotypes that have been found to be involved in cancer and discuss their properties and their involvement in drug resistance.

α-Tubulin Isotypes α-tubulin isotypes have never been extensively investigated, mostly due to the lack of isotype-specific antibodies. Except for α8-tubulin, all isotypes are very similar at the sequence level and are broadly expressed. Specific functions of different α-tubulin isotypes have never been reported. Tubulin alpha 4a and 4b (TUBA4a and TUBA4b) are naturally detyrosinated and detyrosinated microtubules are generally more stable (see below). Since tubulin detyrosination has been associated with breast cancer aggressiveness [200] it would be tempting to speculate that an increased expression of α4 tubulin can contribute to cancer aggressiveness. However, increased TUBA4a or TUBA4b expression have never been reported in tumors. High tubulin alpha 1b (TUBA1B) expression has been correlated with poor prognosis in hepatocellular carcinoma [201] and mantle cell lymphoma [202]. Tubulin alpha 3c (TUBA3C) expression correlates with reduced sensitivity of ovarian cancer to the microtubule stabilizing agent paclitaxel [203]. Nevertheless, at present stage there is no strong link between changes in microtubule dynamics in cancer cells and α-tubulin isotypes.

β-Tubulin Isotypes β-tubulin isotypes have been studied more extensively than α-tubulins, both in healthy tissues and in cancer. Different β-tubulin isotypes have different expression patterns and some of them fulfill specialized functions. Of interest is TUBB3 because of its unique features in the tubulin superfamily. TUBB3 lacks Cys239, which, in other isotypes, is very sensitive to oxidation and alkylation. Oxidation and alkylation of β-tubulin on Cys239 inhibit microtubule polymerization [204, 205]. Also, TUBB3 has an unusual Cyst124, which may act as a scavenger of reactive oxygen species (ROS), forming a disulfide bond with either Cys127 or Cys129. For this unusual cysteine distribution TUBB3 has been proposed to be more resistant to oxidative stress and ROS production [206]. Another feature of TUBB3 is the presence of a phosphorylatable serine at the C-terminus. Phosphorylation of this residue makes microtubules more stable by increasing their assembly rate in presence of MAPs [207, 208]. Moreover, phosphorylated TUBB3 interacts more efficiently with the microtubule stabilizer MAP2 [209]. On the other hand, phosphorylation of TUBB3 on Ser172 by Cdk1 (Cyclin dependent kinase 1) at the onset of mitosis impairs the incorporation of TUBB3 in microtubules, reducing their growth rate [210, 211].

In vitro, microtubules made of TUBB3 are less stable and more dynamic than microtubules made of TUBB2 (A and B) or TUBB4s [212–214], However, in vivo, a high expression of TUBB3 is commonly found in many kinds of cancer [215] reviewed in [8], which would contradict the hypothesis that more stable microtubules can contribute to cancer progression. There may be several, not mutually exclusive, ways in which TUBB3 contributes to cancer. First, TUBB3 is more resistant to oxidative stress because of its Cys 124. This is an advantage in cancer cells, which typically show high ROS concentration (reviewed in [216]). Another interesting hypothesis concerns the phosphorylation of TUBB3 by Cdk1. Since Cyclin A destabilizes kinetochore-microtubule attachments in prometaphase [48] it would be tempting to speculate that Cyclin A is the binding partner of Cdk1 in phosphorylating TUBB3. Cells expressing more TUBB3 may overcome the negative regulation of Cyclin A, stabilizing prematurely kinetochore-microtubule attachments which then results in CIN, as discussed above. In vitro only CyclinB/Cdk1 has been tested and shown to phosphorylate TUBB3 [211] but this does not exclude a contribution of Cyclin A. However, whether TUBB3 overexpression is sufficient to cause CIN is not known. To date, a direct evidence for a contribution to cancer progression is missing. Moreover, how β tubulin isotypes affect microtubule stability is not clear. In the future, it will be interesting to modify the relative contribution of β tubulin isotypes in cultured cells and animal models and assess their role in tumor initiation and progression.

Tubulin Isotypes, Drug Resistance and Future Perspectives In addition to their potential role in modifying microtubule dynamics, tubulin isotypes may contribute to the resistance to tubulin binding agents (TBAs). Different tubulin isotypes bind drugs with different affinities [217–220]. However, the role of tubulin isotypes in drug resistance is not clear. Expression of TUBB, TUBB2 (A and B) and TUBB3 seems to induce resistance to microtubule stabilizing drugs [221–225]. TUBB3 expression confers resistance to platinoids [183, 215, 226–230] reviewed in [8, 224]. Moreover, other studies correlate high expression of TUBB3/TUBB1 to resistance to microtubule stabilizing agents [226, 228, 229, 231]. However, in clear cell ovarian carcinoma, TUBB3 expression is a marker of favorable drug response [232] and in vitro TUBB3 microtubules are more sensitive to Vinca alkaloids [207]. How can we reconcile these data? One possibility is that different tubulin isotypes may contribute to drug resistance via a mechanism independent on their role in microtubule stability. For instance, TUBB3 may protect cancer cells preventing the apoptotic role of ROS, instead of, or in addition to, having a direct role in microtubule stability. Alternatively, the contribution of tubulin isotypes to drug resistance may be highly dependent on the cellular context.

Precisely assessing the role of tubulin isotypes in microtubule stability and drug resistance is difficult. Tubulin is traditionally isolated from cow, sheep and pig brains but such a technique has several drawbacks: (1) due to the antibodies availability tubulin dimers were isolated regardless of the α-isotype. (2) tubulin in nervous tissues carries several PTMs, making difficult to separate the contribution of tubulin PTMs from the one of tubulin isotypes. (3) the purification techniques rely on microtubule polymerization/de-polymerization cycles, which can lead to loss of both hyper-stable and hypo-stable microtubules. Recently developed methods to purify tubulin from different sources [233–235] will provide the solution to have a sufficient amount of specific isotypes of tubulin for in vitro studies. In vivo however it is essential to specifically identify tubulin isotypes. One difficulty is that mRNA levels do not always reflect protein levels [236]. For instance, Kavallaris et al. [231] reported an increase in TUBB2 (A and B) and TUBB4A mRNAs in ovarian tumors resistant to taxol. However proteomic approaches revealed that the proteins were not increased [237]. The fast advance of proteomic techniques at the single cell level may provide a solution to this, although the similarity of tubulin isotypes may constitute a problem for quantitative studies. Alternatively, it may now be possible to attach small tags to different isotypes using the CRIPSR/Cas9 technology. This would allow the in vivo detection of the different isotypes in different cell types. The ability to study tubulin isotypes in vivo in normal and cancer cells combined with the ability to interfere with the levels of these isotypes will help to understand how they affect microtubule stability and cell division.

5.8 Tubulin Post-Translational Modifications and Cancer

As any other protein, after translation, tubulin molecule is subject to post-translational modifications (PTMs; Fig. 5.4). Some of these modifications, such as phosphorylation and acetylation, are common protein modifications. Others, such as polyglutamylation, are unusual in other proteins and a few (tyrosination, polyglycylation) are specific to tubulin. Each subunit of the tubulin dimer ends with a C-terminal tail (10–12 residues for α-tubulin, 16–22 residues for β-tubulin [238] that has been predicted to project outside the globular structure of the protein. The C-terminal tail is the site where most of the PTMs of tubulin occur. Many different modifications have been identified so far and the list may not be complete. Since the C-terminal of tubulin is the region that varies the most between isotypes, different isotypes may carry different PTMs. Changes in the PTMs landscape of microtubules can have dramatic effects on the interaction with MAPs and on microtubule stability (reviewed in [239]).

α-Tubulin Detyrosination One of the most studied tubulin PTM is α-tubulin detyrosination. Most of the α-tubulin isotypes end with a C-terminal tyrosine. This tyrosine can be removed [240] by a so far unknown carboxypeptidase. Tyrosine can also be added back to α-tubulin [241] by a tubulin tyrosine ligase (TTL) [242–246].

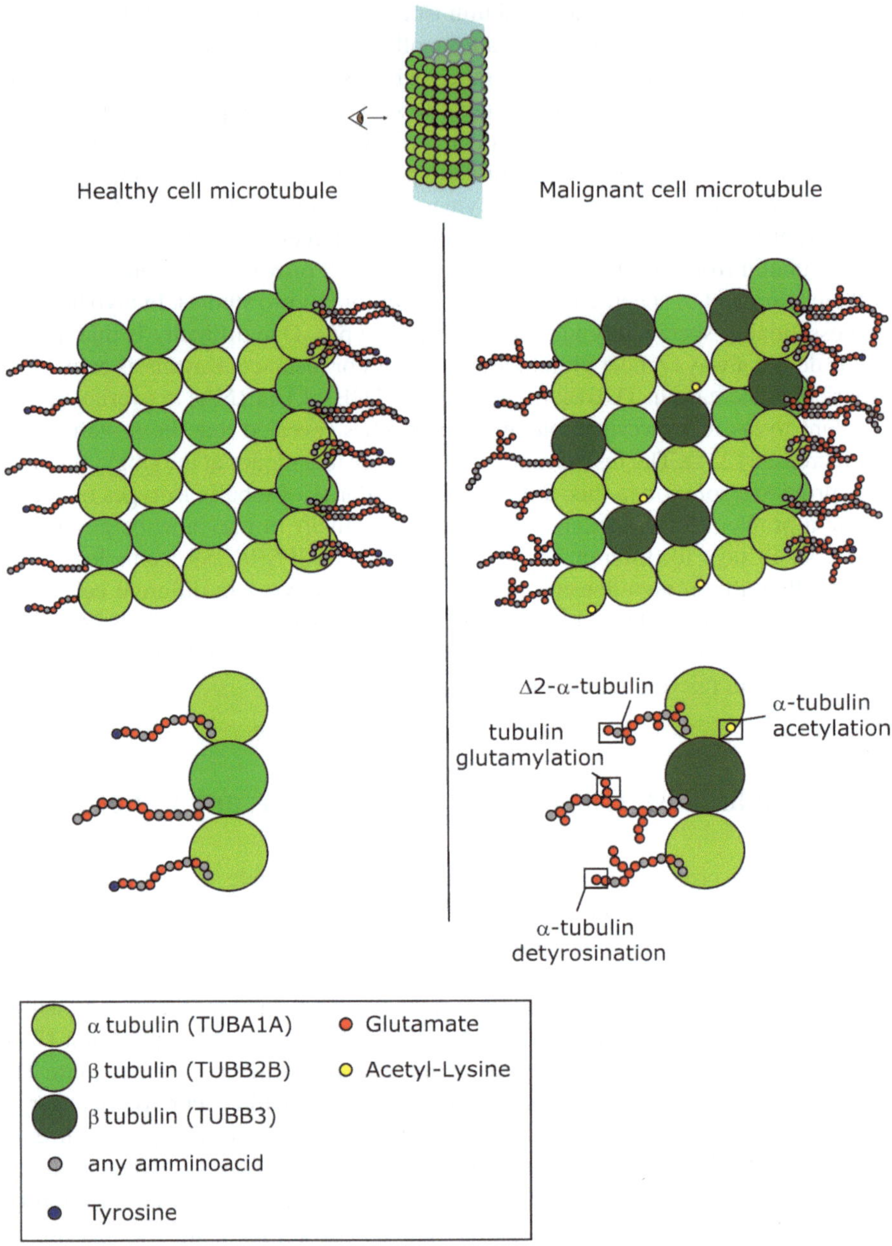

Fig. 5.4 Tubulin post translational modifications and isotypes expression in normal cells versus cancer cells. Shown is a schematic representation of the cross section of microtubules of healthy and malignant cells. Microtubules in cancer cells show higher levels of TUBB3 expression and they carry different PTMs modifications. The depicted modifications, such as α-tubulin acetylation, α-tubulin detyrosination, Δ2-α-tubulin and tubulin glutamylation, have been observed, or proposed to happen, in cancer cells

Detyrosination of tubulin increases microtubule stability by inhibiting the interaction between microtubules and depolymerizing kinesin-13s, such as Kif2s and MCAK [247, 248]. Indeed, detyrosinated tubulin is mostly associated with stable, long lived microtubules [249–251].

In tumors, tubulin detyrosination has been associated with aggressiveness of breast cancer [200]. Consistently, loss of TTL enzyme has been linked to tumor growth and aggressiveness [252] and an active tyrosination machinery correlates with a favorable patient outcome in neuroblastoma [253]. In cancer cells the reduced interaction between kinesin 13s and microtubules due to detyrosinated tubulin could lead to hyperstable kinetochore-microtubule attachments. Recent work has shown that proper chromosome congression depends on de-tyrosination of kinetochore MTs compared to tyrosinated astral microtubules since chromosome transport by CENP-E is strongly enhanced by de-tyrosinated microtubules. Therefore, cells depleted for TTL may display CIN because of altered kinetochore-microtubule stability or because of detyrosination of astral microtubules or both [233].

Δ2-α-Tubulin In Δ2-α-tubulin the penultimate residue of α-tubulin, which is a glutamate, is removed [254]. The formation of Δ2-tubulin requires prior detyrosination and after the removal of glutamate, tubulin cannot be retyrosinated [254]. A specific role of Δ2-tubulin on microtubule dynamics has not been shown, suggesting that its only function is to lock microtubules in a detyrosinated state. High Δ2-α-tubulin expression correlates with poor response to microtubule stabilizers alkaloids derived from Vinca [255].

Tubulin Polyglutamylation Polyglutamylation consists in the addition of up to 20 residues of glutamate to the γ carboxyl group of a C-terminal glutamate of either α or β tubulin. Some tubulin tyrosine ligase like (TTLLs) enzymes can promote this reaction, creating a glutamate chain to the C-terminal tail of tubulin. Polyglutamylation is associated with stable microtubules and several observations suggest that this PTM might increase microtubule stability [256]. Physiologically, glutamylation regulates the beating behavior of cilia [257–259].

Some TTLL enzymes are upregulated in cancer, suggesting a link between tubulin glutamylation and malignancies. Tubulin tyrosine ligase like 12 (Ttll12) is one of the most abundant antigen in prostate cancer [260] and its expression correlates with prostate cancer progression [261]. Overexpression of Ttll12 causes mitotic delay in cancer cells [262] and it leads to aneuploidy [261]. However, although Ttll12 increase correlates with high levels of polyglutamylated tubulin, it has never been shown to directly catalyze such a reaction. Moreover, Ttll12 is known to methylate histones, suggesting that its role in cancer, if any, may not be related to its role on microtubules [261]. Polyglutamylation correlates with drug resistance in cancer. However, the resistance arises against drugs that bind to different sites on tubulin and that have opposite effects on microtubule stability. For example, polyglutamylated tubulin has been implicated in resistance to microtubule destabilizing drugs estramustine and nocodazole [263, 264]. In breast cancer instead, high polyglutamylation correlates with resistance to the microtubule stabilizing drug paclitaxel [265].

We cannot conclude whether microtubule polyglutamylation has a role in cancer progression. It is possible that different extend of the glutamate chain may have an opposite impact on microtubule stability and interaction with MAPs, resulting in a different outcome in term of cell division and cancer progression.

α-Tubulin Acetylation α-Tubulin can be modified via the covalent binding of an acetyl group to a lysine residue. Acetylated tubulin is common in stable microtubules. However, it is unlikely that acetylation on lys40 affects directly microtubule stability since it does not affect microtubule structure [266]. Some line of evidence suggested that lys40 acetylation regulates intracellular transport by increasing the interaction of microtubules with kinesin motors [267, 268]. Acetylation of lys40 of α-tubulin has been reported in cancer cells [269]. Increased histone deacetylase 6 enzyme (HDAC6), which can catalyze tubulin de-acetylation, correlates with better prognosis in breast cancer [270]. Recently, increased levels of acetylated α-tubulin have been found in breast cancer cells and associate with invasive migration [271]. Whether increased acetylation results in CIN is not known.

5.9 Conclusion

In this review, we discussed the role of microtubules in cancer with a focus on mitosis and microtubule dynamics. We highlighted potential mechanisms linking changes in microtubule dynamics and cancer, such as stabilization of kinetochore-microtubule attachment and/or an increased rate of tubulin polymerization, which may cause CIN in mitosis. CIN, in turn, may lead to aneuploidy in the daughter cells, leading to the acquisition of selective advantages that favor cancer progression. However, in contrast to other biological process associated to cancer formation, such as DNA repair or PIK3 kinase signaling, there is lack of clear genetic signature (mutations or deletions) or molecular biomarkers that would link changes in microtubule dynamics to cancer formation. How could it be that such a link has not yet been identified? Formally, one cannot exclude that it does not exist. A second possibility is that the relevant molecular biomarker still needs to be identified. The third possibility, which we favor most, is that changes in microtubule dynamics might be the result of a combination of several minor changes in the expression levels or activity of MAPs or microtubule regulators. Such minor changes, such as epigenetic control of promoters, less than twofold changes in mRNA levels, or minor changes in the turnover of the protein, are not detected by the "omics" technologies, yet might lead to CIN, due to the high sensitivity of chromosome segregation to minor perturbations of kinetochore-microtubule dynamics [49]. Consistent with such an idea, we note how a recent study could correlate minor changes in the expression of kinetochore and centromere protein genes to cancer patient survival and response to chemotherapy when performing a combined analysis on this group of genes, but not when studying single genes [272].

Based on these ideas we predict that in terms of fundamental understanding, we should not restrict our analysis to the effect of single proteins, but rather should aim to understand how a combination of different regulators, MAPs, and post-translational modifications affect microtubule dynamics. Such a more holistic approach might require the development of in vitro recombinant protein systems to study how tubulin isotypes and tubulin post-translational modifications regulate microtubule dynamics on their own and in combination with other MAPs and regulators. This will allow the development of mathematical models capable of predicting the stability of microtubules in different cellular environment (cancer vs. non-cancer cells) and their sensitivity to microtubule-targeting drugs. At the clinical level in contrast, we speculate that instead of focusing on molecular biomarkers, which could be highly variable and difficult to detect, it might be much more effective to use functional biomarkers, such as the relative stability of mitotic microtubules in a given cancer tissue. Although such functional biomarkers are more difficult to measure in tissue biopsies, they might be much more robust in terms of predictive values and might thus allow a much more targeted approach with microtubule-interfering anti-cancer drugs.

Acknowledgements The authors thank Helder Maiato (University of Porto), Carsten Janke (Institut Curie) and the members of the Gotta and Meraldi laboratories for helpful discussions. This work was supported by an SNF-project grant (No 31003A 156013) to M.G., an SNF-project grant (No 31003A_160006) to P.M. and the University of Geneva.

References

1. Mitchison T, Kirschner M (1984) Dynamic instability of microtubule growth. Nature 312:237–242
2. Kirschner M, Mitchison T (1986) Beyond self-assembly: from microtubules to morphogenesis. Cell 45:329–342
3. Boveri T (1902) Über mehrpolige Mitosen als Mittel zur Analyse des Zellkerns. Verhandl Phys-Med Ges Würzburg, 67–90
4. Galeotti G (1893) Beitrag zum Studium des Chromatins in den Epithelzellen der Carcinome. Beitr Pathol Anat Allg Pathol 14:249–271
5. Hansemann D (1890) Ueber asymmetrische Zelltheilung in Epithelkrebsen und deren biologische Bedeutung. Virchows Archiv 119:299–326
6. Mertens F, Johansson B, Höglund M, Mitelman F (1997) Chromosomal imbalance maps of malignant solid tumors: a cytogenetic survey of 3185 neoplasms. Cancer Res 57:2765–2780
7. Bakhoum SF, Thompson SL, Manning AL, Compton DA (2009b) Genome stability is ensured by temporal control of kinetochore-microtubule dynamics. Nat Cell Biol 11:27–35
8. Parker AL, Kavallaris M, McCarroll JA (2014) Microtubules and their role in cellular stress in cancer. Front Oncol 4
9. Etemad B, Kops GJ (2016) Attachment issues: kinetochore transformations and spindle checkpoint silencing. Curr Opin Cell Biol 39:101–108
10. Musacchio A (2015) The molecular Biology of spindle assembly checkpoint signaling dynamics. Curr Biol 25:R1002–R1018

11. Rusan NM, Fagerstrom CJ, Yvon AM, Wadsworth P (2001) Cell cycle-dependent changes in microtubule dynamics in living cells expressing green fluorescent protein-alpha tubulin. Mol Biol Cell 12:971–980
12. Zhai Y, Kronebusch PJ, Borisy GG (1995) Kinetochore microtubule dynamics and the metaphase-anaphase transition. J Cell Biol 131:721–734
13. Bakhoum SF, Genovese G, Compton DA (2009a) Deviant kinetochore microtubule dynamics underlie chromosomal instability. Curr Biol 19:1937–1942
14. Cimini D, Wan X, Hirel CB, Salmon ED (2006) Aurora kinase promotes turnover of kinetochore microtubules to reduce chromosome segregation errors. Curr Biol 16:1711–1718
15. McIntosh JR, Hays T (2016) A brief history of research on mitotic mechanisms. Biology 5:55
16. Cimini D, Howell B, Maddox P, Khodjakov A, Degrassi F, Salmon ED (2001) Merotelic kinetochore orientation is a major mechanism of aneuploidy in mitotic mammalian tissue cells. J Cell Biol 153:517–528
17. Duijf PHG, Schultz N, Benezra R (2013) Cancer cells preferentially lose small chromosomes. Int J Cancer 132:2316–2326
18. Lengauer C, Kinzler KW, Vogelstein B (1997) Genetic instability in colorectal cancers. Nature 386:623–627
19. Lengauer C, Kinzler KW, Vogelstein B (1998) Genetic instabilities in human cancers. Nature 396:643–649
20. Thompson SL, Bakhoum SF, Compton DA (2010) Mechanisms of chromosomal instability. Curr Biol 20:R285–R295
21. Weaver BAA, Cleveland DW (2006) Does aneuploidy cause cancer? Curr Opin Cell Biol 18:658–667
22. Holland AJ, Cleveland DW (2012) Losing balance: the origin and impact of aneuploidy in cancer. EMBO Rep 13:501–514
23. Pfau SJ, Amon A (2012) Chromosomal instability and aneuploidy in cancer: from yeast to man. EMBO Rep 13:515–527
24. Choi C-M, Seo KW, Jang SJ, Oh Y-M, Shim T-S, Kim WS, Lee D-S, Lee S-D (2009) Chromosomal instability is a risk factor for poor prognosis of adenocarcinoma of the lung: fluorescence in situ hybridization analysis of paraffin-embedded tissue from Korean patients. Lung Cancer 64:66–70
25. Gao C, Furge K, Koeman J, Dykema K, Su Y, Cutler ML, Werts A, Haak P, Vande Woude GF (2007) Chromosome instability, chromosome transcriptome, and clonal evolution of tumor cell populations. Proc Natl Acad Sci U S A 104:8995–9000
26. Heilig CE, Löffler H, Mahlknecht U, Janssen JW, Ho AD, Jauch A, Krämer A (2010) Chromosomal instability correlates with poor outcome in patients with myelodysplastic syndromes irrespectively of the cytogenetic risk group. J Cell Mol Med 14:895–902
27. Kuukasjärvi T, Karhu R, Tanner M, Kähkönen M, Schäffer A, Nupponen N, Pennanen S, Kallioniemi A, Kallioniemi OP, Isola J (1997) Genetic heterogeneity and clonal evolution underlying development of asynchronous metastasis in human breast cancer. Cancer Res 57:1597–1604
28. Lee AJX, Endesfelder D, Rowan AJ, Walther A, Birkbak NJ, Futreal PA, Downward J, Szallasi Z, Tomlinson IPM, Howell M et al (2011) Chromosomal instability confers intrinsic multidrug resistance. Cancer Res 71:1858–1870
29. McClelland SE, Burrell RA, Swanton C (2009) Chromosomal instability: a composite phenotype that influences sensitivity to chemotherapy. Cell Cycle (Georgetown, Tex) 8:3262–3266
30. Cimini D, Fioravanti D, Salmon ED, Degrassi F (2002) Merotelic kinetochore orientation versus chromosome mono-orientation in the origin of lagging chromosomes in human primary cells. J Cell Sci 115:507–515
31. Ertych N, Stolz A, Stenzinger A, Weichert W, Kaulfuß S, Burfeind P, Aigner A, Wordeman L, Bastians H (2014) Increased microtubule assembly rates influence chromosomal instability in colorectal cancer cells. Nat Cell Biol 16:779–791

32. Silkworth WT, Nardi IK, Paul R, Mogilner A, Cimini D (2012) Timing of centrosome separation is important for accurate chromosome segregation. Mol Biol Cell 23:401–411
33. Thompson SL, Compton DA (2008) Examining the link between chromosomal instability and aneuploidy in human cells. J Cell Biol 180:665–672
34. Gregan J, Polakova S, Zhang L, Tolic-Norrelykke IM, Cimini D (2011) Merotelic kinetochore attachment: causes and effects. Trends Cell Biol 21:374–381
35. Lampson MA, Grishchuk EL (2017) Mechanisms to avoid and correct erroneous kinetochore-microtubule attachments. Biology 6:1
36. Tanaka TU (2010) Kinetochore-microtubule interactions: steps towards bi-orientation. EMBO J 29:4070–4082
37. Liu D, Vader G, Vromans MJ, Lampson MA, Lens SM (2009) Sensing chromosome bi-orientation by spatial separation of Aurora B kinase from kinetochore substrates. Science 323:1350–1353
38. Welburn JP, Vleugel M, Liu D, Yates JR 3rd, Lampson MA, Fukagawa T, Cheeseman IM (2010) Aurora B phosphorylates spatially distinct targets to differentially regulate the kinetochore-microtubule interface. Mol Cell 38:383–392
39. Foley EA, Kapoor TM (2013) Microtubule attachment and spindle assembly checkpoint signalling at the kinetochore. Nat Rev Mol Cell Biol 14:25–37
40. Godek KM, Kabeche L, Compton DA (2015) Regulation of kinetochore-microtubule attachments through homeostatic control during mitosis. Nat Rev Mol Cell Biol 16:57–64
41. Krenn V, Musacchio A (2015) The Aurora B kinase in chromosome bi-orientation and spindle checkpoint signaling. Front Oncol 5
42. Lampson MA, Cheeseman IM (2011) Sensing centromere tension: Aurora B and the regulation of kinetochore function. Trends Cell Biol 21:133–140
43. Loncarek J, Kisurina-Evgenieva O, Vinogradova T, Hergert P, La Terra S, Kapoor TM, Khodjakov A (2007) The centromere geometry essential for keeping mitosis error free is controlled by spindle forces. Nature 450:745–749
44. Manning AL, Compton DA (2007) Mechanisms of spindle-pole organization are influenced by kinetochore activity in mammalian cells. Curr Biol 17:260–265
45. Barisic M, Aguiar P, Geley S, Maiato H (2014) Kinetochore motors drive congression of peripheral polar chromosomes by overcoming random arm-ejection forces. Nat Cell Biol 16:1249–1256
46. Chmátal L, Yang K, Schultz RM, Lampson MA (2015) Spatial regulation of kinetochore microtubule attachments by destabilization at spindle poles in meiosis I. Curr Biol 25:1835–1841
47. Ye AA, Deretic J, Hoel CM, Hinman AW, Cimini D, Welburn JP, Maresca TJ (2015) Aurora a kinase contributes to a pole-based error correction pathway. Curr Biol 25:1842–1851
48. Kabeche L, Compton DA (2013) Cyclin a regulates kinetochore microtubules to promote faithful chromosome segregation. Nature 502:110–113
49. Manning AL, Bakhoum SF, Maffini S, Correia-Melo C, Maiato H, Compton DA (2010) CLASP1, astrin and Kif2b form a molecular switch that regulates kinetochore-microtubule dynamics to promote mitotic progression and fidelity. EMBO J 29:3531–3543
50. Schmidt JC, Kiyomitsu T, Hori T, Backer CB, Fukagawa T, Cheeseman IM (2010) Aurora B kinase controls the targeting of the Astrin–SKAP complex to bioriented kinetochores. J Cell Biol 191:269–280
51. Domnitz SB, Wagenbach M, Decarreau J, Wordeman L (2012) MCAK activity at microtubule tips regulates spindle microtubule length to promote robust kinetochore attachment. J Cell Biol 197:231–237
52. Wordeman L, Wagenbach M, von Dassow G (2007) MCAK facilitates chromosome movement by promoting kinetochore microtubule turnover. J Cell Biol 179:869–879
53. Andrews PD, Ovechkina Y, Morrice N, Wagenbach M, Duncan K, Wordeman L, Swedlow JR (2004) Aurora B regulates MCAK at the mitotic centromere. Dev Cell 6:253–268

54. Lan W, Zhang X, Kline-Smith SL, Rosasco SE, Barrett-Wilt GA, Shabanowitz J, Hunt DF, Walczak CE, Stukenberg PT (2004) Aurora B phosphorylates centromeric MCAK and regulates its localization and microtubule depolymerization activity. Curr Biol 14:273–286
55. Biggins S, Murray AW (2001) The budding yeast protein kinase Ipl1/Aurora allows the absence of tension to activate the spindle checkpoint. Genes Dev 15:3118–3129
56. Ditchfield C, Johnson VL, Tighe A, Ellston R, Haworth C, Johnson T, Mortlock A, Keen N, Taylor SS (2003) Aurora B couples chromosome alignment with anaphase by targeting BubR1, Mad2, and Cenp-E to kinetochores. J Cell Biol 161:267–280
57. Hauf S, Cole RW, LaTerra S, Zimmer C, Schnapp G, Walter R, Heckel A, van Meel J, Rieder CL, Peters J-M (2003) The small molecule Hesperadin reveals a role for Aurora B in correcting kinetochore-microtubule attachment and in maintaining the spindle assembly checkpoint. J Cell Biol 161:281–294
58. Kallio MJ, McCleland ML, Stukenberg PT, Gorbsky GJ (2002) Inhibition of Aurora B kinase blocks chromosome segregation, overrides the spindle checkpoint, and perturbs microtubule dynamics in mitosis. Curr Biol 12:900–905
59. Lampson MA, Renduchitala K, Khodjakov A, Kapoor TM (2004) Correcting improper chromosome-spindle attachments during cell division. Nat Cell Biol 6:232–237
60. Tanaka TU, Rachidi N, Janke C, Pereira G, Galova M, Schiebel E, Stark MJR, Nasmyth K (2002) Evidence that the Ipl1-Sli15 (Aurora kinase-INCENP) complex promotes chromosome bi-orientation by altering kinetochore-spindle pole connections. Cell 108:317–329
61. Bakhoum SF, Kabeche L, Murnane JP, Zaki BI, Compton DA (2014) DNA-damage response during mitosis induces whole-chromosome missegregation. Cancer Discov 4:1281–1289
62. Sotillo R, Hernando E, Diaz-Rodriguez E, Teruya-Feldstein J, Cordon-Cardo C, Lowe SW, Benezra R (2007) Mad2 overexpression promotes aneuploidy and tumorigenesis in mice. Cancer Cell 11:9–23
63. Ertych N, Stolz A, Valerius O, Braus GH, Bastians H (2016) CHK2-BRCA1 tumor-suppressor axis restrains oncogenic Aurora-a kinase to ensure proper mitotic microtubule assembly. Proc Natl Acad Sci U S A 113:1817–1822
64. Lüddecke S, Ertych N, Stenzinger A, Weichert W, Beissbarth T, Dyczkowski J, Gaedcke J, Valerius O, Braus GH, Kschischo M et al (2016) The putative oncogene CEP72 inhibits the mitotic function of BRCA1 and induces chromosomal instability. Oncogene 35:2398–2406
65. Stolz A, Ertych N, Bastians H (2010) Loss of the tumour-suppressor genes CHK2 and BRCA1 results in chromosomal instability. Biochem Soc Trans 38:1704–1708
66. Antoni L, Sodha N, Collins I, Garrett MD (2007) CHK2 kinase: cancer susceptibility and cancer therapy - two sides of the same coin? Nat Rev Cancer 7:925–936
67. Barr AR, Gergely F (2007) Aurora-a: the maker and breaker of spindle poles. J Cell Sci 120:2987–2996
68. Bartek J, Lukas J (2003) Chk1 and Chk2 kinases in checkpoint control and cancer. Cancer Cell 3:421–429
69. Lens SM, Voest EE, Medema RH (2010) Shared and separate functions of polo-like kinases and Aurora kinases in cancer. Nat Rev Cancer 10:825–841
70. Thompson D, Easton DF, Breast Cancer Linkage C (2002) Cancer incidence in BRCA1 mutation carriers. J Natl Cancer Inst 94:1358–1365
71. Kaseda K, McAinsh AD, Cross RA (2012) Dual pathway spindle assembly increases both the speed and the fidelity of mitosis. Biol Open 1:12–18
72. McHedlishvili N, Wieser S, Holtackers R, Mouysset J, Belwal M, Amaro AC, Meraldi P (2012) Kinetochores accelerate centrosome separation to ensure faithful chromosome segregation. J Cell Sci 125:906–918
73. Janicke MA, Lasko L, Oldenbourg R, LaFountain JR (2007) Chromosome Malorientations after meiosis II arrest cause nondisjunction. Mol Biol Cell 18:1645–1656
74. Thompson SL, Compton DA (2011) Chromosome missegregation in human cells arises through specific types of kinetochore-microtubule attachment errors. Proc Natl Acad Sci U S A 108:17974–17978

75. Torosantucci L, De Santis Puzzonia M, Cenciarelli C, Rens W, Degrassi F (2009) Aneuploidy in mitosis of PtK1 cells is generated by random loss and nondisjunction of individual chromosomes. J Cell Sci 122:3455–3461
76. Crasta K, Ganem NJ, Dagher R, Lantermann AB, Ivanova EV, Pan Y, Nezi L, Protopopov A, Chowdhury D, Pellman D (2012) DNA breaks and chromosome pulverization from errors in mitosis. Nature 482:53–58
77. Zhang R, Alushin GM, Brown A, Nogales E (2015) Mechanistic origin of microtubule dynamic instability and its modulation by EB proteins. Cell 162:849–859
78. Janssen A, Medema RH (2013) Genetic instability: tipping the balance. Oncogene 32:4459–4470
79. Thompson SL, Compton DA (2010) Proliferation of aneuploid human cells is limited by a p53-dependent mechanism. J Cell Biol 188:369–381
80. Ganem NJ, Storchova Z, Pellman D (2007) Tetraploidy, aneuploidy and cancer. Curr Opin Genet Dev 17:157–162
81. Jeganathan K, Malureanu L, Baker DJ, Abraham SC, van Deursen JM (2007) Bub1 mediates cell death in response to chromosome missegregation and acts to suppress spontaneous tumorigenesis. J Cell Biol 179:255–267
82. Michel LS, Liberal V, Chatterjee A, Kirchwegger R, Pasche B, Gerald W, Dobles M, Sorger PK, Murty VVVS, Benezra R (2001) MAD2 haplo-insufficiency causes premature anaphase and chromosome instability in mammalian cells. Nature 409:355–359
83. Ricke RM, Jeganathan KB, van Deursen JM (2011) Bub1 overexpression induces aneuploidy and tumor formation through Aurora B kinase hyperactivation. J Cell Biol 193:1049–1064
84. van Ree JH, Jeganathan KB, Malureanu L, van Deursen JM (2010) Overexpression of the E2 ubiquitin-conjugating enzyme UbcH10 causes chromosome missegregation and tumor formation. J Cell Biol 188:83–100
85. Weaver BA, Silk AD, Montagna C, Verdier-Pinard P, Cleveland DW (2007) Aneuploidy acts both oncogenically and as a tumor suppressor. Cancer Cell 11:25–36
86. Funk LC, Zasadil LM, Weaver BA (2016) Living in CIN: mitotic infidelity and its consequences for tumor promotion and suppression. Dev Cell 39:638–652
87. Giam M, Rancati G (2015) Aneuploidy and chromosomal instability in cancer: a jackpot to chaos. Cell Div 10:3
88. Yuen KWY (2010) Chromosome instability (CIN), aneuploidy and cancer. In: Encyclopedia of life sciences. Wiley, Chichester
89. Torres EM, Sokolsky T, Tucker CM, Chan LY, Boselli M, Dunham MJ, Amon A (2007) Effects of aneuploidy on cellular physiology and cell division in haploid yeast. Science 317:916–924
90. Williams BR, Prabhu VR, Hunter KE, Glazier CM, Whittaker CA, Housman DE, Amon A (2008) Aneuploidy affects proliferation and spontaneous immortalization in mammalian cells. Science 322:703–709
91. Sheltzer JM, Ko JH, Replogle JM, Habibe Burgos NC, Chung ES, Meehl CM, Sayles NM, Passerini V, Storchova Z, Amon A (2017) Single-chromosome gains commonly function as tumor suppressors. Cancer Cell 31:240–255
92. Holland AJ, Cleveland DW (2009) Boveri revisited: chromosomal instability, aneuploidy and tumorigenesis. Nat Rev Mol Cell Biol 10:478–487
93. Levine MS, Bakker B, Boeckx B, Moyett J, Lu J, Vitre B, Spierings DC, Lansdorp PM, Cleveland DW, Lambrechts D et al (2017) Centrosome amplification is sufficient to promote spontaneous tumorigenesis in mammals. Dev Cell 40(313–322):e315
94. Schvartzman J-M, Sotillo R, Benezra R (2010) Mitotic chromosomal instability and cancer: mouse modelling of the human disease. Nat Rev Cancer 10:102–115
95. García-Castillo H, Vásquez-Velásquez AI, Rivera H, Barros-Núñez P (2008) Clinical and genetic heterogeneity in patients with mosaic variegated aneuploidy: delineation of clinical subtypes. Am J Med Genet A 146A:1687–1695

96. Hanks S, Coleman K, Reid S, Plaja A, Firth H, FitzPatrick D, Kidd A, Méhes K, Nash R, Robin N et al (2004) Constitutional aneuploidy and cancer predisposition caused by biallelic mutations in BUB1B. Nat Genet 36:1159–1161
97. Snape K, Hanks S, Ruark E, Barros-Núñez P, Elliott A, Murray A, Lane AH, Shannon N, Callier P, Chitayat D et al (2011) Mutations in CEP57 cause mosaic variegated aneuploidy syndrome. Nat Genet 43:527–529
98. Suijkerbuijk SJE, van Osch MHJ, Bos FL, Hanks S, Rahman N, Kops GJPL (2010) Molecular causes for BUBR1 dysfunction in the human cancer predisposition syndrome mosaic variegated aneuploidy. Cancer Res 70:4891–4900
99. Brito DA, Yang Z, Rieder CL (2008) Microtubules do not promote mitotic slippage when the spindle assembly checkpoint cannot be satisfied. J Cell Biol 182:623–629
100. Hood EA, Kettenbach AN, Gerber SA, Compton DA (2012) Plk1 regulates the kinesin-13 protein Kif2b to promote faithful chromosome segregation. Mol Biol Cell 23:2264–2274
101. Liu D, Davydenko O, Lampson MA (2012) Polo-like kinase-1 regulates kinetochore–microtubule dynamics and spindle checkpoint silencing. J Cell Biol 198:491–499
102. Asteriti IA, Giubettini M, Lavia P, Guarguaglini G (2011) Aurora-a inactivation causes mitotic spindle pole fragmentation by unbalancing microtubule-generated forces. Mol Cancer 10:131
103. Kinoshita K, Noetzel TL, Pelletier L, Mechtler K, Drechsel DN, Schwager A, Lee M, Raff JW, Hyman AA (2005) Aurora a phosphorylation of TACC3/maskin is required for centrosome-dependent microtubule assembly in mitosis. J Cell Biol 170:1047–1055
104. Zhang X, Ems-McClung SC, Walczak CE (2008) Aurora a phosphorylates MCAK to control ran-dependent spindle bipolarity. Mol Biol Cell 19:2752–2765
105. Kabeche L, Compton DA (2012) Checkpoint-independent stabilization of kinetochore-microtubule attachments by Mad2 in human cells. Curr Biol 22:638–644
106. Nakamura Y, Tanaka F, Haraguchi N, Mimori K, Matsumoto T, Inoue H, Yanaga K, Mori M (2007) Clinicopathological and biological significance of mitotic centromere-associated kinesin overexpression in human gastric cancer. Br J Cancer 97:543–549
107. Shimo A, Tanikawa C, Nishidate T, Lin M-L, Matsuda K, Park J-H, Ueki T, Ohta T, Hirata K, Fukuda M et al (2008) Involvement of kinesin family member 2C/mitotic centromere-associated kinesin overexpression in mammary carcinogenesis. Cancer Sci 99:62–70
108. Ishikawa Y, Akishima-Fukasawa Y, Ito K, Akasaka Y, Yokoo T, Ishii T, Behavior TSGfCB (2008) Histopathologic determinants of regional lymph node metastasis in early colorectal cancer. Cancer 112:924–933
109. Ganguly A, Yang H, Cabral F (2011a) Overexpression of mitotic centromere associated kinesin stimulates microtubule detachment and confers resistance to paclitaxel. Mol Cancer Ther 10:929–937
110. Ganguly A, Yang H, Pedroza M, Bhattacharya R, Cabral F (2011b) Mitotic centromere-associated kinesin (MCAK) mediates paclitaxel resistance. J Biol Chem 286:36378–36384
111. Agnese V, Bazan V, Fiorentino FP, Fanale D, Badalamenti G, Colucci G, Adamo V, Santini D, Russo A (2007) The role of Aurora-a inhibitors in cancer therapy. Ann Oncol 18:vi47–vi52
112. Casimiro MC, Crosariol M, Loro E, Li Z, Pestell RG (2012) Cyclins and cell cycle control in cancer and disease. Genes Cancer 3:649–657
113. Fu J, Bian M, Jiang Q, Zhang C (2007) Roles of Aurora kinases in mitosis and tumorigenesis. Mol Cancer Res 5:1–10
114. Hollander Jd, Rimpi S, Doherty JR, Rudelius M, Buck A, Hoellein A, Kremer M, Graf N, Scheerer M, Hall MA et al (2010) Aurora kinases a and B are up-regulated by Myc and are essential for maintenance of the malignant state. Blood 116:1498–1505
115. Malumbres M, Barbacid M (2007) Cell cycle kinases in cancer. Curr Opin Genet Dev 17:60–65
116. Portella G, Passaro C, Chieffi P (2011) Aurora B: a new prognostic marker and therapeutic target in cancer. Curr Med Chem 18:482–496

117. Diaz-Rodríguez E, Sotillo R, Schvartzman J-M, Benezra R (2008) Hec1 overexpression hyperactivates the mitotic checkpoint and induces tumor formation in vivo. Proc Natl Acad Sci U S A 105:16719–16724
118. Kuo T-C, Chang P-Y, Huang S-F, Chou C-K, Chao CC-K (2012) Knockdown of HURP inhibits the proliferation of hepacellular carcinoma cells via downregulation of gankyrin and accumulation of p53. Biochem Pharmacol 83:758–768
119. Nagahara M, Nishida N, Iwatsuki M, Ishimaru S, Mimori K, Tanaka F, Nakagawa T, Sato T, Sugihara K, Hoon DSB et al (2011) Kinesin 18A expression: clinical relevance to colorectal cancer progression. Int J Cancer 129:2543–2552
120. Rucksaken R, Khoontawad J, Roytrakul S, Pinlaor P, Hiraku Y, Wongkham C, Pairojkul C, Boonmars T, Pinlaor S (2012) Proteomic analysis to identify plasma orosomucoid 2 and kinesin 18A as potential biomarkers of cholangiocarcinoma. Cancer Biomarkers Section A Dis Markers 12:81–95
121. Wright C, Brooks WS (2013) Overexpression of the kinetochore-associated proteins SKAP and astrin in human breast cancer. Bios 84:136–141
122. Yuan B, Xu Y, Woo J-H, Wang Y, Bae YK, Yoon D-S, Wersto RP, Tully E, Wilsbach K, Gabrielson E (2006) Increased expression of mitotic checkpoint genes in breast cancer cells with chromosomal instability. Clin Cancer Res 12:405–410
123. Zhang C, Zhu C, Chen H, Li L, Guo L, Jiang W, Lu SH (2010) Kif18A is involved in human breast carcinogenesis. Carcinogenesis 31:1676–1684
124. Thiru P, Kern DM, McKinley KL, Monda JK, Rago F, Su K-C, Tsinman T, Yarar D, Bell GW, Cheeseman IM (2014) Kinetochore genes are coordinately up-regulated in human tumors as part of a FoxM1-related cell division program. Mol Biol Cell 25:1983–1994
125. Dehmelt L, Halpain S (2005) The MAP2/tau family of microtubule-associated proteins. Genome Biol 6:204
126. Folker ES, Baker BM, Goodson HV (2005) Interactions between CLIP-170, tubulin, and microtubules: implications for the mechanism of CLIP-170 plus-end tracking behavior. Mol Biol Cell 16:5373–5384
127. Giodini A, Kallio MJ, Wall NR, Gorbsky GJ, Tognin S, Marchisio PC, Symons M, Altieri DC (2002) Regulation of microtubule stability and mitotic progression by survivin. Cancer Res 62:2462–2467
128. Komarova YA, Akhmanova AS, Kojima S-I, Galjart N, Borisy GG (2002) Cytoplasmic linker proteins promote microtubule rescue in vivo. J Cell Biol 159:589–599
129. Rosa J, Canovas P, Islam A, Altieri DC, Doxsey SJ (2006) Survivin modulates microtubule dynamics and nucleation throughout the cell cycle. Mol Biol Cell 17:1483–1493
130. Sun X, Shi X, Liu M, Li D, Zhang L, Liu X, Zhou J (2011) Mdp3 is a novel microtubule-binding protein that regulates microtubule assembly and stability. Cell Cycle 10:3929–3937
131. Tirnauer JS, Grego S, Salmon ED, Mitchison TJ (2002) EB1-microtubule interactions in Xenopus egg extracts: role of EB1 in microtubule stabilization and mechanisms of targeting to microtubules. Mol Biol Cell 13:3614–3626
132. Kadavath H, Hofele RV, Biernat J, Kumar S, Tepper K, Urlaub H, Mandelkow E, Zweckstetter M (2015) Tau stabilizes microtubules by binding at the interface between tubulin heterodimers. Proc Natl Acad Sci U S A 112:7501–7506
133. Mandelkow E, Mandelkow EM (1995) Microtubules and microtubule-associated proteins. Curr Opin Cell Biol 7:72–81
134. Iqbal K, Liu F, Gong C-X (2016) Tau and neurodegenerative disease: the story so far. Nat Rev Neurol 12:15–27
135. Rouzier R, Rajan R, Wagner P, Hess KR, Gold DL, Stec J, Ayers M, Ross JS, Zhang P, Buchholz TA et al (2005) Microtubule-associated protein tau: a marker of paclitaxel sensitivity in breast cancer. Proc Natl Acad Sci U S A 102:8315–8320
136. Don S, Verrills NM, Liaw TYE, Liu MLM, Norris MD, Haber M, Kavallaris M (2004) Neuronal-associated microtubule proteins class III beta-tubulin and MAP2c in neuroblastoma: role in resistance to microtubule-targeted drugs. Mol Cancer Ther 3:1137–1146

137. Soltani MH, Pichardo R, Song Z, Sangha N, Camacho F, Satyamoorthy K, Sangueza OP, Setaluri V (2005) Microtubule-associated protein 2, a marker of neuronal differentiation, induces mitotic defects, inhibits growth of melanoma cells, and predicts metastatic potential of cutaneous melanoma. Am J Pathol 166:1841–1850
138. Tala N, Xie S, Sun X, Sun X, Ran J, Zhang L, Li D, Liu M, Bao G, Zhou J (2014) Microtubule-associated protein Mdp3 promotes breast cancer growth and metastasis. Theranostics 4:1052–1061
139. Diamantopoulos GS, Perez F, Goodson HV, Batelier G, Melki R, Kreis TE, Rickard JE (1999) Dynamic localization of CLIP-170 to microtubule plus ends is coupled to microtubule assembly. J Cell Biol 144:99–112
140. Bilbe G, Delabie J, Brüggen J, Richener H, Asselbergs FA, Cerletti N, Sorg C, Odink K, Tarcsay L, Wiesendanger W (1992) Restin: a novel intermediate filament-associated protein highly expressed in the reed-Sternberg cells of Hodgkin's disease. EMBO J 11:2103–2113
141. Delabie J, Bilbe G, Brüggen J, Van Leuven F, De Wolf-Peeters C (1993) Restin in Hodgkin's disease and anaplastic large cell lymphoma. Leuk Lymphoma 12:21–26
142. Sun X, Li D, Yang Y, Ren Y, Li J, Wang Z, Dong B, Liu M, Zhou J (2012) Microtubule-binding protein CLIP-170 is a mediator of paclitaxel sensitivity. J Pathol 226:666–673
143. Sun X, Li F, Dong B, Suo S, Liu M, Li D, Zhou J (2013) Regulation of tumor angiogenesis by the microtubule-binding protein CLIP-170. Protein Cell 4:266–276
144. Vitre B, Coquelle FM, Heichette C, Garnier C, Chretien D, Arnal I (2008) EB1 regulates microtubule dynamics and tubulin sheet closure in vitro. Nat Cell Biol 10:415–421
145. Dong X, Liu F, Sun L, Liu M, Li D, Su D, Zhu Z, Dong J-T, Fu L, Zhou J (2010) Oncogenic function of microtubule end-binding protein 1 in breast cancer. J Pathol 220:361–369
146. Kumar M, Mehra S, Thakar A, Shukla NK, Roychoudhary A, Sharma MC, Ralhan R, Chauhan SS (2016) End binding 1 (EB1) overexpression in oral lesions and cancer: a biomarker of tumor progression and poor prognosis. Clin Chim Acta 459:45–52
147. Morimura S, Takahashi K (2011) Rac1 and Stathmin but not EB1 are required for invasion of breast cancer cells in response to IGF-I. Int J Cell Biol 2011:e615912
148. Morrison EE, Wardleworth BN, Askham JM, Markham AF, Meredith DM (1998) EB1, a protein which interacts with the APC tumour suppressor, is associated with the microtubule cytoskeleton throughout the cell cycle. Oncogene 17:3471–3477
149. Nishigaki R, Osaki M, Hiratsuka M, Toda T, Murakami K, Jeang K-T, Ito H, Inoue T, Oshimura M (2005) Proteomic identification of differentially-expressed genes in human gastric carcinomas. Proteomics 5:3205–3213
150. Orimo T, Ojima H, Hiraoka N, Saito S, Kosuge T, Kakisaka T, Yokoo H, Nakanishi K, Kamiyama T, Todo S et al (2008) Proteomic profiling reveals the prognostic value of adenomatous polyposis coli-end-binding protein 1 in hepatocellular carcinoma. Hepatology 48:1851–1863
151. Stypula-Cyrus Y, Mutyal NN, Cruz MAD, Kunte DP, Radosevich AJ, Wali R, Roy HK, Backman V (2014) End-binding protein 1 (EB1) up-regulation is an early event in colorectal carcinogenesis. FEBS Lett 588:829–835
152. Altieri DC (2003) Survivin, versatile modulation of cell division and apoptosis in cancer. Oncogene 22:8581–8589
153. Charrasse S, Mazel M, Taviaux S, Berta P, Chow T, Larroque C (1995) Characterization of the cDNA and pattern of expression of a new gene over-expressed in human hepatomas and colonic tumors. Eur J Biochem 234:406–413
154. Brouhard GJ, Stear JH, Noetzel TL, Al-Bassam J, Kinoshita K, Harrison SC, Howard J, Hyman AA (2008) XMAP215 is a Processive microtubule polymerase. Cell 132:79–88
155. Gutiérrez-Caballero C, Burgess SG, Bayliss R, Royle SJ (2015) TACC3–ch-TOG track the growing tips of microtubules independently of clathrin and Aurora-a phosphorylation. Biol Open 4:170–179
156. Widlund PO, Stear JH, Pozniakovsky A, Zanic M, Reber S, Brouhard GJ, Hyman AA, Howard J (2011) XMAP215 polymerase activity is built by combining multiple tubulin-binding TOG domains and a basic lattice-binding region. Proc Natl Acad Sci 108:2741–2746

157. Booth DG, Hood FE, Prior IA, Royle SJ (2011) A TACC3/ch-TOG/clathrin complex stabilises kinetochore fibres by inter-microtubule bridging. EMBO J 30:906–919
158. Fu W, Tao W, Zheng P, Fu J, Bian M, Jiang Q, Clarke PR, Zhang C (2010) Clathrin recruits phosphorylated TACC3 to spindle poles for bipolar spindle assembly and chromosome alignment. J Cell Sci 123:3645–3651
159. Hood FE, Williams SJ, Burgess SG, Richards MW, Roth D, Straube A, Pfuhl M, Bayliss R, Royle SJ (2013) Coordination of adjacent domains mediates TACC3-ch-TOG-clathrin assembly and mitotic spindle binding. J Cell Biol 202:463–478
160. Lin CH, Hu CK, Shih HM (2010) Clathrin heavy chain mediates TACC3 targeting to mitotic spindles to ensure spindle stability. J Cell Biol 189:1097–1105
161. Royle SJ (2012) The role of clathrin in mitotic spindle organisation. J Cell Sci 125:19–28
162. Thakur HC, Singh M, Nagel-Steger L, Prumbaum D, Fansa EK, Gremer L, Ezzahoini H, Abts A, Schmitt L, Raunser S et al (2013) Role of centrosomal adaptor proteins of the TACC family in the regulation of microtubule dynamics during mitotic cell division. Biol Chem 394:1411–1423
163. Smith CM, Haucke V, McCluskey A, Robinson PJ, Chircop M (2013) Inhibition of clathrin by pitstop 2 activates the spindle assembly checkpoint and induces cell death in dividing HeLa cancer cells. Mol Cancer 12:4
164. Hergovich A, Lisztwan J, Barry R, Ballschmieter P, Krek W (2003) Regulation of microtubule stability by the von Hippel-Lindau tumour suppressor protein pVHL. Nat Cell Biol 5:64–70
165. Stanojevic B, Saenko V, Todorovic L, Petrovic N, Nikolic D, Zivaljevic V, Paunovic I, Nakashima M, Yamashita S, Dzodic R (2014) Low VHL mRNA expression is associated with more aggressive tumor features of papillary thyroid carcinoma. PLoS One 9:e114511
166. Zia MK, Rmali KA, Watkins G, Mansel RE, Jiang WG (2007) The expression of the von Hippel-Lindau gene product and its impact on invasiveness of human breast cancer cells. Int J Mol Med 20:605–611
167. Fodde R, Smits R, Clevers H (2001) APC, signal transduction and genetic instability in colorectal cancer. Nat Rev Cancer 1:55–67
168. Kaplan KB, Burds AA, Swedlow JR, Bekir SS, Sorger PK, Näthke IS (2001) A role for the adenomatous polyposis coli protein in chromosome segregation. Nat Cell Biol 3:429–432
169. Jaulin F, Kreitzer G (2010) KIF17 stabilizes microtubules and contributes to epithelial morphogenesis by acting at MT plus ends with EB1 and APC. J Cell Biol 190:443–460
170. Wen Y, Eng CH, Schmoranzer J, Cabrera-Poch N, Morris EJS, Chen M, Wallar BJ, Alberts AS, Gundersen GG (2004) EB1 and APC bind to mDia to stabilize microtubules downstream of rho and promote cell migration. Nat Cell Biol 6:820–830
171. Zumbrunn J, Kinoshita K, Hyman AA, Näthke IS (2001) Binding of the adenomatous polyposis coli protein to microtubules increases microtubule stability and is regulated by GSK3β phosphorylation. Curr Biol 11:44–49
172. Belmont LD, Mitchison TJ (1996) Identification of a protein that interacts with tubulin dimers and increases the catastrophe rate of microtubules. Cell 84:623–631
173. Hemdan T, Lindén M, Lind SB, Namuduri AV, Sjöstedt E, de Ståhl DT, Asplund A, Malmström P-U, Segersten U (2014) The prognostic value and therapeutic target role of stathmin-1 in urinary bladder cancer. Br J Cancer 111:1180–1187
174. Hsieh S-Y, Huang S-F, Yu M-C, Yeh T-S, Chen T-C, Lin Y-J, Chang C-J, Sung C-M, Lee Y-L, Hsu C-Y (2010) Stathmin1 overexpression associated with polyploidy, tumor-cell invasion, early recurrence, and poor prognosis in human hepatoma. Mol Carcinog 49:476–487
175. Hsu H-P, Li C-F, Lee S-W, Wu W-R, Chen T-J, Chang K-Y, Liang S-S, Tsai C-J, Shiue Y-L (2014) Overexpression of stathmin 1 confers an independent prognostic indicator in nasopharyngeal carcinoma. Tumour Biol 35:2619–2629
176. Li X, Yao R, Yue L, Qiu W, Qi W, Liu S, Yao Y, Liang J (2014) FOXM1 mediates resistance to docetaxel in gastric cancer via up-regulating Stathmin. J Cell Mol Med 18:811–823

177. Wang X, Chen Y, Han Q-b, Chan C-y, Wang H, Liu Z, Cheng CH-k, Yew DT, Lin MCM, He M-l et al (2009) Proteomic identification of molecular targets of gambogic acid: role of stathmin in hepatocellular carcinoma. Proteomics 9:242–253
178. Zheng P, Liu Y-X, Chen L, Liu X-H, Xiao Z-Q, Zhao L, Li G-Q, Zhou J, Ding Y-Q, Li J-M (2010) Stathmin, a new target of PRL-3 identified by proteomic methods, plays a key role in progression and metastasis of colorectal cancer. J Proteome Res 9:4897–4905
179. Aoki K, Taketo MM (2007) Adenomatous polyposis coli (APC): a multi-functional tumor suppressor gene. J Cell Sci 120:3327–3335
180. Fodde R (2002) The APC gene in colorectal cancer. Eur J Cancer 38:867–871
181. Gossage L, Eisen T, Maher ER (2015) VHL, the story of a tumour suppressor gene. Nat Rev Cancer 15:55–64
182. Cheung CHA, Wu S-Y, Lee T-R, Chang C-Y, Wu J-S, Hsieh H-P, Chang J-Y (2010) Cancer cells acquire mitotic drug resistance properties through beta I-tubulin mutations and alterations in the expression of beta-tubulin isotypes. PLoS One 5:e12564
183. Kavallaris M, Tait AS, Walsh BJ, He L, Horwitz SB, Norris MD, Haber M (2001) Multiple microtubule alterations are associated with Vinca alkaloid resistance in human leukemia cells. Cancer Res 61:5803–5809
184. Martello LA, Verdier-Pinard P, Shen H-J, He L, Torres K, Orr GA, Horwitz SB (2003) Elevated levels of microtubule destabilizing factors in a Taxol-resistant/dependent A549 cell line with an alpha-tubulin mutation. Cancer Res 63:1207–1213
185. Mozzetti S, Ferlini C, Concolino P, Filippetti F, Raspaglio G, Prislei S, Gallo D, Martinelli E, Ranelletti FO, Ferrandina G et al (2005) Class III beta-tubulin overexpression is a prominent mechanism of paclitaxel resistance in ovarian cancer patients. Clin Cancer Res 11:298–305
186. Verdier-Pinard P, Wang F, Martello L, Burd B, Orr GA, Horwitz SB (2003b) Analysis of tubulin isotypes and mutations from taxol-resistant cells by combined isoelectrofocusing and mass spectrometry. Biochemistry 42:5349–5357
187. Yin S, Bhattacharya R, Cabral F (2010) Human mutations that confer paclitaxel resistance. Mol Cancer Ther 9:327–335
188. Kelley MJ, Li S, Harpole DH (2001) Genetic analysis of the beta-tubulin gene, TUBB, in non-small-cell lung cancer. J Natl Cancer Inst 93:1886–1888
189. Sale S, Sung R, Shen P, Yu K, Wang Y, Duran GE, Kim J-H, Fojo T, Oefner PJ, Sikic BI (2002) Conservation of the class I beta-tubulin gene in human populations and lack of mutations in lung cancers and paclitaxel-resistant ovarian cancers. Mol Cancer Ther 1:215–225
190. Lamendola DE, Duan Z, Penson RT, Oliva E, Seiden MV (2003) Beta tubulin mutations are rare in human ovarian carcinoma. Anticancer Res 23:681–686
191. Mesquita B, Veiga I, Pereira D, Tavares A, Pinto IM, Pinto C, Teixeira MR, Castedo S (2005) No significant role for beta tubulin mutations and mismatch repair defects in ovarian cancer resistance to paclitaxel/cisplatin. BMC Cancer 5:101
192. Hasegawa S, Miyoshi Y, Egawa C, Ishitobi M, Tamaki Y, Monden M, Noguchi S (2002) Mutational analysis of the class I beta-tubulin gene in human breast cancer. Int J Cancer 101:46–51
193. Maeno K, Ito K-I, Hama Y, Shingu K, Kimura M, Sano M, Nakagomi H, Tsuchiya S-I, Fujimori M (2003) Mutation of the class I beta-tubulin gene does not predict response to paclitaxel for breast cancer. Cancer Lett 198:89–97
194. Urano N, Fujiwara Y, Hasegawa S, Miyoshi Y, Noguchi S, Takiguchi S, Yasuda T, Yano M, Monden M (2003) Absence of beta-tubulin gene mutation in gastric carcinoma. Gastric Cancer 6:108–112
195. Luduena RF, Banerjee A (2008) The isotypes of tubulin: distribution and functional significance. In: Fojo T (ed) Cancer drug discovery and development: the role of microtubules in cell biology. Humana Press, Totowa
196. Luduena RF (1998) Multiple forms of tubulin: different gene products and covalent modifications. Int Rev Cytol 178:207–275
197. Roach MC, Boucher VL, Walss C, Ravdin PM, Luduena RF (1998) Preparation of a monoclonal antibody specific for the class I isotype of beta-tubulin: the beta isotypes of tubulin

differ in their cellular distributions within human tissues. Cell Motil Cytoskeleton 39:273–285
198. Dettman RW, Turner FR, Hoyle HD, Raff EC (2001) Embryonic expression of the divergent drosophila beta3-tubulin isoform is required for larval behavior. Genetics 158:253–263
199. Burgoyne RD, Cambray-Deakin MA, Lewis SA, Sarkar S, Cowan NJ (1988) Differential distribution of beta-tubulin isotypes in cerebellum. EMBO J 7:2311–2319
200. Mialhe A, Lafanechère L, Treilleux I, Peloux N, Dumontet C, Brémond A, Panh MH, Payan R, Wehland J, Margolis RL et al (2001) Tubulin detyrosination is a frequent occurrence in breast cancers of poor prognosis. Cancer Res 61:5024–5027
201. Lu C, Zhang J, He S, Wan C, Shan A, Wang Y, Yu L, Liu G, Chen K, Shi J et al (2013) Increased α-tubulin1b expression indicates poor prognosis and resistance to chemotherapy in hepatocellular carcinoma. Dig Dis Sci 58:2713–2720
202. Blenk S, Engelmann JC, Pinkert S, Weniger M, Schultz J, Rosenwald A, Müller-Hermelink HK, Müller T, Dandekar T (2008) Explorative data analysis of MCL reveals gene expression networks implicated in survival and prognosis supported by explorative CGH analysis. BMC Cancer 8:106
203. Nikas JB, Boylan KLM, Skubitz APN, Low WC (2011) Mathematical prognostic biomarker models for treatment response and survival in epithelial ovarian cancer. Cancer Informat 10:233–247
204. Bai RL, Lin CM, Nguyen NY, Liu TY, Hamel E (1989) Identification of the cysteine residue of beta-tubulin alkylated by the antimitotic agent 2,4-dichlorobenzyl thiocyanate, facilitated by separation of the protein subunits of tubulin by hydrophobic column chromatography. Biochemistry 28:5606–5612
205. Joe PA, Banerjee A, Ludueña RF (2008) The roles of cys124 and ser239 in the functional properties of human betaIII tubulin. Cell Motil Cytoskeleton 65:476–486
206. Guo J, Walss-Bass C, Ludueña RF (2010) The beta isotypes of tubulin in neuronal differentiation. Cytoskeleton 67:431–441
207. Khan IA, Ludueña RF (1996) Phosphorylation of beta III-tubulin. Biochemistry 35:3704–3711
208. Khan IA, Ludueña RF (2003) Different effects of vinblastine on the polymerization of isotypically purified tubulins from bovine brain. Investig New Drugs 21:3–13
209. Littauer UZ, Giveon D, Thierauf M, Ginzburg I, Ponstingl H (1986) Common and distinct tubulin binding sites for microtubule-associated proteins. Proc Natl Acad Sci U S A 83:7162–7166
210. Caudron F, Denarier E, Thibout-Quintana J-C, Brocard J, Andrieux A, Fourest-Lieuvin A (2010) Mutation of Ser172 in yeast β tubulin induces defects in microtubule dynamics and cell division. PLoS One 5:e13553
211. Fourest-Lieuvin A, Peris L, Gache V, Garcia-Saez I, Juillan-Binard C, Lantez V, Job D (2006) Microtubule regulation in mitosis: tubulin phosphorylation by the Cyclin-dependent kinase Cdk1. Mol Biol Cell 17:1041–1050
212. Banerjee A, Roach MC, Trcka P, Luduena RF (1992) Preparation of a monoclonal antibody specific for the class IV isotype of beta-tubulin. Purification and assembly of alpha beta II, alpha beta III, and alpha beta IV tubulin dimers from bovine brain. J Biol Chem 267:5625–5630
213. Hari M, Yang H, Zeng C, Canizales M, Cabral F (2003) Expression of class III beta-tubulin reduces microtubule assembly and confers resistance to paclitaxel. Cell Motil Cytoskeleton 56:45–56
214. Panda D, Miller HP, Banerjee A, Ludueña RF, Wilson L (1994) Microtubule dynamics in vitro are regulated by the tubulin isotype composition. Proc Natl Acad Sci U S A 91:11358–11362
215. McCarroll JA, Sharbeen G, Liu J, Youkhana J, Goldstein D, McCarthy N, Limbri LF, Dischl D, Ceyhan GO, Erkan M et al (2014) βIII-tubulin: a novel mediator of chemoresistance and metastases in pancreatic cancer. Oncotarget 6:2235–2249
216. Liou G-Y, Storz P (2010) Reactive oxygen species in cancer. Free Radic Res 44:479–496

217. Banerjee A, Roach MC, Wall KA, Lopata MA, Cleveland DW, Ludueña RF (1988) A monoclonal antibody against the type II isotype of beta-tubulin. Preparation of isotypically altered tubulin. J Biol Chem 263:3029–3034
218. Kumbhar BV, Borogaon A, Panda D, Kunwar A (2016) Exploring the origin of Differential binding affinities of human tubulin isotypes αβII, αβIII and αβIV for DAMA-colchicine using homology modelling, molecular docking and molecular dynamics simulations. PLoS One 11:e0156048
219. Santoshi S, Naik PK (2014) Molecular insight of isotypes specific β-tubulin interaction of tubulin heterodimer with noscapinoids. J Comput Aided Mol Des 28:751–763
220. Xu K, Schwarz PM, Ludueña RF (2002) Interaction of nocodazole with tubulin isotypes. Drug Dev Res 55:91–96
221. Bernard-Marty C, Treilleux I, Dumontet C, Cardoso F, Fellous A, Gancberg D, Bissery MC, Paesmans M, Larsimont D, Piccart MJ et al (2002) Microtubule-associated parameters as predictive markers of docetaxel activity in advanced breast cancer patients: results of a pilot study. Clin Breast Cancer 3:341–345
222. Haber M, Burkhart CA, Regl DL, Madafiglio J, Norris MD, Horwitz SB (1995) Altered expression of M beta 2, the class II beta-tubulin isotype, in a murine J774.2 cell line with a high level of taxol resistance. J Biol Chem 270:31269–31275
223. Iwamoto Y, Nishio K, Fukumoto H, Yoshimatsu K, Yamakido M, Saijo N (1998) Preferential binding of E7010 to murine β3-tubulin and decreased β3-tubulin in E7010-resistant cell lines. Jpn J Cancer Res 89:954–962
224. Orr GA, Verdier-Pinard P, McDaid H, Horwitz SB (2003) Mechanisms of Taxol resistance related to microtubules. Oncogene 22:7280–7295
225. Sirotnak FM, Danenberg KD, Chen J, Fritz F, Danenberg PV (2000) Markedly decreased binding of vincristine to tubulin in vinca alkaloid-resistant Chinese hamster cells is associated with selective overexpression of alpha and beta tubulin isoforms. Biochem Biophys Res Commun 269:21–24
226. Banerjee A (2002) Increased levels of tyrosinated α-, βIII-, and βIV-tubulin isotypes in paclitaxel-resistant MCF-7 breast cancer cells. Biochem Biophys Res Commun 293:598–601
227. Kavallaris M, Burkhart CA, Horwitz SB (1999) Antisense oligonucleotides to class III beta-tubulin sensitize drug-resistant cells to Taxol. Br J Cancer 80:1020–1025
228. Ranganathan S, Benetatos CA, Colarusso PJ, Dexter DW, Hudes GR (1998) Altered beta-tubulin isotype expression in paclitaxel-resistant human prostate carcinoma cells. Br J Cancer 77:562–566
229. Ranganathan S, Dexter DW, Benetatos CA, Chapman AE, Tew KD, Hudes GR (1996) Increase of beta(III)- and beta(IVa)-tubulin isotopes in human prostate carcinoma cells as a result of estramustine resistance. Cancer Res 56:2584–2589
230. Ranganathan S, McCauley RA, Dexter DW, Hudes GR (2001) Modulation of endogenous beta-tubulin isotype expression as a result of human beta(III)cDNA transfection into prostate carcinoma cells. Br J Cancer 85:735–740
231. Kavallaris M, Kuo DY, Burkhart CA, Regl DL, Norris MD, Haber M, Horwitz SB (1997) Taxol-resistant epithelial ovarian tumors are associated with altered expression of specific beta-tubulin isotypes. J Clin Invest 100:1282–1293
232. Aoki D, Oda Y, Hattori S, Taguchi K-I, Ohishi Y, Basaki Y, Oie S, Suzuki N, Kono S, Tsuneyoshi M et al (2009) Overexpression of class III beta-tubulin predicts good response to taxane-based chemotherapy in ovarian clear cell adenocarcinoma. Clin Cancer Res 15:1473–1480
233. Barisic M, Silva e Sousa R, Tripathy SK, Magiera MM, Zaytsev AV, Pereira AL, Janke C, Grishchuk EL, Maiato H (2015) Mitosis. Microtubule detyrosination guides chromosomes during mitosis. Science 348:799–803
234. Minoura I, Hachikubo Y, Yamakita Y, Takazaki H, Ayukawa R, Uchimura S, Muto E (2013) Overexpression, purification, and functional analysis of recombinant human tubulin dimer. FEBS Lett 587:3450–3455

235. Widlund PO, Podolski M, Reber S, Alper J, Storch M, Hyman AA, Howard J, Drechsel DN (2012) One-step purification of assembly-competent tubulin from diverse eukaryotic sources. Mol Biol Cell 23:4393–4401
236. Gygi SP, Rochon Y, Franza BR, Aebersold R (1999) Correlation between protein and mRNA abundance in yeast. Mol Cell Biol 19:1720–1730
237. Verdier-Pinard P, Wang F, Burd B, Angeletti RH, Horwitz SB, Orr GA (2003a) Direct analysis of tubulin expression in cancer cell lines by electrospray ionization mass spectrometry. Biochemistry 42:12019–12027
238. Roll-Mecak A (2015) Intrinsically disordered tubulin tails: complex tuners of microtubule functions? Semin Cell Dev Biol 37:11–19
239. Janke C (2014) The tubulin code: molecular components, readout mechanisms, and functions. J Cell Biol 206:461–472
240. Hallak ME, Rodriguez JA, Barra HS, Caputto R (1977) Release of tyrosine from tyrosinated tubulin. Some common factors that affect this process and the assembly of tubulin. FEBS Lett 73:147–150
241. Arce CA, Rodriguez JA, Barra HS, Caputto R (1975) Incorporation of l-tyrosine, l-phenylalanine and l-3,4-Dihydroxyphenylalanine as single units into rat brain tubulin. Eur J Biochem 59:145–149
242. Argarana CE, Barra HS, Caputto R (1980) Tubulinyl-tyrosine carboxypeptidase from chicken brain: properties and partial purification. J Neurochem 34:114–118
243. Deanin GG, Gordon MW (1976) The distribution of tyrosyltubulin ligase in brain and other tissues. Biochem Biophys Res Commun 71:676–683
244. Ersfeld K, Wehland J, Plessmann U, Dodemont H, Gerke V, Weber K (1993) Characterization of the tubulin-tyrosine ligase. J Cell Biol 120:725–732
245. Raybin D, Flavin M (1975) An enzyme tyrosylating α-tubulin and its role in microtubule assembly. Biochem Biophys Res Commun 65:1088–1095
246. Schröder HC, Wehland J, Weber K (1985) Purification of brain tubulin-tyrosine ligase by biochemical and immunological methods. J Cell Biol 100:276–281
247. Peris L, Wagenbach M, Lafanechère L, Brocard J, Moore AT, Kozielski F, Job D, Wordeman L, Andrieux A (2009) Motor-dependent microtubule disassembly driven by tubulin tyrosination. J Cell Biol 185:1159–1166
248. Sirajuddin M, Rice LM, Vale RD (2014) Regulation of microtubule motors by tubulin isotypes and post-translational modifications. Nat Cell Biol 16:335–344
249. Brown A, Li Y, Slaughter T, Black MM (1993) Composite microtubules of the axon: quantitative analysis of tyrosinated and acetylated tubulin along individual axonal microtubules. J Cell Sci 104:339–352
250. Cambray-Deakin MA, Burgoyne RD (1987) Posttranslational modifications of alpha-tubulin: acetylated and detyrosinated forms in axons of rat cerebellum. J Cell Biol 104:1569–1574
251. Robson SJ, Burgoyne RD (1989) Differential localisation of tyrosinated, detyrosinated, and acetylated α-tubulins in neurites and growth cones of dorsal root ganglion neurons. Cell Motil Cytoskeleton 12:273–282
252. Lafanechère L, Courtay-Cahen C, Kawakami T, Jacrot M, Rüdiger M, Wehland J, Job D, Margolis RL (1998) Suppression of tubulin tyrosine ligase during tumor growth. J Cell Sci 111(Pt 2):171–181
253. Kato C, Miyazaki K, Nakagawa A, Ohira M, Nakamura Y, Ozaki T, Imai T, Nakagawara A (2004) Low expression of human tubulin tyrosine ligase and suppressed tubulin tyrosination/detyrosination cycle are associated with impaired neuronal differentiation in neuroblastomas with poor prognosis. Int J Cancer 112:365–375
254. Paturle-Lafanechere L, Manier M, Trigault N, Pirollet F, Mazarguil H, Job D (1994) Accumulation of delta 2-tubulin, a major tubulin variant that cannot be tyrosinated, in neuronal tissues and in stable microtubule assemblies. J Cell Sci 107:1529–1543
255. Sève P, Isaac S, Trédan O, Souquet P-J, Pachéco Y, Pérol M, Lafanéchère L, Penet A, Peiller E-L, Dumontet C (2005) Expression of class III β-tubulin is predictive of patient outcome in

patients with non–small cell lung cancer receiving Vinorelbine-based chemotherapy. Clin Cancer Res 11:5481–5486
256. Bonnet C, Boucher D, Lazereg S, Pedrotti B, Islam K, Denoulet P, Larcher JC (2001) Differential binding regulation of microtubule-associated proteins MAP1A, MAP1B, and MAP2 by tubulin polyglutamylation. J Biol Chem 276:12839–12848
257. Ikegami K, Sato S, Nakamura K, Ostrowski LE, Setou M (2010) Tubulin polyglutamylation is essential for airway ciliary function through the regulation of beating asymmetry. Proc Natl Acad Sci 107:10490–10495
258. Janke C, Rogowski K, Wloga D, Regnard C, Kajava AV, Strub J-M, Temurak N, Dijk Jv, Boucher D, Dorsselaer AV et al (2005) Tubulin Polyglutamylase enzymes are members of the TTL domain protein family. Science 308:1758–1762
259. Pathak N, Obara T, Mangos S, Liu Y, Drummond IA (2007) The zebrafish fleer Gene encodes an essential regulator of cilia tubulin polyglutamylation. Mol Biol Cell 18:4353–4364
260. Massoner P, Lueking A, Goehler H, Höpfner A, Kowald A, Kugler KG, Amersdorfer P, Horninger W, Bartsch G, Schulz-Knappe P et al (2012) Serum-autoantibodies for discovery of prostate cancer specific biomarkers. Prostate 72:427–436
261. Wasylyk C, Zambrano A, Zhao C, Brants J, Abecassis J, Schalken JA, Rogatsch H, Schaefer G, Pycha A, Klocker H et al (2010) Tubulin tyrosine ligase like 12 links to prostate cancer through tubulin posttranslational modification and chromosome ploidy. Int J Cancer 127:2542–2553
262. Brants J, Semenchenko K, Wasylyk C, Robert A, Carles A, Zambrano A, Pradeau-Aubreton K, Birck C, Schalken JA, Poch O et al (2012) Tubulin tyrosine ligase like 12, a TTLL family member with SET- and TTL-like domains and roles in histone and tubulin modifications and mitosis. PLoS One 7:e51258
263. Sangrajrang S, Denoulet P, Laing NM, Tatoud R, Millot G, Calvo F, Tew KD, Fellous A (1998) Association of estramustine resistance in human prostatic carcinoma cells with modified patterns of tubulin expression. Biochem Pharmacol 55:325–331
264. Wloga D, Dave D, Meagley J, Rogowski K, Jerka-Dziadosz M, Gaertig J (2010) Hyperglutamylation of tubulin can either stabilize or destabilize microtubules in the same cell. Eukaryot Cell 9:184–193
265. Sangrajrang S, Fellous A (2000) Taxol resistance. Chemotherapy 46:327–334
266. Howes SC, Alushin GM, Shida T, Nachury MV, Nogales E (2014) Effects of tubulin acetylation and tubulin acetyltransferase binding on microtubule structure. Mol Biol Cell 25:257–266
267. Dompierre JP, Godin JD, Charrin BC, Cordelières FP, King SJ, Humbert S, Saudou F (2007) Histone deacetylase 6 inhibition compensates for the transport deficit in Huntington's disease by increasing tubulin acetylation. J Neurosci 27:3571–3583
268. Reed NA, Cai D, Blasius TL, Jih GT, Meyhofer E, Gaertig J, Verhey KJ (2006) Microtubule acetylation promotes kinesin-1 binding and transport. Curr Biol 16:2166–2172
269. Giustiniani J, Daire V, Cantaloube I, Durand G, Poüs C, Perdiz D, Baillet A (2009) Tubulin acetylation favors Hsp90 recruitment to microtubules and stimulates the signaling function of the Hsp90 clients Akt/PKB and p53. Cell Signal 21:529–539
270. Zhang Z, Yamashita H, Toyama T, Sugiura H, Omoto Y, Ando Y, Mita K, Hamaguchi M, Hayashi S-i, Iwase H (2004) HDAC6 expression is correlated with better survival in breast cancer. Clin Cancer Res 10:6962–6968
271. Boggs AE, Vitolo MI, Whipple RA, Charpentier MS, Goloubeva OG, Ioffe OB, Tuttle KC, Slovic J, Lu Y, Mills GB et al (2015) α-tubulin acetylation elevated in metastatic and basal-like breast cancer cells promotes microtentacle formation, adhesion and invasive migration. Cancer Res 75:203–215
272. Zhang W, Mao JH, Zhu W, Jain AK, Liu K, Brown JB, Karpen GH (2016) Centromere and kinetochore gene misexpression predicts cancer patient survival and response to radiotherapy and chemotherapy. Nat Commun 7:12619

Chapter 6
Clinical Development of Anti-mitotic Drugs in Cancer

Anna-Maria Olziersky and S. Intidhar Labidi-Galy

Abstract Mitosis is one of the most fundamental processes of life by which a mammalian cell divides into two daughter cells. Mitosis has been an attractive target for anticancer therapies since fast proliferation was identified as one of the hallmarks of cancer cells. Despite efforts into developing specific inhibitors for mitotic kinases and kinesins, very few drugs have shown the efficiency of microtubule targeting-agents in cancer cells with paclitaxel being the most successful. A deeper translational research accompanying clinical trials of anti-mitotic drugs will help in identifying potent biomarkers predictive for response. Here, we review the current knowledge of mitosis targeting agents that have been tested so far in the clinics.

Keywords Mitosis • Cancer • Anti-mitotic drugs • PLK-1 • Microtubules • Aurora kinase • Antibody-drug conjugates • Kinesin-5 • CENP-E

6.1 Introduction

Cell division is a fundamental process driven by several mitotic kinases and phosphatases. There are three types of cell division: Binary Fission, Mitosis and Meiosis. In this review, we will focus on mitosis, as this is how mammalian somatic cells divide. Mitosis proceeds in five phases: prophase, prometaphase, metaphase, anaphase and telophase. A successful mitosis depends on the equal segregation of genetic material in the two daughter cells [1]. During prophase, the two centrosomes migrate to the opposite sides of the mitotic cell. After nuclear envelope

A.-M. Olziersky
Department of Cell Physiology and Metabolism, Faculty of Medicine, Geneva University, Geneva, Switzerland

S.I. Labidi-Galy (✉)
Department of Oncology, Geneva University Hospitals, Rue Gabrielle Perret-Gentil 4, Geneva 1205, Switzerland
e-mail: Intidhar.labidi-galy@hcuge.ch

M. Gotta, P. Meraldi (eds.), *Cell Division Machinery and Disease*, Advances in Experimental Medicine and Biology 1002, DOI 10.1007/978-3-319-57127-0_6

breakdown, microtubules emanating from centrosomes grow and shrink in search of chromosomes in the cytoplasm. The bipolar spindle forms and chromosomes align in the middle of it. Chromosomes ought to bipolarly attach to microtubules to ensure faithful segregation of sister chromatids. If the attachments established are erroneous or unstable, chromosome missegregation might cause chromosomal instability (CIN) and aneuploidy [2, 3], a key hallmark of cancer cells [4, 5]. Cells have evolved different mechanisms to ensure proper chromosome segregation, in particular the spindle assembly checkpoint (SAC), which delays anaphase in the presence of erroneous chromosome-microtubule attachments or absence of attachments [6].

The idea of targeting mitosis for cancer treatment was first initiated when scientists discovered that one of the hallmarks of cancer cells is uncontrolled cell proliferation irresponsive to anti-growth signals [4, 5]. All proliferative cells will eventually go through mitosis, making it an ideal target for cancer cells. Extensive efforts have lead to identification of several compounds that specifically block factors important for mitosis.

One important structure for mitosis is the mitotic spindle, which is composed of constantly polymerizing and depolymerizing microtubules. Microtubule-targeting compounds were introduced in the clinics in the 1990's. The clinical success of these agents and their classification as mitotic poisons spurred the development of small molecules that inhibit mitotic kinases important for mitotic entry, progression and exit [7]. The goal behind these efforts was to identify compounds with similar clinical efficacy than microtubule-targeting agents while avoiding some of their dose-limiting toxicities. None of these new agents has shown so far in solid tumors the clinical efficacy of microtubule-targeting agents, such as taxanes. Targeted anti-mitotic drugs showed their highest response rates in hematologic diseases [8, 9], which are characterized by shorter doubling time when compared to solid tumors. However, doubling time measurement of tumor cells at the patient level is quite vague, as it is determined by the measurement of tumor volume between two time intervals [10].

After reviewing the mechanisms of action of the different anti-mitotic drugs, we will focus on agents that have been approved for marketing or investigated in clinical trials.

6.2 Microtubule Targeting Agents

Microtubules (MT) are protein polymers made up of alternating subunits of α and β-tubulin [11], that play a key role in mitosis but also for intracellular trafficking, signaling and motility. MT are present in all eukaryotes, during interphase and cell division. Formation of microtubules involves a highly dynamic process of polymerization and depolymerization of α and β-tubulin heterodimers. This dynamic property is crucial for the assembly of mitotic spindle and precise segregation of chromosomes during cell division [11, 12]. Disruption of MT dynamics, either by inhibiting their polymerization or depolymerization, perturbs the function of the

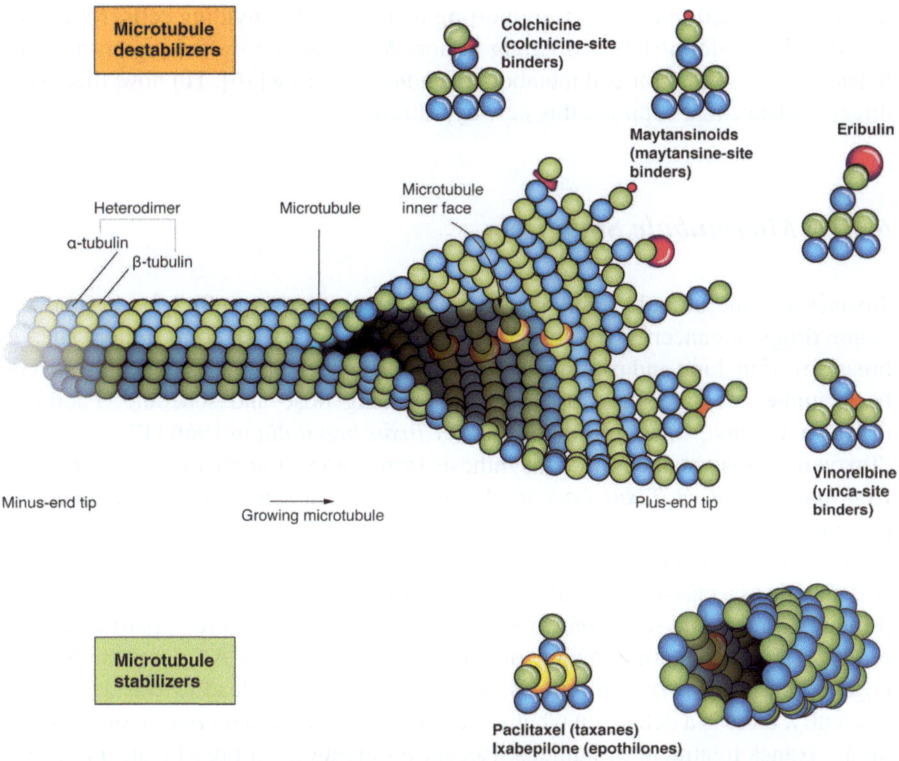

Fig. 6.1 Microtubule-targeting agents binding sites on the microtubule. The microtubule is a polymer that goes through cycles of polymerization and depolymerization during mitosis. This growth and shrinkage capacity is very important for mitosis. Microtubule-targeting agents bind on microtubules and either stabilize or destabilize microtubules. Adapted by permission from Macmillan Publishers Ltd. [Nature Reviews Molecular Cell Biology] [13], copyright (2015)

mitotic spindle and consequently cell division. This results in either mitotic arrest or mitotic progression with segregation errors during anaphase. Both situations will lead to cell death [11, 12].

MTs are one of the most validated intracellular targets in oncology. Microtubule-targeting agents (MTAs) are widely used in cancer chemotherapy as monotherapy and combination therapy. Currently there are two conventional classes of MTAs: those that **stabilize** MTs such as taxanes (paclitaxel and docetaxel) and epothilones (ixabepilone) and those that **destabilize** MTs such as vinca alkaloids (vinblastine, vincristine, vinorelbine, vindesine and vinflunine), colchicine and eribulin (Fig. 6.1). A new class of anticancer drugs that conjugate microtubule inhibitors (maytansinoids or auristatins) and monoclonal antibodies, namely antibody-drug conjugate (ADC)—recently emerged with outstanding clinical efficacy in some tumors [14]. From studies on cells in culture and animals, it seems clear that mitosis is the main target of MTAs in these rapidly-dividing cells [15]. However, microtubules are also important in interphase cells. There are currently some speculation on the exact mechanism of action of MTAs, supporting that in rapidly-dividing cells, MTAs target mitosis, while in slowly-dividing tumors MTAs act on trafficking on microtubules that is essential for cell metabolism and/or signaling [16]. Till now, there is no direct evidence that supports this new hypothesis.

6.2.1 Microtubule Stabilizers

Taxanes are microtubule stabilizers that are among the most efficient chemotherapeutic drugs in cancer. They are backbone therapy in several solid tumors such as breast, ovarian, lung and prostate cancers. Few single chemotherapeutic agents have been studied as rigorously as taxanes regarding dose and schedule. Paclitaxel (Taxol®) was first isolated from the bark of *Taxus brevifolia* in 1966 [17]. Docetaxel (Taxotere®) was obtained by semisynthesis from a noncytotoxic precursor extracted from the needles of *Taxus baccata L.* Docetaxel is a semi-synthetic analogue of paclitaxel [18].

Based on the pivotal role of MTs dynamics in mitosis, extensive research identified mitosis as a classic target of taxanes. Paclitaxel binds to β-tubulin, bridging two β-tubulin dimers and stabilizes microtubules. This leads to severe impairment of the mitotic spindle, creating many unattached chromosomes. As a result, SAC is engaged and cells arrest in mitosis, which will ultimately lead to cell death [19]. Currently, there is a debate on whether mitotic arrest is the main reason of cell death upon taxanes treatment in humans. Recent observations in breast cancers in vitro and in vitro suggest that mitotic arrest is not necessary for tumor regression in patients. Instead, the authors demonstrate that the efficacy of paclitaxel rather depends on chromosome mis-segregation in highly abnormal, multipolar spindles (Fig. 6.2) [20]. This results into daughter cells with abnormal number of chromosomes that will undergo cell death in the subsequent interphase.

MTs are also very important for the directional intracellular transport of vesicles, proteins and messenger ribonucleic acid. Another way by which paclitaxel could induce cell death is blocking intracellular transport of vesicles which depends on microtubules in interphase. A recently described example of such trafficking is the transport of the androgen receptor (AR) in castration-resistant prostate cancer. Prostate cancer growth relies on AR signaling, as it stimulates proliferation. This requires the binding of androgens to the AR and consequently its translocation from the cell surface to the nucleus, where it transcriptionally activates different genes. This nuclear translocation depends on MT-dynein-dependent intracellular trafficking [21]. One of the most common therapies for prostate cancer requires androgen ablation. As a result, this leads to tumor regression. Unfortunately, patients develop secondary resistance to these treatments after 12–24 months, pointing out the necessity of finding alternative ways to block AR signalling. To date, taxanes (docetaxel and cabazitaxel) are the most efficient chemotherapy and standard of care therapy in castration-resistant prostate cancer, a disease with indolent behavior. Taxanes block the AR nuclear translocation by stabilizing the MTs, inhibiting the androgen-mediated

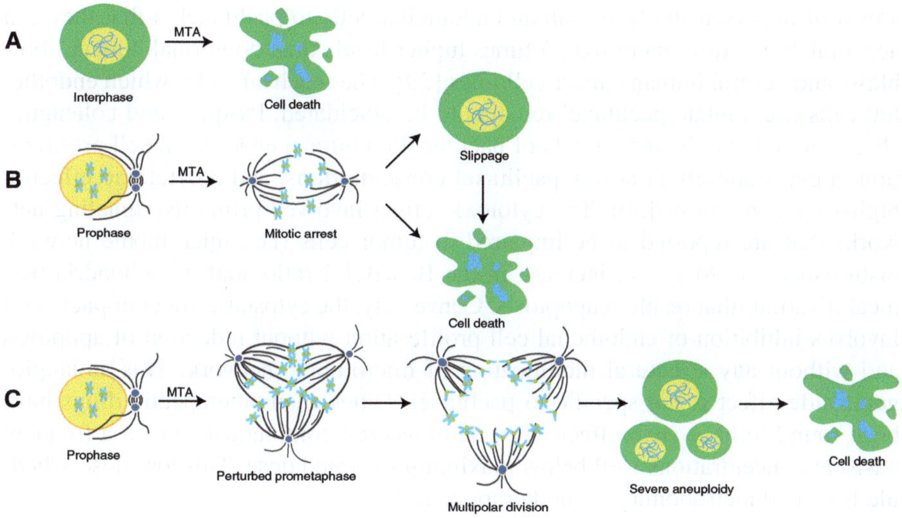

Fig. 6.2 Different hypotheses on how microtubule-targeting agents kill cancer cells. (**A**) It has been hypothesized that Microtubule-Targeting Agents (MTAs) can kill cancer cells in interphase due to perturbed signaling. (**B**) MTAs kill cancer cells by arresting them in mitosis. The MTAs induced mitotic arrest can either lead to mitotic cell death or slippage (exit of mitosis without division). In the latter case, after slippage cells most likely will undergo cell death. (**C**) MTAs can lead to divisions with severe abnormalities that can give rise to daughter cells with an unequal number of chromosomes and eventually die

signaling and the subsequent inhibition of androgen receptor activity in castration-resistant prostate cancer [22–24]. These studies indicate that the activity of taxanes in castration-resistant prostate cancer is mediated, at least in part, by inhibiting AR nuclear transport and signaling via microtubules (review in [25]). Additionally, it has been shown that MTAs (paclitaxel and vincristine) can synergize with DNA damage agents by interfering with cellular trafficking of DNA damage and repair proteins such as RAD51, ATM and ATR on interphase microtubules [26]. One limit of this study is that it relied on doses that are much higher than the clinically relevant doses, making it difficult to evaluate the physiological relevance. Although the hypothesis of cell death due to perturbed intracellular transport upon taxanes treatment is attractive, more studies are required to show a direct link between the two.

In the mid-1990s, researchers began to investigate antiangiogenic activity of paclitaxel as an additional mechanism contributing to its antineoplastic activity (review in [27]). In vitro, it was shown that endothelial cells are 10–100-fold more sensitive to paclitaxel than are tumor cells. The drug inhibits proliferation, migration, and differentiation of cultured human umbilical vein endothelial cells [27]. In a highly metastatic breast cancer mouse model, it was shown that low-dose of paclitaxel displayed a stronger anti-tumor activity with less side effects and a stronger anti-angiogenic activity than the maximum tolerated dose [28]. One possible mechanism of hypersensitivity of human endothelial cells to paclitaxel is that they can accumulate the drug more than 5 times higher levels than do normal human fibroblasts and several human cancer cell lines [29]. The mechanism by which endothelial cells accumulate paclitaxel remains to be elucidated. Pasquier and colleagues characterized two distinct effects of paclitaxel on human endothelial cell proliferation: a cytostatic effect at low paclitaxel concentrations, and a cytotoxic effect at higher concentrations [30]. The cytotoxic effect involved primarily signaling networks that are reported to be impaired in tumor cells (i.e., microtubule network disturbance, G2-M arrest, increase in the Bax/Bcl-2 ratio, and mitochondria permeabilization) that result in apoptosis. Conversely, the cytostatic effect of paclitaxel involves inhibition of endothelial cell proliferation without induction of apoptosis and without any structural modification of microtubule network. This antiangiogenic side-effect is not specific to paclitaxel. Other chemotherapeutic drugs have been found to have such effect when administered continuously, or on a frequent basis, at concentrations well below maximum tolerated dose. This low-dose schedule is named metronomic chemotherapy [31, 32].

In the clinics, paclitaxel was initially evaluated and approved in the 1990s on every- 3-week's schedule. However, it was shown in the last 10 years superiority of weekly paclitaxel over every-3-weeks schedule in both metastatic [33] and early breast cancers [34] with longer disease-free survival and overall survival. In ovarian cancer, sub-group analysis of GOG-262 phase III trial showed that ovarian cancer patients who received weekly paclitaxel seemed to do as well as those receiving every-3-weeks paclitaxel and bevacizumab, a monoclonal antibody anti-VEGF [35].

Although paclitaxel is the best selling chemotherapy in history and the most efficient MTA, most of the patients ultimately develop secondary resistance after several months of therapy. Mechanisms of resistance to microtubule stabilizers and especially paclitaxel have been studied for very long and different hypotheses have been reported. However, it is difficult to define what is clinically relevant or not. The most accepted mechanism of resistance is upregulation of the multidrug resistance pumps, which are p-glycoproteins on the cell surface with the capacity to pump out drug molecules that enter the cell [36]. Another well-studied way that cells use to resist death upon paclitaxel is their capacity to destabilize microtubules by changing the ratio of β-tubulin isoforms [37].

Paclitaxel has a markedly hydrophobic profile. To improve its delivery to tumor tissues, a new formulation of paclitaxel has been developed using nanotechnologies by encapsulating it with albumin nanoparticles. This albumin-drug complex named nab-paclitaxel, is distributed exactly as endogenous albumin. Whereas preclinical data suggest nab-paclitaxel to be more efficient than paclitaxel [38], conflicting results were observed in patients treated for metastatic breast cancers [39, 40]. In weekly schedule, nab-paclitaxel did not show superiority to paclitaxel [39]. Currently, nab-paclitaxel is used in the clinics mainly in metastatic pancreatic cancer, where it showed prolonged survival when associated with antifolate drug gemcitabine [41].

A new class of microtubules stabilizers, epothilones, is currently under clinical investigation in several cancers. Epothilones are macrolides and have similar mechanism of action than taxanes [42]. They compete with paclitaxel to the same binding pocket on β-tubulin, while they have different characteristics for microtubule binding [43, 44]. In vitro, they showed efficacy in taxanes resistant cell lines, where resistance is the result of an upregulation of multidrug resistance pumps [45]. Ixabepilone is the most clinically advanced epothilone. Ixabepilone is approved for metastatic breast cancers resistant to anthracyclins and taxanes. Recent trials suggest inferiority of ixabepilone when compared to weekly paclitaxel in metastatic breast cancers [39] or relapsing endometrial cancers [46].

To resume, MT stabilizers are still the most commonly used drugs in cancer treatment. They lead to cell death by inducing errors during mitosis and by inhibiting cellular trafficking especially in hormone dependent prostate cancer. Even though they are efficient, patients ultimately develop secondary resistance. More studies are required to find out how to overcome such resistance.

6.2.2 Microtubule Destabilizers

The so-called destabilizing agents inhibit microtubule polymerization, thus perturbing microtubule dynamicity, leading to a mitotic arrest which will eventually lead to cell death. Most of these agents bind to one of two binding domains of tubulin, either vinca domain or colchicine domain. Vinca-site binders include vinca alkaloids (vinblastine, vincristine, vinorelbine, vindesine and vinflunine), cryptophycins and dolastatins (spongistatin, rhizoxin, maytansinoids and tasidotin). Colchicine-site binders include colchicine and its analogues [47]. Vinca alkaloids, originally isolated from Madagascar periwinkle plant *Catharanthus rosea*, are the oldest and to date the most diversified family in terms of number of approved compound in a given family [47]. Vinblastine and vincristine are integrated in combination curative chemotherapy regimens in patients with lymphomas, sarcomas and several pediatric tumors. Vinorelbine is administered in a large range of solid tumors including breast, lung and sarcomas. Vinflunine has been recently approved for bladder cancers [48]. It is unclear why some drugs act preferentially on certain subgroup of tumors. As for paclitaxel, vinorelbine showed anti-angiogenic effect at non-cytotoxic concentrations by reducing proliferation of human umbilical vein endothelial cells [49] and blocking intracellular trafficking of DNA repair proteins [26].

Eribulin is a synthetic analogue of marine sponge product halichondrin B. It inhibits polymerization by binding the plus-end tip of the polymerizing microtubules and leads to a robust mitotic arrest [50, 51]. Eribulin mesylate is the one among halichondrin B derivatives that reached the most advanced clinical development. It showed prolonged survival in relapsing metastatic breast cancers [52] and soft tissue sarcomas (liposarcomas and leiomyosarcomas) [53], and is approved in both indications.

Secondary resistance to microtubule destabilizers is observed faster compared to paclitaxel and this is one of the reasons why paclitaxel remains the drug of reference.

Multidrug resistance pumps are responsible for resistance against microtubule destabilizers, as well as changes in microtubule stability [54–58]. So far, there is no biomarker that has been identified to predict resistance or sensitivity to microtubule destabilizers. Although, microtubule destabilizers have shown good clinical response for some tumors, they are still not as efficient as paclitaxel for the treatment of solid tumors highlighting the need for predictive biomarkers.

6.2.3 Antibody-Drug Conjugate

Antibody-drug conjugates (ADC) are tripartite drugs comprising a tumor-specific monoclonal antibody (mAb) conjugated to a potent cytotoxin via a stable linker. Development of ADC takes advantage of the specificity of a mAb and potent cytotoxic effect of chemotherapy, leading to enhanced cytotoxicity in target cells and limiting toxicity in normal tissue (review in [14]). Design of ADC relies on the proper selection of a tumor-specific antigen that is accessible for antibody binding and subsequent delivery of ADC to its pharmacologic target. The target antigen has to be expressed at high levels at the surface of tumor cells. When ADC bound to the antigen, it has to undergo efficient internalization (Fig. 6.3). Once inside the cell, the conjugate will be degraded by the lysosome releasing the cytotoxic compound in the cytoplasm [59]. The cytotoxic drug should have prolonged stability and be potent at low concentrations, since only a small proportion of administered ADC reaches tumor cells [59]. Currently, the majority of ADCs in clinical development employ either maytansinoids or auristatins [14], which are microtubule targeting agents. Maytansinoids interfere with microtubule assembly. Upon maytasine treatment, microtubules lose their dynamicity, as it both inhibits their growth and shrinkage. Therefore, it induces a mitotic arrest which leads to cell death. Maytasinoids bind to the same site as vinca alkaloids [60], but their cytotoxic effects is almost 100-fold higher than vinca alcaloids [61–63]. This is the reason why they were chosen for ADCs. When they are administered as standard chemotherapy, they have toxic effects that would not let them go further in the clinics. However, in the case ADCs, one can take advantage of the potency of these drugs. Auristatins, monomethyl auristin E (MMAE) and monomethyl auristin F (MMAF) are microtubule targeting agents that depolymerize microtubules.

There are currently two approved ADCs-trastuzumab emtansine and bretuximab vedotin-both humanized antibodies conjugated to MTAs. Trastuzumab emtansine (TDM-1) is an ADC that comprises trastuzumab (a monoclonal antibody) linked to a tubulin polymerization inhibitor, mertansine (a maytansine derivative; also known as DM-1). Trastuzumab targets human epidermal growth factor receptor 2 (HER-2), an oncoprotein overexpressed in 15–20% of breast cancers and associated with poor outcome [64]. TDM-1 is approved as second-line monotherapy for HER-2 overexpressing metastatic breast cancers [65]. Brentuximab vedotin, an ADC composed of anti-CD30 antibody linked to MMAE is approved for the treatment of refractory CD30-positive lymphomas, i.e. Hodgkin lymphoma and anaplastic large-cell

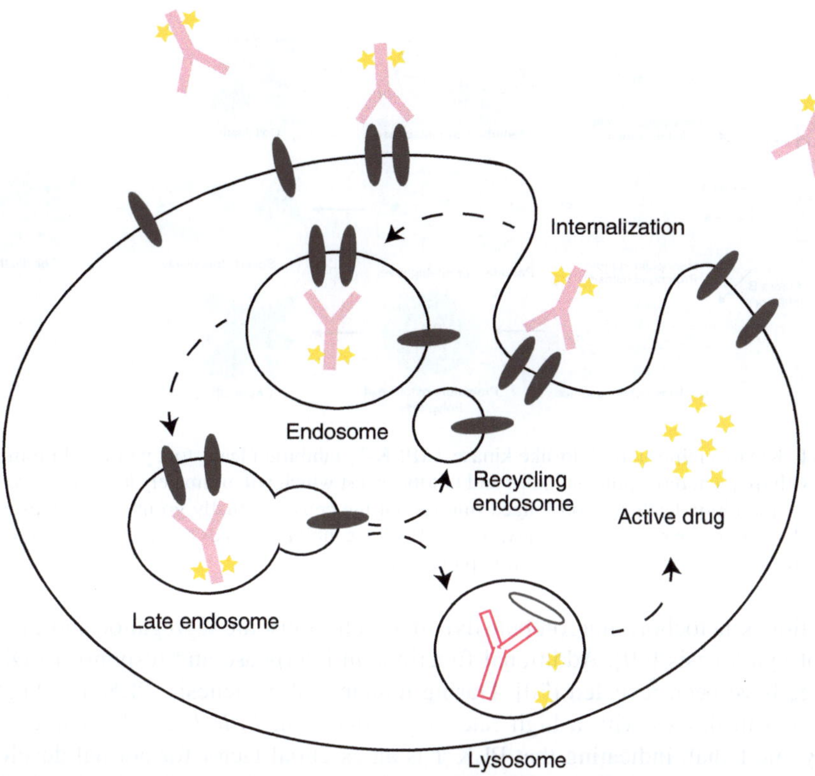

Fig. 6.3 Cell entry of the antibody-drug conjugates and activation of the cytotoxic compound. The antibody-drug conjugates binds the epitope on cell surface. The complex of the antibody with the epitope gets internalized and degraded by the lysosome. The cytotoxic compound is released in the cytoplasm

lymphomas [66, 67]. The discovery and synthesis of ADCs was a major advancement in the fight against cancer. Using ADCs, one can now specifically target cancer cells while avoiding damage of healthy tissues.

6.3 Polo-Like Kinase-1

Nowadays, there are efforts to specifically target mitotic proteins, such as PLK-1. Five members of the polo-like kinase family have been identified in humans (PLK1–5) and PLK-1 stands out as a promising target in oncology [68]. PLK-1 is a conserved serine/threonine protein kinase that has several functions in the cell cycle. Expression and activity of PLK-1 begins to increase from S/G2 phase, and its activity peaks during mitosis, identifying it as an important mitotic player. PLK-1 facilitates entry in mitosis and is involved in centrosome maturation, bipolar spindle

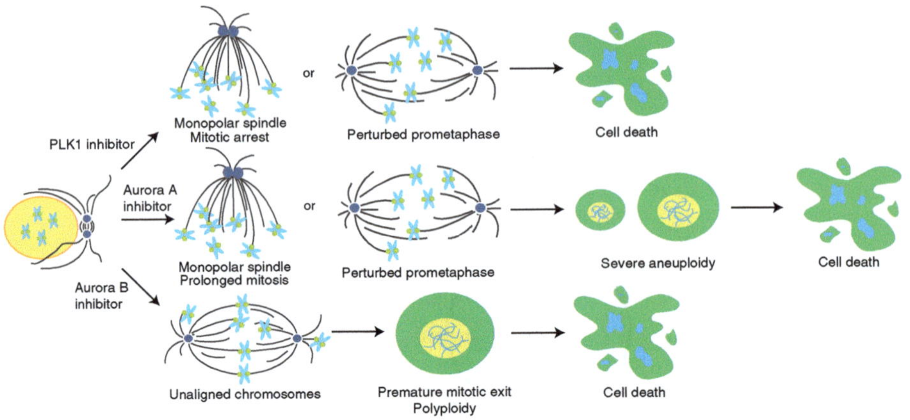

Fig. 6.4 Kinases inhibition. Polo-like kinase-1 (PLK-1) inhibition leads to a perturbed prometaphase with frequent monopolar spindles and mitotic arrest which will ultimately lead to cell death. Aurora A inhibition leads to a prolonged mitosis but the cells eventually go into erroneous anaphase which will give rise to aneuploidy. Aurora B kinase inhibition leads to a premature exit from mitosis resulting to the formation of a polyploid cell

formation, kinetochore-microtubule dynamics, chromosome segregation and execution of cytokinesis [69]. Additional functions in interphase and response to DNA damage have been revealed [70]. During human embryogenesis, PLK-1 is highly expressed in tissues with a high rate of proliferation. *PLK-1* knockout mice are embryonic lethal, indicating that PLK-1 is an essential factor for normal development [71]. In adults, PLK-1 is often detected in testis, spleen and bone marrow but not in non-proliferative tissues [72].

PLK-1 is found overexpressed in a variety of human cancers and this has been associated with poor outcome [73–76]. Conversely, *PLK-1*$^{+/-}$ mice have threefold increase in spontaneous tumor formation compared with control [71], indicating that (at least in mice) PLK-1 functions as an oncoprotein but also as a tumor suppressor. These observations lead to recognizing PLK-1 as a promising target for cancer therapy, through the effects of PLK-1 inhibitors. Downregulation of PLK-1 inhibits the growth of cancer cells in vitro and in vivo, due to mitotic defects [77]. Inhibitors of PLK-1 elicit a typical "Polo" phenotype: cells arrest in mitosis with a monopolar or disorganized spindle and eventually undergo cell death (Fig. 6.4) [78].

Another function of PLK-1 that is probably targeted by PLK-1 inhibitors is its contribution in establishing correct spindle orientation [79–81] and maintaining polarity [82–84]. The inhibition of this function of PLK-1 could be important for the treatment of tumors that contain stem-like cancer cells. In accordance to this hypothesis, Robin G. Lerner and colleagues found that CD133^{+} tumor propagating cells isolated from glioblastoma primary cultures go through polarized divisions, show high levels of PLK-1 expression and are resistant to BRAF inhibitors (inhibitor targeting the MAPK pathway). Inhibition of PLK-1 abolished polarity in these cells and enhanced the antiproliferative and proapoptotic effects of

BRAF inhibitor [85]. It would be interesting to question if similar observations would occur in other types of tumors that are enriched in stem-cell like cancer cells.

Currently, several PLK-1 inhibitors are available and have been tested in the clinics (BI2536, GSK461364A, TAK-960 and BI6727). The most frequent and dose-limiting side effect is hematological toxicity, mainly neutropenia and leukopenia [86]. So far, the most effective molecule is BI6727 (volasertib). It showed significant clinical benefit in ovarian cancers and acute myeloid leukemia. Volasertib has been tested in a phase II randomized trial that compared chemotherapy to volasertib (1:1) in platinum resistant or refractory ovarian cancers. Volasertib showed low response rate (13%) but this response was durable (>1 year) [87]. Volasertib has shown a significant response rate (31%) and prolonged survival when combined with low-dose cytarabine in relapsing acute myeloid leukemia [8].

Unfortunately, even though PLK–1 inhibitors showed promising preclinical results in several tumors, they demonstrated minimal clinical activity. Identifying biomarkers predictive for response to PLK-1 inhibitors would definitely accelerate their development. While one would expect that these drugs would be more efficient in tumors with overexpression of PLK-1 or high proliferative ones, no correlation was found between response to volasertib and expression of PLK-1, Ki-67 or phospho-Histone 3 (proliferation markers) in ovarian cancer samples. The six patients that showed long-term disease control under volasertib (>1 year) had serous histologic subtype, which are characterized by *TP53* mutation in virtually all cases [88]. This suggests a potential correlation between *TP53* mutation and sensitivity to PLK-1 inhibitor in solid tumors. Indeed, PLK-1 overexpression correlates with *TP53* mutation in human cancers [74, 89]. There is evidence that while p53 regulates directly and indirectly the expression of PLK-1, PLK-1 inhibits the function of p53, through activation of MDM-2, the negative regulator of p53 [90, 91]. Therefore, PLK-1 and p53 regulation seem to be tightly connected and this could explain why *TP53* mutated tumors respond well to PLK-1 inhibitors. It is important to mention that at the tumor level, it is difficult to investigate PLK-1 levels, as it accumulates only during very specific phases of the cell cycle.

In pre-clinical models, another factor that makes tumor cells more vulnerable to PLK-1 inhibitor (BI2536) is activating mutation of *KRAS* oncogene. Ji Luo and colleagues have shown that *KRAS* mutant cells depend more on PLK-1 for mitotic progression compared to wild-type. After nocodazole release in PLK-1 inhibitor, *KRAS* mutant cells arrest for longer than wild type cells [92], but data are lacking on patient samples'. Another potential mechanism of sensitivity of *KRAS* mutant cells to PLK-1 inhibitors is their increased proliferation rate.

6.4 Aurora Kinases

Aurora kinases comprise a family of highly conserved serine-threonine protein kinases that play a pivotal role in regulation of the cell cycle [93]. Three members of this family, Aurora A, B and C are encoded in humans by the genes *AURKA*, *AURKB* and *AURKC*. The three kinases are expressed in a cell specific pattern in rapidly dividing tissues; for instance in hematopoietic cells Aurora A and B are highly expressed whereas in germ cells only Aurora C is expressed [94]. Conversely, Aurora kinases expression is low or absent in most adult tissues due to their lower rates of proliferation [95].

The three Aurora kinases are involved in cell division but they have very distinct expression and localization patterns as well as functions [96]. Aurora A kinase (AURKA) regulates mitotic entry, centrosome maturation and spindle formation. AURKA is located at the centrosome since its duplication time and on both spindle poles and spindle MTs during early mitosis (prophase and prometaphase). Aurora B kinase (AURKB) plays a critical role in chromosome condensation and cohesion, chromosome biorientation and cytokinesis, as a member of chromosome passenger complex. AURKB is expressed in proliferating cells during G2 and mitosis and shows chromosomal-passenger localization. *AURKA* or *AURKB* knockout mice revealed the importance of their function for normal development [97–99].

Less is known about Aurora kinase C (AURKC) compared to the other Aurora kinases. AURKC is required for spermatogenesis and the first divisions of mouse embryogenesis. It is mainly expressed in testis and has a localization pattern and interaction partners similar to AURKB [100].

6.4.1 Aurora A Kinase Inhibitors

AURKA is overexpressed and/or amplified in several subtypes of solid tumors that include breast [101, 102], ovarian [103, 104], colon [95], lung [105] and pancreas [106]. High levels of AURKA are associated with poor outcome [95, 101, 104]. However, the contribution of overexpression of AURKA to tumorigenesis on its own is controversial. Some studies found that overexpression of AURKA can drive transformation of immortalized rodent cell lines and tumors in nude mice [95, 107]. Other studies suggest that overexpression of AURKA alone is not sufficient to transform primary mouse embryonic fibroblasts, even after *TP53* mutation [108]. In vivo, both p53-proficient and p53-deficient transgenic mice that overexpress AURKA in mammary glands do not develop malignant tumors [109, 110]. This controversy is probably due to the different models used for every study; in the studies that support that overexpression of AURKA is sufficient to transform cells, Rat-1 and immortalized rodent cells were used and they seem to be more prone to transformation than primary cells [111]. Instead, malignant transformation by AURKA is potentiated by additional oncogenic events, such as activating mutation of RAS

[112]. This probably happens because overexpression of AURKA induces increased segregation errors during anaphase, which in combination with other spontaneous mutations that might arise could lead to tumorigenesis. This was very elegantly shown by Enrique C. Torchia and colleagues. They introduced a conditionally expressed variant of Aurora A that after induction leads to its overexpression only on the skin. They noticed that after treatment with TPA (which induces proliferation) neither control nor the mice expressing the variant developed tumors. On the contrary, when they additionally treated the mice with the carcinogen/mutagenizing agent DMBA, while both the control mice and the ones expressing the variant developed papillomas with the same frequency, the papillomas of the mice overexpressing the variant progressed to a more aggressive metastatic squamous cell carcinoma with a higher frequency than control mice [113]. Conversely, heterozygous deletion of AURKA in mice can increase incidence of spontaneous tumor formation, indicating that AURKA would function more as a haploinsufficient tumor suppressor than as an oncogene [97]. It is important to mention that this is also true for PLK-1 [71], maybe because it shares the same pathway with AURKA [114–117].

Selective inhibition of AURKA leads to mitotic defects such as abnormal mitotic spindles and chromosome segregation defects, and cells consequently become aneuploid [118]. However, it does not seem that AURKA inhibitors cause permanent mitotic arrest (Fig. 6.4). Cells rather go trough an abnormal division that leads to the formation of aneuploid daughter cells which will eventually die in interphase.

Except for its role in mitotic entry, centrosome maturation, spindle formation and chromosome segregation, Aurora A has been shown to be implicated in the establishment of polarity in asymmetrically diving Drosophila neuroblasts [119] and spindle positioning in *C. elegans* and human cells [120, 121]. Similarly, it regulates spindle orientation and cell fate in mouse mammary epithelial cells, promoting asymmetric cell divisions [122]. Therefore, inhibition of AURKA could perturb spindle orientation in asymmetrically dividing cells, inducing differentiation. This seems to be the case for acute megakaryocytic leukemia, a rare subtype of leukemia [123]. In a seminal work, it was shown that AURKA inhibitors are effective therapies in multiple pre-clinical models of acute megakaryocytic leukemia [124]. In this case, AURKA inhibitors act by inducing terminal differentiation of acute megakaryocytic leukemia blasts [124]. Similar effects of AURKA inhibitors on differentiation of megakaryocytes have been recently shown in primary myelofibrosis, another myeloproliferative neoplasm [125].

Several AURKA inhibitors have been investigated in humans as potential anticancer therapy (MLN8237/alisertib, MLN8054, MK-5108/VX-689 and ENMD-2076) but clinical development of most of these molecules has been stopped early (Phase I/II trials) due to lack of efficacy and/or toxicity (review in [126]). Alisertib has by far the most advanced clinical development program with up to 40 clinical trials. In a large phase I/II trial (n = 249 patients), alisertib showed an interesting response rate of 18% in metastatic breast cancers and 21% in small cell lung cancers [127]. When combined with paclitaxel, alisertib showed prolonged progression-free survival in platinum-resistant or refractory ovarian cancer patients in a large randomized phase II trial [128]. Alisertib also showed promising response rate of 30% in patients with peripheral T cell lymphomas in a phase II trial [9], but this benefit was not confirmed in the phase III study (NCT01482962). Overall, it seems that alisertib has anti-tumoral effect on a subset of solid and hematologic tumors and this effect does not seem to correlate with the level of expression of AURKA on tumor samples [9, 129].

Identifying predictive biomarker for response to alisertib can accelerate its clinical development. As for PLK-1, a potential candidate is p53, as amplification of AURKA shows a clear correlation with *TP53* mutational status [98, 130]. However, it is important to mention that the crosstalk between these two proteins is not really well understood. The lack of p53 prolonges mitotic timing upon aurora A inhibition but it does not necessarily lead to cell death. Conversely, when p53 is present, the cells rather exit mitosis, being aneuploid and as a result die in interphase [131]. Mechanistically, p53 negatively regulates AURKA at transcriptional and post-translational level, while AURKA can phosphorylate p53 at three different sites. While Ser-315-phosphorylated p53 undergoes ubiquitination and subsequent degradation and Ser-215-phosphorylated p53 shows a reduced DNA binding activity, phosphorylated Ser106 on p53 inhibits its interaction with MDM2, its negative regulator. Besides p53, p21 and retinoblastoma protein are also important for post-mitotic checkpoint and may influence cell's response to Aurora inhibition [132]. To date, no predictive biomarker for response to Aurora A inhibitors has been identified.

6.4.2 Aurora B Kinase Inhibitors

Overexpression of AURKB has been detected in multiple human tumors [133–135] and it correlates with genomic instability, high tumor grade and/or poor prognosis [133, 134, 136]. However, no cancer- associated mutations of *AURKB* have been identified, and the locus was not found to be amplified [132]. Overexpression of AURKB in Chinese hamster embryo cell can lead to tumor formation in nude mice [137] and generates aneuploidy in human cancer cell lines [138].

Selective inhibition of Aurora B leads to a premature mitotic exit and aneuploidy [139] (Fig. 6.4). Several AURKB inhibitors have been tested in the clinics (AZD1152/barasertib, ENMD-2076, ABT-348/ilorasertib and AT9283). AZD1152 is highly selective for AURKB, 1000-fold higher than for AURKA [140]. AZD1152 showed best efficiency in hematologic malignancies, with up to 25% of response rate in acute myeloid leukemia [141] and did better than low-dose cytarabine in relapsing patients [142]. However, its clinical development is limited by the hematological toxicities. In solid tumors which do not proliferate rapidly when compared to hematological malignancies, AZD1152 did not show any objective response at any dose or schedule in the phase I trial [143].

Due to complex structure-activity relationships and similarities among different kinase families, a property of many AURKB inhibitors is concomitant inhibition of tyrosine kinase relevant in angiogenesis such as KDR/VEGFR2, Flt3 and BCR-Abl [144]. This beneficial property has been exploited in clinical development of ilorasertib/ABT-348 [145], AT9283 [146] and ENMD-2076 [147]. Indeed ENMD-2076 showed a response rate of 8% in recurrent platinum-resistant ovarian cancer [148], a disease that highly benefits from anti-angiogenic therapies [149]. However, this response rate is lower than single agent bevacizumab (anti-VEGF monoclonal antibody) in this setting [150].

6.4.3 Dual Aurora Inhibitors

Several inhibitors that target both AURKA and AURKB have been developed. VX-680/MK-0457 [151] and PHA-739358/danusertib [152]. MK-0457 demonstrated minimal efficacy in hematological and solid malignancies [153, 154]. Danusertib showed clinical efficacy in patients with chronic myeloid leukemia that harbor T315I Abl kinase mutations [155].

6.5 Kinesins

Kinesins are motor proteins that use ATP energy to translocate, often with cargo, along microtubules. So far, 16 different kinesins have been shown to have mitotic functions in spindle assembly, chromosome alignment, chromosome segregation and cytokinesis [132].

6.5.1 Kinesin-5 Inhibitors

Kinesin-5 (Eg5/Kif11) family members were the first identified to be essential in mitosis in frog extracts [156]. The name Eg points to the fact that it was identified in xenopus eggs [157, 158]. Kinesin-5 motors assemble into a bipolar homotetrameric structure that is capable of modulating dynamics and organization of microtubules [159]. In mammalian cells, they are the most important motors for centrosome separation during prophase by creating pushing forces between the centrosomes. They are plus-end microtubule motors that are able to slide on antiparallel microtubules emanating from the two opposite poles [160].

Selective inhibition of kinesin-5 leads to the formation of monopolar spindles. As a result, spindle assembly checkpoint cannot get satisfied and cells get arrested in this monopolar configuration eventually leading to cell death. However, kinesin-5 inhibition does not have a direct effect on microtubule dynamics (Fig. 6.5) [161].

Given the central role of kinesin-5 in mitosis, over 100 small chemical inhibitors of human kinesin-5 have been developed as potent anticancer therapies. Monastrol

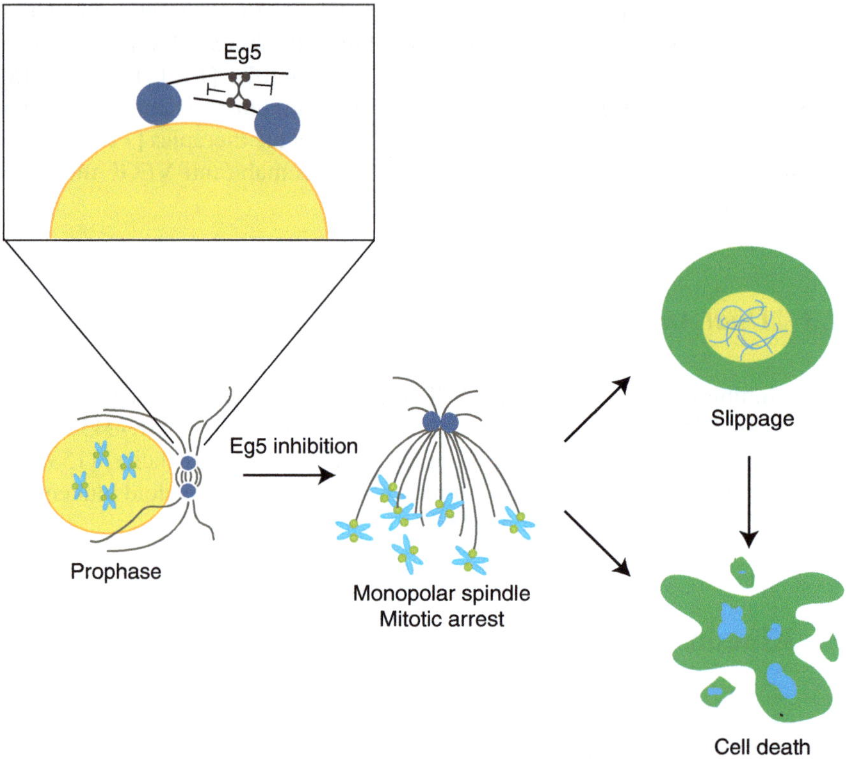

Fig. 6.5 Eg5 inhibition. Eg5 inhibition blocks centrosome separation and cells are arrested in a monopolar configuration

was the proof-of principle molecule targeting kinesin-5 [162]. Several target therapies against kinesin-5 have shown promise in preclinical studies [163–165]. In the clinics, kinesin-5 inhibitors got disappointing results in cancer patients with very few if any objective responses obtained so far [166–168]. Toxicities were limited with this class of drugs. One potential explanation for the disappointing results in the clinics could be the necessity for kinesin-5 inhibitors to accumulate inside cancer cells at very high concentration. Additionally, cells become resistant to Eg-5 inhibitors extremely fast by overexpressing alternative kinases for centrosome separation such as Kif15 [169, 170]. In cancer patients, it might be difficult to assess an objective response to eg-5 inhibitors as resistant cells appear very fast, probably between two evaluations of response.

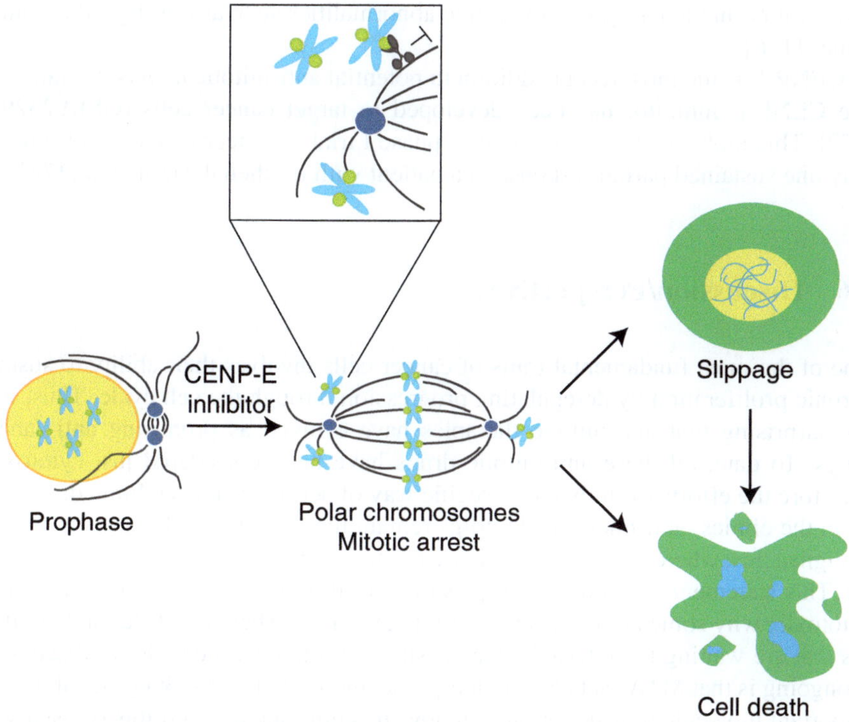

Fig. 6.6 CENP-E inhibition. CENP-E is a motor protein that is able to carry polar chromosomes to the metaphase plate. When it is blocked, the cells form a metaphase plate with polar chromosomes and cells are arrested in this configuration

6.5.2 CENP-E Inhibitor

The kinesin motor protein, Centromere protein E (CENP-E) is a dimeric kinesin (Kinesin-7 family) with critical roles in mitosis. CENP-E is a plus-end motor protein that is able to slide on microtubules helping at the movement of mono-oriented chromosomes for proper alignment at metaphase. CENP-E exhibits a periodic accumulation and loss, with maximal levels found during late G2 and M-phases of cell cycle, and minimum levels in early G1 [171]. Consistently, in human tissues, mRNA expression pattern of CENP-E shows a strong association with proliferation. CENP-E expression is undetectable or low in most of the tissues except those with significant cellular turnover such as gastrointestinal epithelia, hematopoetic sites or testis [172]. Lung adenocarcinomas and squamous cell carcinomas showed increase levels of CENP-E mRNA whereas prostate cancers have very low [172].

Depletion or selective inhibition of CENP-E results in mitotic arrest due to polar chromosomes [173, 174]. This mitotic arrest is followed by apoptotic cell death (Fig. 6.6) [175]. Homozygous disruption of CENP-E in developing mice leads to early embryonic lethality, due to mitotic abnormalities such as misaligned chromosomes [176].

CENP-E is the most recent addition to potential anti-mitotic targets. So far, only one CENP-E inhibitor has been developed to target cancer cells (GSK923295) [177]. This molecule has been tested in phase I trial in cancer patients and showed only one sustained partial response in a patient with urothelial carcinoma [178].

6.6 Discussion/Perspectives

One of the most fundamental traits of cancer cells involves their ability to sustain chronic proliferation by deregulating progression through the cell cycle. Thus, it is not surprising that anti-mitotic therapies have evolved as promising anti-cancer drugs. To date, all these anti-mitotic drugs have been considered just cytotoxic, therefore the efforts to study their specific way of action have been limited.

In the clinics, anti-microtubule drugs remain the first choice therapy for several malignancies, whereas none of the kinase or kinesin inhibitors has been approved so far. This statement raises different questions; if all mitosis targeting drugs are just cytotoxic, why some are more successful than others? There are different hypotheses that are waiting to be tested. One possible reason, on which a lot of discussion is ongoing is that MTAs act also on interphase microtubules, blocking signaling and important molecule transportation. Till now, it is not established if this is just a secondary effect or it could be the main cause of cancer cell death upon treatment with MTAs. For some tumors highly dependent on receptor transportation such as prostate cancer, signaling blockage can have a major effect on cell survival. Another possibility is that MTAs do not inhibit a single molecule, as inhibitors of kinases or kinesins, but the microtubules, which form complex structures with different participating proteins. Consequently, MTAs can be more efficient at a lower concentration in exerting their anti-mitotic activity compared to drugs that target a single kinesin or kinase.

Another question that remains to be answered is why some chemotherapeutic drugs are more efficient in certain types of cancer than others. This is based on clinical observations but it would be interesting to understand the molecular mechanism that lays behind. What is the common molecular characteristic of cancers that respond to taxanes or vinca-alkaloids? In the era of precise medicine, answering this question will be of a great benefit with improving the clinical efficiency of MTAs while avoiding toxicity to patients that are unlikely to respond.

Finally, understanding the mechanisms of resistance to chemotherapeutic drugs is a major area of interest. Resistance can be "intrinsic" (primary resistance), which means that treatment has no clinical benefit on the patient from the beginning. It can also be "acquired" (secondary resistance), which means that resistance is developed after or under therapy. These questions have been in part elucidated in DNA damage agents such as platinum and PARP inhibitors [179]. For anti-mitotic drugs, the question of resistance remains a major issue and there is a need for identifying reproducible biomarkers of resistance in preclinical and clinical models.

References

1. Walczak CE, Cai S, Khodjakov A (2010) Mechanisms of chromosome behaviour during mitosis. Nat Rev Mol Cell Biol 11(2):91–102
2. Cimini D (2008) Merotelic kinetochore orientation, aneuploidy, and cancer. Biochim Biophys Acta 1786(1):32–40
3. Pfau SJ, Amon A (2012) Chromosomal instability and aneuploidy in cancer: from yeast to man. EMBO Rep 13(6):515–527
4. Hanahan D, Weinberg RA (2000) The hallmarks of cancer. Cell 100(1):57–70
5. Hanahan D, Weinberg RA (2011) Hallmarks of cancer: the next generation. Cell 144(5):646–674
6. Foley EA, Kapoor TM (2013) Microtubule attachment and spindle assembly checkpoint signalling at the kinetochore. Nat Rev Mol Cell Biol 14(1):25–37
7. Rohena CC, Mooberry SL (2014) Recent progress with microtubule stabilizers: new compounds, binding modes and cellular activities. Nat Prod Rep 31(3):335–355
8. Dohner H, Lubbert M, Fiedler W, Fouillard L, Haaland A, Brandwein JM et al (2014) Randomized, phase 2 trial of low-dose cytarabine with or without volasertib in AML patients not suitable for induction therapy. Blood 124(9):1426–1433
9. Barr PM, Li H, Spier C, Mahadevan D, LeBlanc M, Ul Haq M et al (2015) Phase II intergroup trial of alisertib in relapsed and refractory peripheral T-cell lymphoma and transformed mycosis fungoides: SWOG 1108. J Clin Oncol Off J Am Soc Clin Oncol 33(21):2399–2404
10. Mehrara E, Forssell-Aronsson E, Ahlman H, Bernhardt P (2007) Specific growth rate versus doubling time for quantitative characterization of tumor growth rate. Cancer Res 67(8):3970–3975
11. Desai A, Mitchison TJ (1997) Microtubule polymerization dynamics. Annu Rev Cell Dev Biol 13:83–117
12. Mistry SJ, Atweh GF (2002) Role of stathmin in the regulation of the mitotic spindle: potential applications in cancer therapy. Mt Sinai J Med 69(5):299–304
13. Akhmanova A, Steinmetz MO (2015) Control of microtubule organization and dynamics: two ends in the limelight. Nat Rev Mol Cell Biol 16(12):711–726
14. Klute K, Nackos E, Tasaki S, Nguyen DP, Bander NH, Tagawa ST (2014) Microtubule inhibitor-based antibody-drug conjugates for cancer therapy. Onco Targets Ther 7:2227–2236
15. Jordan MA, Kamath K (2007) How do microtubule-targeted drugs work? An overview. Curr Cancer Drug Targets 7(8):730–742
16. Komlodi-Pasztor E, Sackett DL, Fojo AT (2012) Inhibitors targeting mitosis: tales of how great drugs against a promising target were brought down by a flawed rationale. Clin Cancer Res 18(1):51–63
17. Wani MC, Taylor HL, Wall ME, Coggon P, McPhail AT (1971) Plant antitumor agents. VI. The isolation and structure of taxol, a novel antileukemic and antitumor agent from *Taxus brevifolia*. J Am Chem Soc 93(9):2325–2327
18. Bissery MC, Guenard D, Gueritte-Voegelein F, Lavelle F (1991) Experimental antitumor activity of taxotere (RP 56976, NSC 628503), a taxol analogue. Cancer Res 51(18):4845–4852
19. Jordan MA, Toso RJ, Thrower D, Wilson L (1993) Mechanism of mitotic block and inhibition of cell proliferation by taxol at low concentrations. Proc Natl Acad Sci U S A 90(20):9552–9556

20. Zasadil LM, Andersen KA, Yeum D, Rocque GB, Wilke LG, Tevaarwerk AJ et al (2014) Cytotoxicity of paclitaxel in breast cancer is due to chromosome missegregation on multipolar spindles. Sci Transl Med 6(229):229–243
21. Feldman BJ, Feldman D (2001) The development of androgen-independent prostate cancer. Nat Rev Cancer 1(1):34–45
22. Zhu ML, Horbinski CM, Garzotto M, Qian DZ, Beer TM, Kyprianou N (2010) Tubulin-targeting chemotherapy impairs androgen receptor activity in prostate cancer. Cancer Res 70(20):7992–8002
23. Darshan MS, Loftus MS, Thadani-Mulero M, Levy BP, Escuin D, Zhou XK et al (2011) Taxane-induced blockade to nuclear accumulation of the androgen receptor predicts clinical responses in metastatic prostate cancer. Cancer Res 71(18):6019–6029
24. Thadani-Mulero M, Nanus DM, Giannakakou P (2012) Androgen receptor on the move: boarding the microtubule expressway to the nucleus. Cancer Res 72(18):4611–4615
25. Mistry SJ, Oh WK (2013) New paradigms in microtubule-mediated endocrine signaling in prostate cancer. Mol Cancer Ther 12(5):555–566
26. Poruchynsky MS, Komlodi-Pasztor E, Trostel S, Wilkerson J, Regairaz M, Pommier Y et al (2015) Microtubule-targeting agents augment the toxicity of DNA-damaging agents by disrupting intracellular trafficking of DNA repair proteins. Proc Natl Acad Sci U S A 112(5):1571–1576
27. Bocci G, Di Paolo A, Danesi R (2013) The pharmacological bases of the antiangiogenic activity of paclitaxel. Angiogenesis 16(3):481–492
28. Jiang H, Tao W, Zhang M, Pan S, Kanwar JR, Sun X (2010) Low-dose metronomic paclitaxel chemotherapy suppresses breast tumors and metastases in mice. Cancer Investig 28(1):74–84
29. Merchan JR, Jayaram DR, Supko JG, He X, Bubley GJ, Sukhatme VP (2005) Increased endothelial uptake of paclitaxel as a potential mechanism for its antiangiogenic effects: potentiation by Cox-2 inhibition. Int J Cancer 113(3):490–498
30. Pasquier E, Carre M, Pourroy B, Camoin L, Rebai O, Briand C et al (2004) Antiangiogenic activity of paclitaxel is associated with its cytostatic effect, mediated by the initiation but not completion of a mitochondrial apoptotic signaling pathway. Mol Cancer Ther 3(10):1301–1310
31. Hanahan D, Bergers G, Bergsland E (2000) Less is more, regularly: metronomic dosing of cytotoxic drugs can target tumor angiogenesis in mice. J Clin Invest 105(8):1045–1047
32. Kerbel RS, Kamen BA (2004) The anti-angiogenic basis of metronomic chemotherapy. Nat Rev Cancer 4(6):423–436
33. Seidman AD, Berry D, Cirrincione C, Harris L, Muss H, Marcom PK et al (2008) Randomized phase III trial of weekly compared with every-3-weeks paclitaxel for metastatic breast cancer, with trastuzumab for all HER-2 overexpressors and random assignment to trastuzumab or not in HER-2 nonoverexpressors: final results of cancer and leukemia group B protocol 9840. J Clin Oncol 26(10):1642–1649
34. Sparano JA, Wang M, Martino S, Jones V, Perez EA, Saphner T et al (2008) Weekly paclitaxel in the adjuvant treatment of breast cancer. N Engl J Med 358(16):1663–1671
35. Chan JK, Brady MF, Penson RT, Huang H, Birrer MJ, Walker JL et al (2016) Weekly vs. every-3-week paclitaxel and carboplatin for ovarian cancer. N Engl J Med 374(8):738–748
36. Yusuf RZ, Duan Z, Lamendola DE, Penson RT, Seiden MV (2003) Paclitaxel resistance: molecular mechanisms and pharmacologic manipulation. Curr Cancer Drug Targets 3(1):1–19
37. Orr GA, Verdier-Pinard P, McDaid H, Horwitz SB (2003) Mechanisms of Taxol resistance related to microtubules. Oncogene 22(47):7280–7295
38. Desai N, Trieu V, Yao Z, Louie L, Ci S, Yang A et al (2006) Increased antitumor activity, intratumor paclitaxel concentrations, and endothelial cell transport of cremophor-free, albumin-bound paclitaxel, ABI-007, compared with cremophor-based paclitaxel. Clin Cancer Res 12(4):1317–1324

39. Rugo HS, Barry WT, Moreno-Aspitia A, Lyss AP, Cirrincione C, Leung E et al (2015) Randomized phase III trial of paclitaxel once per week compared with nanoparticle albumin-bound nab-paclitaxel once per week or ixabepilone with bevacizumab as first-line chemotherapy for locally recurrent or metastatic breast cancer: CALGB 40502/NCCTG N063H (alliance). J Clin Oncol 33(21):2361–2369
40. Gradishar WJ, Tjulandin S, Davidson N, Shaw H, Desai N, Bhar P et al (2005) Phase III trial of nanoparticle albumin-bound paclitaxel compared with polyethylated castor oil-based paclitaxel in women with breast cancer. J Clin Oncol 23(31):7794–7803
41. Von Hoff DD, Ervin T, Arena FP, Chiorean EG, Infante J, Moore M et al (2013) Increased survival in pancreatic cancer with nab-paclitaxel plus gemcitabine. N Engl J Med 369(18):1691–1703
42. Cortes J, Baselga J (2007) Targeting the microtubules in breast cancer beyond taxanes: the epothilones. Oncologist 12(3):271–280
43. Nettles JH, Li H, Cornett B, Krahn JM, Snyder JP, Downing KH (2004) The binding mode of epothilone A on alpha,beta-tubulin by electron crystallography. Science 305(5685):866–869
44. Bode CJ, Gupta ML Jr, Reiff EA, Suprenant KA, Georg GI, Himes RH (2002) Epothilone and paclitaxel: unexpected differences in promoting the assembly and stabilization of yeast microtubules. Biochemistry 41(12):3870–3874
45. Kowalski RJ, Giannakakou P, Hamel E (1997) Activities of the microtubule-stabilizing agents epothilones A and B with purified tubulin and in cells resistant to paclitaxel (Taxol(R)). J Biol Chem 272(4):2534–2541
46. McMeekin S, Dizon D, Barter J, Scambia G, Manzyuk L, Lisyanskaya A et al (2015) Phase III randomized trial of second-line ixabepilone versus paclitaxel or doxorubicin in women with advanced endometrial cancer. Gynecol Oncol 138(1):18–23
47. Dumontet C, Jordan MA (2010) Microtubule-binding agents: a dynamic field of cancer therapeutics. Nat Rev Drug Discov 9(10):790–803
48. Bellmunt J, Fougeray R, Rosenberg JE, von der Maase H, Schutz FA, Salhi Y et al (2013) Long-term survival results of a randomized phase III trial of vinflunine plus best supportive care versus best supportive care alone in advanced urothelial carcinoma patients after failure of platinum-based chemotherapy. Ann Oncol 24(6):1466–1472
49. Biziota E, Briasoulis E, Mavroeidis L, Marselos M, Harris AL, Pappas P (2016) Cellular and molecular effects of metronomic vinorelbine and 4-O-deacetylvinorelbine on human umbilical vein endothelial cells. Anti-Cancer Drugs 27(3):216–224
50. Jordan MA, Kamath K, Manna T, Okouneva T, Miller HP, Davis C et al (2005) The primary antimitotic mechanism of action of the synthetic halichondrin E7389 is suppression of microtubule growth. Mol Cancer Ther 4(7):1086–1095
51. Okouneva T, Azarenko O, Wilson L, Littlefield BA, Jordan MA (2008) Inhibition of centromere dynamics by eribulin (E7389) during mitotic metaphase. Mol Cancer Ther 7(7):2003–2011
52. Cortes J, O'Shaughnessy J, Loesch D, Blum JL, Vahdat LT, Petrakova K et al (2011) Eribulin monotherapy versus treatment of physician's choice in patients with metastatic breast cancer (EMBRACE): a phase 3 open-label randomised study. Lancet 377(9769):914–923
53. Schoffski P, Chawla S, Maki RG, Italiano A, Gelderblom H, Choy E et al (2016) Eribulin versus dacarbazine in previously treated patients with advanced liposarcoma or leiomyosarcoma: a randomised, open-label, multicentre, phase 3 trial. Lancet 387(10028):1629–1637
54. Lautier D, Canitrot Y, Deeley RG, Cole SP (1996) Multidrug resistance mediated by the multidrug resistance protein (MRP) gene. Biochem Pharmacol 52(7):967–977
55. Oba T, Izumi H, Ito KI (2016) ABCB1 and ABCC11 confer resistance to eribulin in breast cancer cell lines. Oncotarget 7(43):70011–70027
56. Wu J, He Z, Wang DL, Sun FL (2016) Depletion of JMJD5 sensitizes tumor cells to microtubule-destabilizing agents by altering microtubule stability. Cell Cycle 15(21):2980–2991

57. Kanakkanthara A, Northcote PT, Miller JH (2012) betaI-tubulin and betaIII-tubulin mediate sensitivity to peloruside A and laulimalide, but not paclitaxel or vinblastine, in human ovarian carcinoma cells. Mol Cancer Ther 11(2):393–404
58. Kanakkanthara A, Wilmes A, O'Brate A, Escuin D, Chan A, Gjyrezi A et al (2011) Peloruside- and laulimalide-resistant human ovarian carcinoma cells have betaI-tubulin mutations and altered expression of betaII- and betaIII-tubulin isotypes. Mol Cancer Ther 10(8):1419–1429
59. Peters C, Brown S (2015) Antibody-drug conjugates as novel anti-cancer chemotherapeutics. Biosci Rep 35(4)
60. Hamel E (1992) Natural products which interact with tubulin in the vinca domain: maytansine, rhizoxin, phomopsin A, dolastatins 10 and 15 and halichondrin B. Pharmacol Ther 55(1):31–51
61. Erickson HK, Park PU, Widdison WC, Kovtun YV, Garrett LM, Hoffman K et al (2006) Antibody-maytansinoid conjugates are activated in targeted cancer cells by lysosomal degradation and linker-dependent intracellular processing. Cancer Res 66(8):4426–4433
62. Oroudjev E, Lopus M, Wilson L, Audette C, Provenzano C, Erickson H et al (2010) Maytansinoid-antibody conjugates induce mitotic arrest by suppressing microtubule dynamic instability. Mol Cancer Ther 9(10):2700–2713
63. Huang AB, Lin CM, Hamel E (1985) Maytansine inhibits nucleotide binding at the exchangeable site of tubulin. Biochem Biophys Res Commun 128(3):1239–1246
64. Slamon DJ, Godolphin W, Jones LA, Holt JA, Wong SG, Keith DE et al (1989) Studies of the HER-2/neu proto-oncogene in human breast and ovarian cancer. Science 244(4905):707–712
65. Verma S, Miles D, Gianni L, Krop IE, Welslau M, Baselga J et al (2012) Trastuzumab emtansine for HER2-positive advanced breast cancer. N Engl J Med 367(19):1783–1791
66. Younes A, Bartlett NL, Leonard JP, Kennedy DA, Lynch CM, Sievers EL et al (2010) Brentuximab vedotin (SGN-35) for relapsed CD30-positive lymphomas. N Engl J Med 363(19):1812–1821
67. Moskowitz CH, Nademanee A, Masszi T, Agura E, Holowiecki J, Abidi MH et al (2015) Brentuximab vedotin as consolidation therapy after autologous stem-cell transplantation in patients with Hodgkin's lymphoma at risk of relapse or progression (AETHERA): a randomised, double-blind, placebo-controlled, phase 3 trial. Lancet 385(9980):1853–1862
68. Lens SM, Voest EE, Medema RH (2010) Shared and separate functions of polo-like kinases and aurora kinases in cancer. Nat Rev Cancer 10(12):825–841
69. Strebhardt K, Ullrich A (2006) Targeting polo-like kinase 1 for cancer therapy. Nat Rev Cancer 6(4):321–330
70. Hyun SY, Hwang HI, Jang YJ (2014) Polo-like kinase-1 in DNA damage response. BMB Rep 47(5):249–255
71. Lu LY, Wood JL, Minter-Dykhouse K, Ye L, Saunders TL, Yu X et al (2008) Polo-like kinase 1 is essential for early embryonic development and tumor suppression. Mol Cell Biol 28(22):6870–6876
72. Winkles JA, Alberts GF (2005) Differential regulation of polo-like kinase 1, 2, 3, and 4 gene expression in mammalian cells and tissues. Oncogene 24(2):260–266
73. Weichert W, Denkert C, Schmidt M, Gekeler V, Wolf G, Kobel M et al (2004) Polo-like kinase isoform expression is a prognostic factor in ovarian carcinoma. Br J Cancer 90(4):815–821
74. King SI, Purdie CA, Bray SE, Quinlan PR, Jordan LB, Thompson AM et al (2012) Immunohistochemical detection of polo-like kinase-1 (PLK1) in primary breast cancer is associated with TP53 mutation and poor clinical outcom. Breast Cancer Res 14(2):R40
75. Triscott J, Lee C, Foster C, Manoranjan B, Pambid MR, Berns R et al (2013) Personalizing the treatment of pediatric medulloblastoma: polo-like kinase 1 as a molecular target in high-risk children. Cancer Res 73(22):6734–6744

76. Wolf G, Elez R, Doermer A, Holtrich U, Ackermann H, Stutte HJ et al (1997) Prognostic significance of polo-like kinase (PLK) expression in non-small cell lung cancer. Oncogene 14(5):543–549
77. Spankuch-Schmitt B, Wolf G, Solbach C, Loibl S, Knecht R, Stegmuller M et al (2002) Downregulation of human polo-like kinase activity by antisense oligonucleotides induces growth inhibition in cancer cells. Oncogene 21(20):3162–3171
78. Steegmaier M, Hoffmann M, Baum A, Lenart P, Petronczki M, Krssak M et al (2007) BI 2536, a potent and selective inhibitor of polo-like kinase 1, inhibits tumor growth in vivo. Curr Biol 17(4):316–322
79. Kiyomitsu T, Cheeseman IM (2012) Chromosome- and spindle-pole-derived signals generate an intrinsic code for spindle position and orientation. Nat Cell Biol 14(3):311–317
80. Yan M, Chu L, Qin B, Wang Z, Liu X, Jin C et al (2015) Regulation of NDR1 activity by PLK1 ensures proper spindle orientation in mitosis. Sci Rep 5:10449
81. Hanafusa H, Kedashiro S, Tezuka M, Funatsu M, Usami S, Toyoshima F et al (2015) PLK1-dependent activation of LRRK1 regulates spindle orientation by phosphorylating CDK5RAP2. Nat Cell Biol 17(8):1024–1035
82. Shrestha R, Little KA, Tamayo JV, Li W, Perlman DH, Devenport D (2015) Mitotic control of planar cell polarity by polo-like kinase 1. Dev Cell 33(5):522–534
83. Noatynska A, Panbianco C, Gotta M (2010) SPAT-1/bora acts with polo-like kinase 1 to regulate PAR polarity and cell cycle progression. Development 137(19):3315–3325
84. Rivers DM, Moreno S, Abraham M, Ahringer J (2008) PAR proteins direct asymmetry of the cell cycle regulators polo-like kinase and Cdc25. J Cell Biol 180(5):877–885
85. Lerner RG, Grossauer S, Kadkhodaei B, Meyers I, Sidorov M, Koeck K et al (2015) Targeting a Plk1-controlled polarity checkpoint in therapy-resistant glioblastoma-propagating cells. Cancer Res 75(24):5355–5366
86. Yim H (2013) Current clinical trials with polo-like kinase 1 inhibitors in solid tumors. Anti-Cancer Drugs 24(10):999–1006
87. Pujade-Lauraine E, Selle F, Weber B, Ray-Coquard IL, Vergote I, Sufliarsky J et al (2016) Volasertib versus chemotherapy in platinum-resistant or -refractory ovarian cancer: a randomized phase II groupe des investigateurs nationaux pour l'Etude des cancers de l'Ovaire study. J Clin Oncol Off J Am Soc Clin Oncol 34(7):706–713
88. Cancer Genome Atlas Research N (2011) Integrated genomic analyses of ovarian carcinoma. Nature 474(7353):609–615
89. Watanabe G, Ishida T, Furuta A, Takahashi S, Watanabe M, Nakata H et al (2015) Combined immunohistochemistry of PLK1, p21, and p53 for predicting TP53 status: an independent prognostic factor of breast cancer. Am J Surg Pathol 39(8):1026–1034
90. Louwen F, Yuan J (2013) Battle of the eternal rivals: restoring functional p53 and inhibiting polo-like kinase 1 as cancer therapy. Oncotarget 4(7):958–971
91. Dias SS, Hogan C, Ochocka AM, Meek DW (2009) Polo-like kinase-1 phosphorylates MDM2 at Ser260 and stimulates MDM2-mediated p53 turnover. FEBS Lett 583(22):3543–3548
92. Luo J, Emanuele MJ, Li D, Creighton CJ, Schlabach MR, Westbrook TF et al (2009) A genome-wide RNAi screen identifies multiple synthetic lethal interactions with the Ras oncogene. Cell 137(5):835–848
93. Carmena M, Earnshaw WC, Glover DM (2015) The dawn of Aurora kinase research: from fly genetics to the clinic. Front Cell Dev Biol 3:73
94. Su AI, Wiltshire T, Batalov S, Lapp H, Ching KA, Block D et al (2004) A gene atlas of the mouse and human protein-encoding transcriptomes. Proc Natl Acad Sci U S A 101(16):6062–6067
95. Bischoff JR, Anderson L, Zhu Y, Mossie K, Ng L, Souza B et al (1998) A homologue of Drosophila aurora kinase is oncogenic and amplified in human colorectal cancers. EMBO J 17(11):3052–3065

96. Carmena M, Earnshaw WC (2003) The cellular geography of aurora kinases. Nat Rev Mol Cell Biol 4(11):842–854
97. Lu LY, Wood JL, Ye L, Minter-Dykhouse K, Saunders TL, Yu X et al (2008) Aurora A is essential for early embryonic development and tumor suppression. J Biol Chem 283(46):31785–31790
98. Katayama H, Sasai K, Kawai H, Yuan ZM, Bondaruk J, Suzuki F et al (2004) Phosphorylation by aurora kinase A induces Mdm2-mediated destabilization and inhibition of p53. Nat Genet 36(1):55–62
99. Fernandez-Miranda G, Trakala M, Martin J, Escobar B, Gonzalez A, Ghyselinck NB et al (2011) Genetic disruption of aurora B uncovers an essential role for aurora C during early mammalian development. Development 138(13):2661–2672
100. Malumbres M, Perez de Castro I (2014) Aurora kinase A inhibitors: promising agents in antitumoral therapy. Expert Opin Ther Targets 18(12):1377–1393
101. Nadler Y, Camp RL, Schwartz C, Rimm DL, Kluger HM, Kluger Y (2008) Expression of Aurora A (but not Aurora B) is predictive of survival in breast cancer. Clin Cancer Res 14(14):4455–4462
102. Staff S, Isola J, Jumppanen M, Tanner M (2010) Aurora-A gene is frequently amplified in basal-like breast cancer. Oncol Rep 23(2):307–312
103. Gritsko TM, Coppola D, Paciga JE, Yang L, Sun M, Shelley SA et al (2003) Activation and overexpression of centrosome kinase BTAK/Aurora-A in human ovarian cancer. Clin Cancer Res 9(4):1420–1426
104. Landen CN Jr, Lin YG, Immaneni A, Deavers MT, Merritt WM, Spannuth WA et al (2007) Overexpression of the centrosomal protein Aurora-A kinase is associated with poor prognosis in epithelial ovarian cancer patients. Clin Cancer Res 13(14):4098–4104
105. Lo Iacono M, Monica V, Saviozzi S, Ceppi P, Bracco E, Papotti M et al (2011) Aurora kinase A expression is associated with lung cancer histological-subtypes and with tumor de-differentiation. J Transl Med 9:100
106. Li D, Zhu J, Firozi PF, Abbruzzese JL, Evans DB, Cleary K et al (2003) Overexpression of oncogenic STK15/BTAK/Aurora A kinase in human pancreatic cancer. Clin Cancer Res 9(3):991–997
107. Zhou H, Kuang J, Zhong L, Kuo WL, Gray JW, Sahin A et al (1998) Tumour amplified kinase STK15/BTAK induces centrosome amplification, aneuploidy and transformation. Nat Genet 20(2):189–193
108. Anand S, Penrhyn-Lowe S, Venkitaraman AR (2003) AURORA-A amplification overrides the mitotic spindle assembly checkpoint, inducing resistance to Taxol. Cancer Cell 3(1):51–62
109. Zhang D, Shimizu T, Araki N, Hirota T, Yoshie M, Ogawa K et al (2008) Aurora A overexpression induces cellular senescence in mammary gland hyperplastic tumors developed in p53-deficient mice. Oncogene 27(31):4305–4314
110. Zhang D, Hirota T, Marumoto T, Shimizu M, Kunitoku N, Sasayama T et al (2004) Cre-loxP-controlled periodic Aurora-A overexpression induces mitotic abnormalities and hyperplasia in mammary glands of mouse models. Oncogene 23(54):8720–8730
111. Reynolds VL, DiPietro M, Lebovitz RM, Lieberman MW (1987) Inherent tumorigenic and metastatic properties of rat-1 and rat-2 cells. Cancer Res 47(23):6384–6387
112. Tatsuka M, Sato S, Kitajima S, Suto S, Kawai H, Miyauchi M et al (2005) Overexpression of Aurora-A potentiates HRAS-mediated oncogenic transformation and is implicated in oral carcinogenesis. Oncogene 24(6):1122–1127
113. Torchia EC, Chen Y, Sheng H, Katayama H, Fitzpatrick J, Brinkley WR et al (2009) A genetic variant of Aurora kinase A promotes genomic instability leading to highly malignant skin tumors. Cancer Res 69(18):7207–7215
114. Asteriti IA, De Mattia F, Guarguaglini G (2015) Cross-talk between AURKA and Plk1 in mitotic entry and spindle assembly. Front Oncol 5:283

115. Archambault V, Carmena M (2012) Polo-like kinase-activating kinases: Aurora A, Aurora B and what else? Cell Cycle 11(8):1490–1495
116. De Luca M, Lavia P, Guarguaglini G (2006) A functional interplay between Aurora-A, Plk1 and TPX2 at spindle poles: Plk1 controls centrosomal localization of Aurora-A and TPX2 spindle association. Cell Cycle 5(3):296–303
117. Bruinsma W, Macurek L, Freire R, Lindqvist A, Medema RH (2014) Bora and Aurora-A continue to activate Plk1 in mitosis. J Cell Sci 127(Pt 4):801–811
118. Manfredi MG, Ecsedy JA, Meetze KA, Balani SK, Burenkova O, Chen W et al (2007) Antitumor activity of MLN8054, an orally active small-molecule inhibitor of Aurora A kinase. Proc Natl Acad Sci U S A 104(10):4106–4111
119. Wirtz-Peitz F, Nishimura T, Knoblich JA (2008) Linking cell cycle to asymmetric division: Aurora-A phosphorylates the par complex to regulate numb localization. Cell 135(1):161–173
120. Kotak S, Afshar K, Busso C, Gonczy P (2016) Aurora A Kinase regulates proper spindle positioning in *C. elegans* and in human cells. J Cell Sci 129(15):3015–3025
121. Gallini S, Carminati M, De Mattia F, Pirovano L, Martini E, Oldani A et al (2016) NuMA phosphorylation by Aurora-A orchestrates spindle orientation. Curr Biol 26(4):458–469
122. Regan JL, Sourisseau T, Soady K, Kendrick H, McCarthy A, Tang C et al (2013) Aurora A kinase regulates mammary epithelial cell fate by determining mitotic spindle orientation in a notch-dependent manner. Cell Rep 4(1):110–123
123. Goldenson B, Crispino JD (2015) The aurora kinases in cell cycle and leukemia. Oncogene 34(5):537–545
124. Wen Q, Goldenson B, Silver SJ, Schenone M, Dancik V, Huang Z et al (2012) Identification of regulators of polyploidization presents therapeutic targets for treatment of AMKL. Cell 150(3):575–589
125. Jeremy Wen Q, Yang Q, Goldenson B, Malinge S, Lasho T, Schneider RK et al (2015) Targeting megakaryocytic-induced fibrosis in myeloproliferative neoplasms by AURKA inhibition. Nat Med 21(12):1473–1480
126. Malumbres M, Perez de Castro I (2014) Aurora kinase A inhibitors: promising agents in antitumoral therapy. Expert Opin Ther Targets 18(12):1–17
127. Melichar B, Adenis A, Lockhart AC, Bennouna J, Dees EC, Kayaleh O et al (2015) Safety and activity of alisertib, an investigational aurora kinase A inhibitor, in patients with breast cancer, small-cell lung cancer, non-small-cell lung cancer, head and neck squamous-cell carcinoma, and gastro-oesophageal adenocarcinoma: a five-arm phase 2 study. Lancet Oncol 16(4):395–405
128. Coleman R, Roszak A, Behbakht K, Ray-Coquard IL, Matulonis U, Liu H et al (eds) (2014) Randomized phase 2 study of Iinvestigational, selective Aurora A kinase inhibitor Alisertib (MLN8237) with weekly paclitaxel vs paclitaxel alone in patients (pts) with recurrent ovarian cancer (OC). annual ESMO congress; Madrid
129. Matulonis UA, Sharma S, Ghamande S, Gordon MS, Del Prete SA, Ray-Coquard I et al (2012) Phase II study of MLN8237 (alisertib), an investigational Aurora A kinase inhibitor, in patients with platinum-resistant or -refractory epithelial ovarian, fallopian tube, or primary peritoneal carcinoma. Gynecol Oncol 127(1):63–69
130. Mao JH, Wu D, Perez-Losada J, Jiang T, Li Q, Neve RM et al (2007) Crosstalk between Aurora-A and p53: frequent deletion or downregulation of aurora-A in tumors from p53 null mice. Cancer Cell 11(2):161–173
131. Nair JS, Ho AL, Schwartz GK (2012) The induction of polyploidy or apoptosis by the Aurora A kinase inhibitor MK8745 is p53-dependent. Cell Cycle 11(4):807–817
132. Salmela AL, Kallio MJ (2013) Mitosis as an anti-cancer drug target. Chromosoma 122(5):431–449
133. Chen YJ, Chen CM, Twu NF, Yen MS, Lai CR, Wu HH et al (2009) Overexpression of Aurora B is associated with poor prognosis in epithelial ovarian cancer patients. Virchows Arch 455(5):431–440

134. Garcia-Fernandez E, De Diego JI, Collantes-Bellido E, Mendiola M, Prim MP, Perez-Fernandez E et al (2011) Aurora B kinase expression in laryngeal squamous cell carcinoma and its prognostic implications. Histopathology 58(3):368–376
135. Chieffi P, Cozzolino L, Kisslinger A, Libertini S, Staibano S, Mansueto G et al (2006) Aurora B expression directly correlates with prostate cancer malignancy and influence prostate cell proliferation. Prostate 66(3):326–333
136. Goos JA, Coupe VM, Diosdado B, Delis-Van Diemen PM, Karga C, Belien JA et al (2013) Aurora kinase A (AURKA) expression in colorectal cancer liver metastasis is associated with poor prognosis. Br J Cancer 109(9):2445–2452
137. Ota T, Suto S, Katayama H, Han ZB, Suzuki F, Maeda M et al (2002) Increased mitotic phosphorylation of histone H3 attributable to AIM-1/Aurora-B overexpression contributes to chromosome number instability. Cancer Res 62(18):5168–5177
138. Gonzalez-Loyola A, Fernandez-Miranda G, Trakala M, Partida D, Samejima K, Ogawa H et al (2015) Aurora B overexpression causes aneuploidy and p21Cip1 repression during tumor development. Mol Cell Biol 35(20):3566–3578
139. Trakala M, Fernandez-Miranda G, Perez de Castro I, Heeschen C, Malumbres M (2013) Aurora B prevents delayed DNA replication and premature mitotic exit by repressing p21(Cip1). Cell Cycle 12(7):1030–1041
140. Walsby E, Walsh V, Pepper C, Burnett A, Mills K (2008) Effects of the aurora kinase inhibitors AZD1152-HQPA and ZM447439 on growth arrest and polyploidy in acute myeloid leukemia cell lines and primary blasts. Haematologica 93(5):662–669
141. Lowenberg B, Muus P, Ossenkoppele G, Rousselot P, Cahn JY, Ifrah N et al (2011) Phase 1/2 study to assess the safety, efficacy, and pharmacokinetics of barasertib (AZD1152) in patients with advanced acute myeloid leukemia. Blood 118(23):6030–6036
142. Kantarjian HM, Martinelli G, Jabbour EJ, Quintas-Cardama A, Ando K, Bay JO et al (2013) Stage I of a phase 2 study assessing the efficacy, safety, and tolerability of barasertib (AZD1152) versus low-dose cytosine arabinoside in elderly patients with acute myeloid leukemia. Cancer 119(14):2611–2619
143. Schwartz GK, Carvajal RD, Midgley R, Rodig SJ, Stockman PK, Ataman O et al (2013) Phase I study of barasertib (AZD1152), a selective inhibitor of Aurora B kinase, in patients with advanced solid tumors. Investig New Drugs 31(2):370–380
144. Bamborough P, Drewry D, Harper G, Smith GK, Schneider K (2008) Assessment of chemical coverage of kinome space and its implications for kinase drug discovery. J Med Chem 51(24):7898–7914
145. Garcia-Manero G, Tibes R, Kadia T, Kantarjian H, Arellano M, Knight EA et al (2015) Phase 1 dose escalation trial of ilorasertib, a dual Aurora/VEGF receptor kinase inhibitor, in patients with hematologic malignancies. Investig New Drugs 33(4):870–880
146. Arkenau HT, Plummer R, Molife LR, Olmos D, Yap TA, Squires M et al (2012) A phase I dose escalation study of AT9283, a small molecule inhibitor of aurora kinases, in patients with advanced solid malignancies. Ann Oncol ESMO 23(5):1307–1313
147. Fletcher GC, Brokx RD, Denny TA, Hembrough TA, Plum SM, Fogler WE et al (2011) ENMD-2076 is an orally active kinase inhibitor with antiangiogenic and antiproliferative mechanisms of action. Mol Cancer Ther 10(1):126–137
148. Matulonis UA, Lee J, Lasonde B, Tew WP, Yehwalashet A, Matei D et al (2013) ENMD-2076, an oral inhibitor of angiogenic and proliferation kinases, has activity in recurrent, platinum resistant ovarian cancer. Eur J Cancer 49(1):121–131
149. Pujade-Lauraine E, Hilpert F, Weber B, Reuss A, Poveda A, Kristensen G et al (2014) Bevacizumab combined with chemotherapy for platinum-resistant recurrent ovarian cancer: the AURELIA open-label randomized phase III trial. J Clin Oncol Off J Am Soc Clin Oncol 32(13):1302–1308
150. Cannistra SA, Matulonis UA, Penson RT, Hambleton J, Dupont J, Mackey H et al (2007) Phase II study of bevacizumab in patients with platinum-resistant ovarian cancer or peritoneal serous cancer. J Clin Oncol Off J Am Soc Clin Oncol 25(33):5180–5186

151. Harrington EA, Bebbington D, Moore J, Rasmussen RK, Ajose-Adeogun AO, Nakayama T et al (2004) VX-680, a potent and selective small-molecule inhibitor of the Aurora kinases, suppresses tumor growth in vivo. Nat Med 10(3):262–267
152. Carpinelli P, Ceruti R, Giorgini ML, Cappella P, Gianellini L, Croci V et al (2007) PHA-739358, a potent inhibitor of Aurora kinases with a selective target inhibition profile relevant to cancer. Mol Cancer Ther 6(12 Pt 1):3158–3168
153. Seymour JF, Kim DW, Rubin E, Haregewoin A, Clark J, Watson P et al (2014) A phase 2 study of MK-0457 in patients with BCR-ABL T315I mutant chronic myelogenous leukemia and philadelphia chromosome-positive acute lymphoblastic leukemia. Blood Cancer J 4:e238
154. Traynor AM, Hewitt M, Liu G, Flaherty KT, Clark J, Freedman SJ et al (2011) Phase I dose escalation study of MK-0457, a novel Aurora kinase inhibitor, in adult patients with advanced solid tumors. Cancer Chemother Pharmacol 67(2):305–314
155. Borthakur G, Dombret H, Schafhausen P, Brummendorf TH, Boissel N, Jabbour E et al (2015) A phase I study of danusertib (PHA-739358) in adult patients with accelerated or blastic phase chronic myeloid leukemia and Philadelphia chromosome-positive acute lymphoblastic leukemia resistant or intolerant to imatinib and/or other second generation c-ABL therapy. Haematologica 100(7):898–904
156. Sawin KE, LeGuellec K, Philippe M, Mitchison TJ (1992) Mitotic spindle organization by a plus-end-directed microtubule motor. Nature 359(6395):540–543
157. Le Guellec R, Paris J, Couturier A, Roghi C, Philippe M (1991) Cloning by differential screening of a Xenopus cDNA that encodes a kinesin-related protein. Mol Cell Biol 11(6):3395–3398
158. Liao H, Li G, Yen TJ (1994) Mitotic regulation of microtubule cross-linking activity of CENP-E kinetochore protein. Science 265(5170):394–398
159. Kapitein LC, Peterman EJ, Kwok BH, Kim JH, Kapoor TM, Schmidt CF (2005) The bipolar mitotic kinesin Eg5 moves on both microtubules that it crosslinks. Nature 435(7038):114–118
160. Ferenz NP, Gable A, Wadsworth P (2010) Mitotic functions of kinesin-5. Semin Cell Dev Biol 21(3):255–259
161. Blangy A, Lane HA, d'Herin P, Harper M, Kress M, Nigg EA (1995) Phosphorylation by p34cdc2 regulates spindle association of human Eg5, a kinesin-related motor essential for bipolar spindle formation in vivo. Cell 83(7):1159–1169
162. Mayer TU, Kapoor TM, Haggarty SJ, King RW, Schreiber SL, Mitchison TJ (1999) Small molecule inhibitor of mitotic spindle bipolarity identified in a phenotype-based screen. Science 286(5441):971–974
163. Ye XS, Fan L, Van Horn RD, Nakai R, Ohta Y, Akinaga S et al (2015) A novel Eg5 inhibitor (LY2523355) causes mitotic arrest and apoptosis in cancer cells and shows potent antitumor activity in xenograft tumor models. Mol Cancer Ther 14(11):2463–2472
164. Chen Y, Chow JP, Poon RY (2012) Inhibition of Eg5 acts synergistically with checkpoint abrogation in promoting mitotic catastrophe. Mol Cancer Res 10(5):626–635
165. Nakai R, Iida S, Takahashi T, Tsujita T, Okamoto S, Takada C et al (2009) K858, a novel inhibitor of mitotic kinesin Eg5 and antitumor agent, induces cell death in cancer cells. Cancer Res 69(9):3901–3909
166. Wakui H, Yamamoto N, Kitazono S, Mizugaki H, Nakamichi S, Fujiwara Y et al (2014) A phase 1 and dose-finding study of LY2523355 (litronesib), an Eg5 inhibitor, in Japanese patients with advanced solid tumors. Cancer Chemother Pharmacol 74(1):15–23
167. Kantarjian HM, Padmanabhan S, Stock W, Tallman MS, Curt GA, Li J et al (2012) Phase I/II multicenter study to assess the safety, tolerability, pharmacokinetics and pharmacodynamics of AZD4877 in patients with refractory acute myeloid leukemia. Investig New Drugs 30(3):1107–1115
168. Gerecitano JF, Stephenson JJ, Lewis NL, Osmukhina A, Li J, Wu K et al (2013) A phase I trial of the kinesin spindle protein (Eg5) inhibitor AZD4877 in patients with solid and lymphoid malignancies. Investig New Drugs 31(2):355–362

169. Sturgill EG, Norris SR, Guo Y, Ohi R (2016) Kinesin-5 inhibitor resistance is driven by kinesin-12. J Cell Biol 213(2):213–227
170. Dumas ME, Sturgill EG, Ohi R (2016) Resistance is not futile: surviving Eg5 inhibition. Cell Cycle 15(21):2845–2847
171. Brown KD, Coulson RM, Yen TJ, Cleveland DW (1994) Cyclin-like accumulation and loss of the putative kinetochore motor CENP-E results from coupling continuous synthesis with specific degradation at the end of mitosis. J Cell Biol 125(6):1303–1312
172. Wood KW, Chua P, Sutton D, Jackson JR (2008) Centromere-associated protein E: a motor that puts the brakes on the mitotic checkpoint. Clin Cancer Res 14(23):7588–7592
173. Yao X, Abrieu A, Zheng Y, Sullivan KF, Cleveland DW (2000) CENP-E forms a link between attachment of spindle microtubules to kinetochores and the mitotic checkpoint. Nat Cell Biol 2(8):484–491
174. Tanudji M, Shoemaker J, L'Italien L, Russell L, Chin G, Schebye XM (2004) Gene silencing of CENP-E by small interfering RNA in HeLa cells leads to missegregation of chromosomes after a mitotic delay. Mol Biol Cell 15(8):3771–3781
175. Wood KW, Lad L, Luo L, Qian X, Knight SD, Nevins N et al (2010) Antitumor activity of an allosteric inhibitor of centromere-associated protein-E. Proc Natl Acad Sci U S A 107(13):5839–5844
176. Putkey FR, Cramer T, Morphew MK, Silk AD, Johnson RS, McIntosh JR et al (2002) Unstable kinetochore-microtubule capture and chromosomal instability following deletion of CENP-E. Dev Cell 3(3):351–365
177. Qian X, McDonald A, Zhou HJ, Adams ND, Parrish CA, Duffy KJ et al (2010) Discovery of the first potent and selective inhibitor of centromere-associated protein E: GSK923295. ACS Med Chem Lett 1(1):30–34
178. Chung V, Heath EI, Schelman WR, Johnson BM, Kirby LC, Lynch KM et al (2012) First-time-in-human study of GSK923295, a novel antimitotic inhibitor of centromere-associated protein E (CENP-E), in patients with refractory cancer. Cancer Chemother Pharmacol 69(3):733–741
179. Sakai W, Swisher EM, Karlan BY, Agarwal MK, Higgins J, Friedman C et al (2008) Secondary mutations as a mechanism of cisplatin resistance in BRCA2-mutated cancers. Nature 451(7182):1116–1120

Chapter 7
Mitotic Dysfunction Associated with Aging Hallmarks

Joana Catarina Macedo, Sara Vaz, and Elsa Logarinho

Abstract Aging is a biological process characterized by the progressive deterioration of physiological functions known to be the main risk factor for chronic diseases and declining health. There has been an emerging connection between aging and aneuploidy, an aberrant number of chromosomes, even though the molecular mechanisms behind age-associated aneuploidy remain largely unknown. In recent years, several genetic pathways and biochemical processes controlling the rate of aging have been identified and proposed as aging hallmarks. Primary hallmarks that cause the accumulation of cellular damage include genomic instability, telomere attrition, epigenetic alterations and loss of proteostasis (López-Otín et al., Cell 153:1194–1217, 2013). Here we review the provocative link between these aging hallmarks and the loss of chromosome segregation fidelity during cell division, which could support the correlation between aging and aneuploidy seen over the past decades. Secondly, we review the systemic impacts of aneuploidy in cell physiology and emphasize how these include some of the primary hallmarks of aging. Based on the evidence, we propose a mutual causality between aging and aneuploidy, and suggest modulation of mitotic fidelity as a potential means to ameliorate healthy lifespan.

Keywords Aging • Mitosis • Aneuploidy • Genomic instability • Telomeres • Epigenetics • Proteostasis

---

Authors Joana Catarina Macedo and Sara Vaz contributed equally in this chapter.

J.C. Macedo • S. Vaz
Aging and Aneuploidy Laboratory, Instituto de Biologia Molecular e Celular, Instituto de Investigação e Inovação em Saúde – i3S, Universidade do Porto, Rua Alfredo Allen 208, 4200-135 Porto, Portugal

E. Logarinho (✉)
Aging and Aneuploidy Laboratory, Instituto de Biologia Molecular e Celular, Instituto de Investigação e Inovação em Saúde – i3S, Universidade do Porto, Rua Alfredo Allen 208, 4200-135 Porto, Portugal

Cell Division Unit, Department of Experimental Biology, Faculdade de Medicina, Universidade do Porto, Alameda Prof. Hernâni Monteiro, 4200-319 Porto, Portugal
e-mail: elsa.logarinho@ibmc.up.pt

M. Gotta, P. Meraldi (eds.), *Cell Division Machinery and Disease*, Advances in Experimental Medicine and Biology 1002, DOI 10.1007/978-3-319-57127-0_7

7.1 Age-Associated Aneuploidy

Aging has been linked to an increase in aneuploidy[1] for the past several decades [1]. This association has been well documented for oocytes and is considered to be the main cause of female reproductive infertility as well as of mis-carriage and birth defects in humans [2]. However, aneuploidy can also arise in somatic cells, and a number of studies have reported age-dependent increases in aneuploidy. Men are for long known to be subject to age-related loss of the Y chromosome (LOY) [3–5], and recent studies have suggested that LOY is associated with shorter survival and higher risk of cancer [6]. Also, loss of an X chromosome with advancing age has been reported for females [7, 8], though the physiological consequence of this phenomenon remains unclear. Similarly to sexual chromosomes, an increase in autosomal aneuploidy has been observed in elderly peripheral blood lymphocytes, bone marrow cells, myeloid cells and fibroblasts, using different techniques such as metaphase spreads and fluorescence *in situ* hybridization in both interphase nuclei and cytokinesis-blocked binucleated cells [9–11]. The results obtained have shown that aging is positively correlated with the incidence of chromosome mis-segregation,[2] even though different chromosomes might have distinct susceptibility to mis-segregation [11, 12], or alternatively generate aneusomies[3] more compatible with cell survival [9, 10]. Chromosome-specific aneuploidies were also found in the aging brain as well as in buccal cells from both older patients and patients with Alzheimer's disease (AD), suggesting they might contribute to neurodegeneration [13–15]. However, this became recently controversial with the advent of single-cell whole genome sequencing to measure aneuploidy [16, 17].

The relative high frequencies of multiple mis-segregation events (involving more than one chromosome) have raised the question whether there is a general dysfunction of the mitotic apparatus in aged cells [12]. Oligonucleotide microarrays in a panel of fibroblast and lymphocyte cultures from young and old individuals were used to determine changes in gene expression specific for increased aneuploidy with age [18, 19]. These analyses revealed an association between age-related aneuploidy and the expression levels of genes involved in centromere[4] and kinetochore[5] function and in the microtubule and spindle assembly apparatus [18]. This opens the question whether dividing cells of elderly proliferative tissues exhibit any of the mitotic defects known to lead to aneuploidy. These include defective sister chromatid cohesion, weakened spindle assembly checkpoint (SAC),[6] supernumerary

[1] Chromosome content that is not an exact multiple of the haploid complement; imbalanced karyotype.

[2] Segregation of a whole chromosome to the incorrect daughter cell during mitosis.

[3] Gains or losses of entire chromosomes arising from mis-segregation events during cell division; generally denotes a diploid organism with subpopulations of aneuploid somatic cells.

[4] Part of the chromosome comprised of repetitive DNA where the sister-chromatids are connected and the kinetochores assembled.

[5] Large protein complex that allows the attachment of chromosomes to spindle microtubules.

[6] Surveillance mechanism that halts cell cycle progression until all chromosomes are correctly attached to the spindle.

centrosomes[7] and abnormal dynamics of kinetochore-microtubule attachments, as reviewed [20, 21]. Proper chromosome segregation requires the maintenance of cohesion between replicated chromosomes (sister chromatids) through the G2 and M phases and then the sudden disruption at anaphase onset. The SAC acts to prevent the destruction of sister chromatid cohesion as long there are chromosomes not properly attached to spindle microtubules. Even though the back-to-back geometry of sister kinetochores favors chromosome bi-orientation (attachment to microtubules from opposite poles) [22], the stochastic nature of the attachments might result in errors such as one single kinetochore attached to microtubules from different poles (merotely). Merotely occurs naturally in early mitosis but is corrected before anaphase by specific molecular mechanisms [23]. If left uncorrected, merotely results in anaphase lagging[8] chromosomes that might generate aneuploid daughter cells. The prevalence of merotelic attachments increases with cohesion defects that impair the orderly packing of centromeric chromatin and the typical back-to-back orientation of sister kinetochores. The incidence of merotely also increases in cells with attenuated SAC activity likely due to insufficient time for correction prior to anaphase onset. Cells with extra centrosomes induce transient multipolar spindles before the coalescence of centrosomes into bipolar spindles, and this event was shown to promote merotely and lagging chromosomes [24, 25]. Finally, insults that stabilize kinetochore-microtubule attachments prevent efficient correction of wrong attachments (which acts by releasing inappropriately attached microtubules) and promote chromosome mis-segregation [26].

Interestingly, the analysis of the mitotic process in aging models and diseases, even though limited, has often revealed the presence of dysfunctional phenotypes. Thus, we next review the data supporting that the primary causes of cellular damage during aging can induce mitotic defects and aneuploidization.

7.2 Mitotic Defects Associated with Primary Hallmarks of Aging

Genomic instability, telomere loss, epigenetic drift and defective proteostasis have been classified as primary hallmarks of aging, which act as initiating triggers whose damaging consequences progressively accumulate with time leading to secondary hallmarks [27]. In this section, and for each one of these types of damage, we will (1) briefly summarize how it has been linked to aging, and (2) highlight how often it leads to mitotic defects. By compiling these data, we argue for the existence of an age-associated mitotic decline.

[7] The main microtubule-organizing center of the cell, which contains the centrioles (in animal cells) and duplicates before mitosis to form a bipolar spindle.

[8] Chromosome that is left behind at the spindle equator when all the other chromosomes have segregated to opposite spindle poles.

7.2.1 *Genomic Instability*

Recent evidence points to DNA damage accumulation as an important driver of the aging process [28]. DNA integrity is constantly challenged by exogenous agents such as radiation, environmental chemicals, endogenous reactive oxygen and nitrogen species. These genotoxic agents induce altogether an enormous variety of DNA lesions that include point mutations, single- or double-stranded DNA breaks, and chromosomal rearrangements, losses and gains. To counteract DNA damage, the organisms have evolved a powerful network of repair mechanisms that jointly correct most of the nuclear DNA lesions. Aging arises from a wide range of phenotypic changes at the cellular level once the repair mechanisms are not sufficient to cope with a given level of damage [29]. In addition to the direct lesions in DNA, defects in nuclear architecture, known as laminopathies,[9] also lead to genome instability and premature aging[10] [30].

7.2.1.1 Nuclear DNA Damage

Nuclear DNA Damage and Aging

A prominent role of genome integrity in aging has been emphasized by the perception that most multi-system premature aging syndromes are caused by mutations in DNA repair genes. Examples are Werner syndrome (WS), Bloom syndrome (BS), Rothmund–Thomson (RTS), Fanconi's anemia (FA), Cockayne syndrome (CS), Trichothiodystrophy (TTD), and XFE syndrome [28, 31, 32]. Mice with defects in specific DNA repair mechanisms also display many progeroid phenotypes[11] [33]. Integrative analysis of human syndromes and mouse models has tentatively assigned aging phenotypes to specific DNA lesions, although the impact of specific lesions depends additionally on the underlying DNA repair mechanism and cellular context (stage in the cell cycle, proliferation and differentiation status, overall condition) [34]. In proliferative tissues and cell compartments, such as the hematopoietic and gonadal systems, double strand breaks (DSBs) and interstrand crosslinks (ICLs) are the types of lesions translating into segmental aging features.[12] DSBs and ICLs block DNA replication, activating the DNA damage response (DDR). When not repaired by the error-prone mechanisms, these lesions lead to cell death or senescence[13] [35]. Cell death induces loss of tissue homeostasis or depletion of somatic

[9] Group of rare diseases caused by mutations in genes functionally linked to formation/maintenance of nuclear lamina.

[10] A condition in which aging features arise early in life.

[11] Features of late-life aging in young individuals, as for instance grey hair, cataracts and body mass loss.

[12] Partially mimic an aging phenotype; do not include all signs of aging.

[13] Refers to an essentially irreversible growth arrest that occurs when cells that can divide encounter oncogenic stress (for instance, strong mitogenic signals); it is a secondary aging hallmark or compensatory response to primary causes of cellular damage.

stem cell pools, whereas cellular senescence induces a secretory phenotype including several pro-inflammatory cytokines, leading to tissue functional decline [36]. An example of a human syndrome caused by defects in ICL repair is FA, whereas the syndromes WS, RTS and BS are all caused by defective DSB repair due to mutations in RecQ helicases. These syndromes exhibit cancer and/or aging phenotypes depending whether ICLs/DSBs are repaired or not by error-prone mechanisms, respectively. In post-mitotic tissues, such as the neuronal system, helix-distorting lesions are the type of lesions most commonly leading to tissue decline and aging. Helix-disruptive lesions block RNA polymerase elongation and repair is initiated by transcription-coupled repair (TCR), a multi-step 'cut-and-patch'-type excision reaction that uses many core nucleotide-excision repair (NER) factors. Defective TCR repair is the main driver of premature aging phenotypes in the human disorders TTD, CS and XFE, and corresponding mouse models [37].

Finally, not only defects in DNA maintenance may lead to accelerated aging, but also there is sufficient evidence that all pathways of DNA repair become less efficient with age [38]. Even though the causes behind this decline remain elusive, several studies have identified an age-related decrease in the expression of DNA repair enzymes or their activities [39].

Mitotic Defects Induced by DNA Damage

The DNA damage response (DDR) is a complex signaling cascade that leads to cell cycle arrest in the presence of DNA lesions including those abovementioned that stall replication, transcription and that translate into aging phenotypes. DDR has been extensively investigated in interphase and shown to comprise two pathways, ATR/CHK1 and ATM/CHK2, that inhibit mitotic entry in order to provide sufficient time for DNA repair [40, 41]. However, recent studies have found that mitotic cells can also elicit a 'primary DDR' comprised of early events such as ATM activation, but then are unable to repair DNA damage due to the inhibition of 53BP1 recruitment to DNA lesions by mitotic kinase activity [42]. The rationale behind this impaired recruitment is to restrain end-to-end chromosome fusions that interfere with chromosome segregation [43] (see Sect. 7.2.2.2). Nevertheless, as lately shown, partial activation of DDR during mitosis increases the frequency of lagging chromosomes during anaphase by selectively stabilizing kinetochore-microtubule attachments and, thereby, preventing efficient correction of erroneous attachments [44] (Fig. 7.1a).

Moreover, mutations in several DNA repair genes have been frequently reported to induce chromosome segregation errors. This is the case of FA, WS, BS and RTS human diseases in which DNA damage interferes with DNA replication. FA is caused by mutations in genes of the FANC pathway and cells from FA patients typically exhibit gross aneuploidy [45, 46]. Recent studies have evaluated the role of FA proteins in chromosome segregation. FANCD2 localizes to discrete sites on mitotic chromosomes to promote anaphase resolution of chromosome entanglements induced by replication fork stalling [47] (Fig. 7.1b). In addition, FA proteins

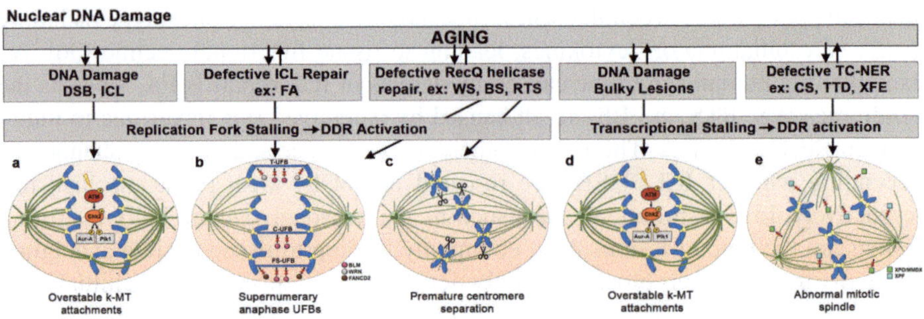

Fig. 7.1 Age-associated nuclear DNA Damage leads to mitotic defects. (**a**–**e**). DNA damage accumulation or defective DNA repair mechanisms are important drivers of aging. (**a**) DSB and ICL lesions induce replication fork stalling and elicit a primary DDR (ATM/CHK2 activation) that leads to the phosphorylation of Aur-A and Plk-1 mitotic kinases, stabilization of kinetochore-microtubule attachments and increased frequency of lagging chromosomes during anaphase. (**b**) Defective ICL repair (FA) and defective WRN and BLM helicase activity result in limited resolution of anaphase ultrafine bridges (UFBs) emerging from different chromosomal loci, centromeres (C-UFBs), telomeres (T-UFBs) and fragile sites (FS-UFBs). (**c**) Moreover, defective Recql4 helicase (RTS) leads to premature centromere separation. (**d**) Bulky lesions induce transcriptional stalling and primary DDR that leads to overstable kinetochore-microtubule attachments. (**e**) Mutations in NER core factors, as for instance XPD and XPF, which interfere with their localization to the mitotic spindle, result in spindle defects and chromosome mis-segregation

differentially localize to structures of the mitotic apparatus generating a signal essential for SAC activity [48]. Three of the five human RecQ homologs have been shown to be associated with the autosomal recessive syndromes, WS, BS and RTS. The RecQ protein family is a highly conserved group of DNA helicases involved in recombination-related processes, the non-homologous end-joining and homologous recombination, that function in the repair of DSBs, ICLs and recovery of stalled or broken replication forks [49]. Importantly, these helicases are required to protect the genome from illegitimate recombination during chromosome segregation. Analyses of patient-derived cells have demonstrated unusually high frequencies of chromosomal abnormalities including aneuploidy [50–54]. Chromosomal instability is also found in the mouse models of these human diseases [55–57]. Cells from Recql4 mutant mice (Type II RTS model) have high frequencies of premature centromere separation and aneuploidy, suggesting a role for Recql4 in sister-chromatid cohesion [57] (Fig. 7.1c). Both WRN and BLM helicases function to resolve stalled DNA replication forks and preclude chromatin mis-segregation [58–60]. By unwinding various DNA structures, BLM not only prevents elevated frequency of sister chromatid homologous recombination, but also resolves ultrafine anaphase bridges[14] during the later stages of mitosis [61]. BLM localizes to all types of ultrafine anaphase bridges (UFBs) emerging from different chromosomal loci, centromeres, telomeres and fragile sites. Centromere-UFBs are likely due to double strand DNA catenation, whereas telomere-UFBs and fragile site-UFBs most likely

[14] Stretched chromatin structure in between two daughter cells.

contain incompletely replicated DNA or hemicatenates [62]. Localization of BLM to UBFs depends on the Plk1-interacting checkpoint kinase PICH and is required to prevent the formation of supernumerary UFBs [63] (Fig. 7.1b).

Studies of NER mutations in human patients and mouse models suggest that, like replication stalling, also transcriptional blocking might compromise chromosome segregation (Fig. 7.1d). Whereas clinical phenotypes of highly elevated cancer predisposition, such as in Xeroderma pigmentosa (XP), are caused by mutations that affect the global genome NER pathway, those of accelerated aging, such as in Cockayne Syndrome (CS), are caused by mutations that affect the transcription-coupled NER pathway [37]. However, following DNA damage recognition, both pathways converge into common NER factors that unwind the helix (XPB, XPD and XPA) and remove the damaged DNA strand (ERCC1, XPF, and XPG). Mutations in NER core factors, for example XPD, cause diseases with combined phenotypes of XP and CS, such as XP/CS and Trichothiodystrophy (TTD). Evidence for a role of transcription-coupled NER in mitotic fidelity have come from reports showing increased aneuploidy in CS cells [53] as well as in XP-D and XP-D/CS cells [64]. XPD forms a specific complex called MMXD that localizes to the mitotic spindle and is required for proper chromosome segregation [64] (Fig. 7.1e). XPF, mutated in XFE syndrome, also localizes to the mitotic spindle [65] and, in addition, co-localizes with FANCD2 on mitotic chromosomes playing a role in processing of replication stress at fragile sites until mitosis [66].

It is possible that the distinct mitotic defects found in association with mutations in different DNA repair enzymes, do actually concur in an age-associated mitotic decline, if we consider that all pathways of DNA repair become less efficient during natural aging [38].

7.2.1.2 Nuclear Lamina Defects

Defects in the nuclear lamina, a structure near the inner nuclear membrane and the peripheral chromatin, compromise nuclear architecture and can cause genomic instability [67]. *LMNA* is a gene differentially expressed and spliced to produce the nuclear intermediate filament proteins lamins A and C, the major components of the nuclear lamina that direct and indirectly act in the maintenance of nuclear structure, gene expression, chromatin organization, cell cycle regulation and apoptosis [68].

Nuclear Structure and Aging

In the past 20 years, an increasing number of mutations in lamins and lamin-binding proteins was found to be linked to at least 15 different diseases, called laminopathies, of which Hutchinson-Gilford Progeria Syndrome (HGPS), Restrictive Dermopathy (RD) and Néstor-Guillermo Progeria Syndrome (NPGS) exhibit segmental premature aging features [69]. Classic HGPS is caused by a point mutation in the *LMNA* gene that activates a cryptic donor splice site leading to partial deletion of exon 11 and generation of a mutant lamin A (LA) protein termed "progerin/LAΔ50". Normally, mature lamin A is generated from a prelamin A precursor through substantial post-translational modification of its C-terminal CaaX motif, which includes cysteine farnesylation, cleavage of the aaX amino acids by the FACE-1/Zmspte24 metalloproteinase, carboxymethylation of the farnesylated cysteine and, finally, a second cleavage of the 15 terminal residues also by FACE-1 [70]. The progerin-truncated isoform remains permanently farnesylated and carboxymethylated since it cannot be secondarily cleaved by FACE-1. This leads to toxic accumulation of progerin at the nuclear envelope of HGPS patient cells causing nuclear shape abnormalities and chromatin stress [71]. Several studies have supported the dominant-negative disruption of lamin-related functions by progerin. Expression of LAΔ50 in normal cells recapitulates the nuclear abnormalities of HGPS cells [71], whereas expression of nonfarnesylated and carboxymethylated versions of progerin [72] does not show the classical HGPS phenotypes. Moreover, administration of farnesyl transferase inhibitors rescues some of the HGPS phenotypes [73, 74]. Another progeroid syndrome, RD, is caused by homozygous loss of FACE-1/ZMPSTE24 [75, 76]. Again accumulation of farnesylated prelamin A associates with the disease phenotypes, and mice deficient in Zmpste24 extensively phenocopy HGPS. The NGPS is caused by mutations in the *barrier to autointegration factor BAF/BANF1* gene which codes for an essential DNA-binding protein involved in many pathways, namely the maintenance of nuclear structure through interaction with lamins and nuclear membrane proteins [77].

Remarkably, studies of progeroid laminopathies and lamin A-related mouse models have generated significant insight into the normal aging process. Fibroblast samples from elderly wild type individuals have been shown to exhibit increased DNA damage, morphological abnormalities and changes in histone modifications comparable to those of HGPS patients [71, 78]. In addition, the appearance of progerin in fibroblasts and skin samples from elderly individuals [79, 80], further suggests that progerin may play a role in normal aging. Progerin synthesis due to sporadic use of the cryptic splice site in the *LMNA* gene and decreased levels of lamin B1 are putative determinants of the age-related nuclear defects [80–82].

Mitotic Defects in Progeroid Laminopathies

Early studies measuring aneuploidy in cultured cells from human progeria syndromes found significantly increased levels compared to controls [53]. More recently, expression of progerin/LAΔ50 in human cells was shown to cause chromosome segregation defects [83, 84]. Stable farnesylation and carboxymethylation of the mutant LAΔ50 cause an abnormal association with membranes during mitosis as well as formation of insoluble cytoplasmic aggregates. The abnormal dynamic behavior of progerin interferes with the mitotic membrane network morphogenesis, leading to a significant increase of lagging chromosomes at anaphase, a delay in the onset and progression of cytokinesis, and often binucleation (Fig. 7.2a). Moreover, similar mitotic defects correlating with the presence of progerin membrane-like

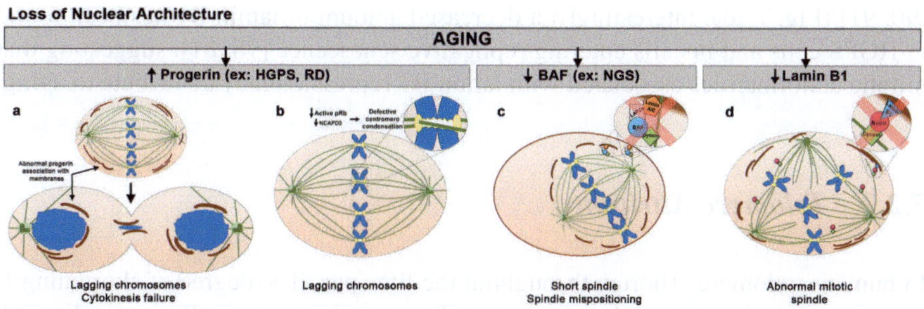

Fig. 7.2 Mitotic defects associated with loss of nuclear architecture (**a–d**). (**a**) Accumulation of mutant lamin A or progerin in HGPS or RD patients as well as in elderly individuals, interferes with the mitotic membrane network morphogenesis causing chromosome segregation and cytokinesis defects. (**b**) Progerin also leads to reduced levels of active pRB1 and its transcriptional target, centromere condensin subunit NCAP-D3, causing merotelic attachments (*inset*). (**c**) In NGS, disruption of the lamin A/C-LAP2α-BAF complex, that binds the actin filaments in the cell cortex and the membranous spindle matrix to the spindle-associated dynein (*inset*), prevents proper spindle assembly and positioning. (**d**) Decreased levels of lamin B1 in HGPS or elderly cells lead to spindle defects due to disruption of the spindle matrix architecture that requires the interaction between lamin B1 bound to membrane vesicles and the NudEL/dynein complex at spindle microtubules (*inset*)

aggregates and increasing with cell culture passage number were found in both HGPS and normal cells. In addition, cells from *Lmna/Disheveled hair and ears(Dhe)* heterozygous mice exhibit many phenotypes of human HGPS cells, including perturbations of the nuclear shape and lamina, increased DNA damage, and slow growth rates due to mitotic delay. Reduced levels of active hypophosphorylated RB1 and of its target NCAP-D3, a mitosis-specific centromere condensin subunit, were suggested to account for the chromosome segregation defects and consistent aneuploidy in the *Lmna(Dhe/+)* fibroblasts [85] (Fig. 7.2b). Lack of FACE-1/Zmspte24 metalloproteinase, also induces formation of lobulated nuclei and micronuclei[15] [86, 87]. Recently, it was shown that lamin-A/C are part of a stable complex with LAP2α and BAF that binds the actin filaments in the cell cortex and membranous spindle matrix[16] to the spindle-associated dynein, thereby ensuring proper spindle assembly and positioning [88] (Fig. 7.2c). However, the existence and functional role of a spindle matrix remain controversial as well as its molecular and structural composition [89]. Regarding BAF, the proper control of its association with other proteins and with DNA seems to be critical at multiple mitotic stages, being its dynamic phosphorylation and dephosphorylation particularly essential to drive nuclear disassembly and reassembly, respectively [77].

Finally, because B-type lamins lack a C-terminal CaaX motif, they remain bound to membrane vesicles through their farnesyl anchor during mitosis, and possibly make part of the spindle matrix controlling mitotic spindle assembly and orientation

[15] Small nucleus in a daughter cell generated by chromosome mis-segregation.

[16] Vesicular membraneous matrix that embeds the microtubule spindle apparatus during mitosis.

[90, 91] (Fig. 7.2d). Interestingly, a decreased amount of lamin B1 has been found in HGPS cells and in cells entering replicative senescence [78, 81], suggesting that mitotic abnormalities associated with lamin B1 repression may contribute to aging.

7.2.2 *Telomere Attrition*

In humans, telomeres shorten throughout the life span. The degree of shortening is roughly proportionate to risks of common diseases of aging as well as mortality risk [92]. Telomeres are dynamic complexes at chromosome ends containing tandem short DNA repeats and associated proteins [93]. The telomeres protect the genomic DNA by two means. First, telomeres are bound by the multiprotein complex shelterin to prevent the end of the linear chromosomal DNA from being recognized as DNA breaks [94]. Whereas this deflects the action of DNA repair mechanisms that would lead to chromosome rearrangements and instability, it turns DNA damage more persistent at telomeres contributing to cellular senescence and/or apoptosis [95, 96]. Second, the ribonucleoprotein enzyme telomerase adds telomeric repeat sequences to the chromosome ends to prevent the attrition arising from the inability of the DNA polymerases to completely replicate the extreme ends of linear chromosomes [97]. However, the levels of telomerase (or of its action on telomeres) are limiting in most mammalian somatic cells, causing progressive loss of telomere-protective sequences.

7.2.2.1 Age-Related Telomere Erosion

Cells with critically short or sufficiently damaged telomeres elicit a sustained DNA signaling, which leads to the loss of proliferative capacity known as replicative senescence (or Hayflick limit) [98, 99], unless telomerase is ectopically expressed [100]. However, the idea of telomere length as a mitotic clock ticking during normative aging[17] is too simplistic as telomerase enrichment in stem cells ensures their capacity to constantly renew somatic tissue cells in vivo [101] and it is unknown how much cellular senescence or death can arise from other causes than telomere erosion [102]. Also, extrapolating findings from aging studies using short-lived animal models has been limited because of the much longer time frame of human aging. Unless telomere maintenance is experimentally repressed genetically, laboratory animal models normally die of old age with relatively long telomeres [103]. But the experimental deletion of a telomerase component or a telomere protective protein does cause accelerated aging phenotypes in short-lived animals. The challenge has been to establish the extent by which telomere attrition contributes to normative aging phenotypes in the human. The study of monogenetic disorders of telomere maintenance has been valuable in this regard. Inactivating mutations are

[17] Natural or chronological aging.

known in 11 genes that encode either a telomerase component or a telomere-binding protein [92]. These mutations are associated with premature development of diseases, such as pulmonary fibrosis, dyskeratosis congenita, and aplastic anemia, which involve the loss of the regenerative capacity of different tissues [104]. They parallel many phenotypes of experimental mouse models that are null for a telomere maintenance gene [105]. Further supporting telomere loss as a hallmark of aging, evidence indicates that aging can be reverted by telomerase activation [106].

7.2.2.2 Mitotic Defects Associated with Telomere Dysfunction

Similarly to the naturally occurring chromosome ends where DSB repair must be prevented by the telomere capping, mitosis is another condition under which DSB repair must be silenced [43, 107]. The mechanisms behind DSB repair silencing in telomeres and during mitosis both act to block downstream ubiquitin signaling in the DDR after initial upstream phosphorylation signaling occurs [108]. In the human, cellular aging leads to spontaneous accumulation of shortened intermediate-state telomeres, which activate an ATM-dependent DDR but still have sufficient levels of shelterin proteins (i.e., TRF2) to suppress non-homologous end joining DNA repair mechanisms [109]. Intermediate-state telomeres routinely pass through mitosis in increasing abundance during cellular aging, until sufficient numbers accumulate to induce senescence [110] (Fig. 7.3a). If DSB repair were active during mitosis, the transit of intermediate-state telomeres through cell division would drive genome instability in pre-senescent cells. However, under certain conditions such as excessive telomere shortening and defects in telomere-associated proteins, fully uncapped-state telomeres might occur which lack sufficient TRF2 to inhibit DSB repair [111–113]. Impairment of telomere function together with a compromised senescence/apoptosis response leads to chromosomal instability (CIN)[18] through end-to-end chromosome fusions entering BFB (breakage-fusion-bridge) cycles [114]. Interchromosome dicentrics or isodicentric chromatids are prone to bridge in anaphase [115] (Fig. 7.3b). If the chromatin bridges are unequally broken, the new broken ends generated will perpetuate the BFB cycle leading to structural CIN [116]. Alternatively, if chromatin bridges are not resolved by breaking, whole chromosome mis-segregation will happen [117]. Detachment of dicentric chromatids from microtubules of one or both poles during anaphase was originally proposed as the mechanism underlying chromosome bridge-induced aneuploidy [117]. However, one recent study has elucidated that dicentric chromatid bridges rarely break during mitosis and exhibit persistently bound microtubule k-fibers[19] that hardly shorten during anaphase [118]. This causes the bridged chromosomes to lag behind in anaphase leading to chromosome non-disjunction or alternatively, micronuclei formation (Fig. 7.3b). Physical isolation of chromosomes in micronuclei can lead to

[18] Increased frequency of chromosome mis-segregation events.

[19] Microtubule bundles that attach sister kinetochores to spindle poles and power chromosome movement during mitosis.

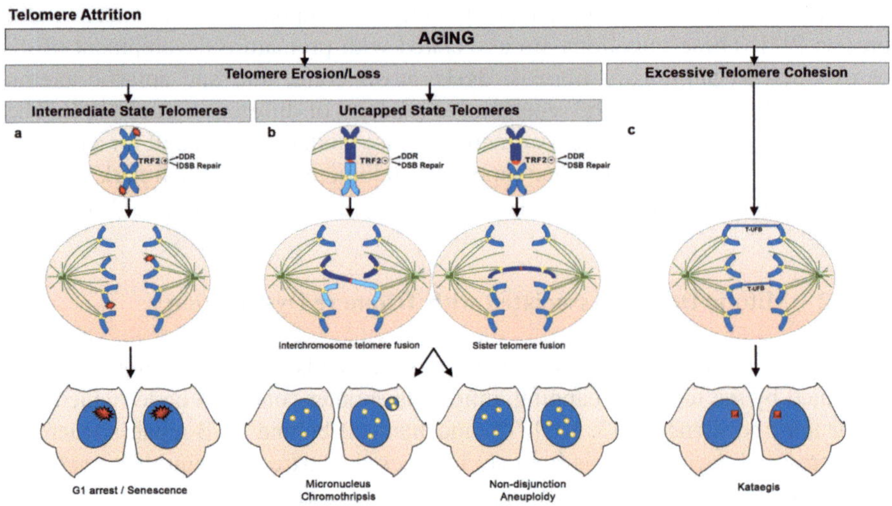

Fig. 7.3 Telomere attrition during natural aging compromises mitotic fidelity. (**a**) Cellular aging leads to spontaneous accumulation of shortened intermediate-state telomeres, which activate DDR but still have sufficient levels of shelterin proteins (i.e., TRF2) to suppress DSB repair. Intermediate-state telomeres routinely pass through mitosis, until sufficient numbers accumulate to induce cell cycle arrest and senescence. (**b**) However, excessive telomere shortening and defects in telomere-associated proteins generate uncapped-state telomeres that lack sufficient TRF2 to inhibit DSB repair and lead to telomere fusions. Interchromosome or sister telomere fusions cause bridged chromosomes to lag in anaphase. These lagging chromosomes either can lead to micronucleus formation and consequently chromothripsis, or can lead to chromosome non-disjunction and aneuploidy. (**c**) Excessive telomere cohesion leads to formation of anaphase ultrafine bridges (T-UBFs), which can cause micronuclei formation and chromothripsis or alternatively be resolved by the TREX1/APOBEC mechanism that generates clusters of point mutations in the DNA bridge known as kataegis

chromothripsis, the presence of massive chromosome rearrangements in confined genomic regions of one or a few chromosomes [119]. Fragmentation and subsequent reassembly events occur restrictively in the micronucleus chromatin in one cell division, and after mitosis the mutated chromosome can be incorporated into daughter nuclei [120]. In addition, one recent study has shown that a pattern of localized hypermutation known as kataegis may also arise during telomere crisis from the fragmentation of dicentric chromosome bridges during late mitosis [121]. The base mutations in kataegistic clusters are mainly cytosine to thymidine transitions in the context of a TpC nucleotide induced by AID/APOBEC editing deaminases in single strand DNA [122]. The mechanism by which dicentric chromosome bridges lead to kataegis involves a transient nuclear envelope breakdown at the bridge late in telophase, which makes DNA accessible to the cytoplasmic 3′ nuclease TREX1. This nuclease generates single stranded DNA that becomes a target for AID/APOBEC deamination and kataegis [121] (Fig. 7.3c).

Anaphase chromatin bridges might arise not only from end-to-end chromosome fusions, but also from defects in sister chromatid resolution. Sister chromatid resolution is spatial-temporally coordinated at the centromeres, arms and telomeres. Resolution of telomere cohesion occurs in early mitosis and it is distinctly regulated from centromere and arm cohesion as it requires the poly(ADP-ribose) polymerase (PARP) tankyrase 1 [123, 124] and the isoform SA1 of the cohesin subunit Scc3 [125]. SA1 associates with the shelterin subunit TRF1 and its partner TIN2 [126], and TRF1 PARsylation by tankyrase 1 [127] releases SA1 from telomeres in prophase [128, 129]. Inhibition of tankyrase 1 and overexpression of SA1 or TIN2 induce excess cohesion at telomeres in mitosis preventing a robust and efficient anaphase [130]. Interestingly, excessive telomere cohesion was found during replicative aging of primary fibroblasts [131] (Fig. 7.3c).

7.2.3 *Epigenetic Alterations*

Recent evidence indicates that several of the conserved longevity pathways mediated by signaling pathways (e.g. insulin/IGF1, TOR, AMPK) and downstream transcription factors (e.g., FOXA, FOXO, NRF2) can modulate chromatin states [132]. Chromatin state is governed by a series of epigenetic modifications that include DNA methylation, histone modification and chromatin remodeling.

7.2.3.1 Epigenetic Modifications and Aging

DNA methylation represents the addition of methyl groups to cytosine residues in the context of CG dinucleotides, referred to as 'CpG site'. The relationship between DNA methylation (DNAm) and aging remains elusive as there is global hypomethylation concurrently with loci-specific hypermethylation [133, 134], and no evidence exists thus far for lifespan extension through modulation of DNAm. Nevertheless, DNAm is perhaps the best-characterized epigenetic modification, and 'epigenetic-signatures' at specific CpG sites have been reported as quantitative models of both in vitro and natural aging [135–137]. Epigenetic drift at specific CpG sites in the genome is mediated, among other factors, by changes in the histone code. Age-associated histone modifications include increased H4K16 acetylation, H4K20 trimethylation, or H3K4 trimethylation, as well as decreased H3K9 methylation or H3K27 trimethylation [132, 138]. Multiple enzymes including acetyltransferases, deacetylases, methyltransferases and demethylases reversibly catalyze these modifications. Studies on chromatin regulators and lifespan have focused mostly on histone acetylases and deacetylases, in particular the sirtuin family. However, in the past few years, histone methyltransferases and demethylases have been also described to affect lifespan [132], even though unclear if through epigenetic mechanisms or transcriptional changes impacting on longevity signaling pathways [139]. Members of the sirtuin family of NAD-dependent protein deacetylases and ADP ribosyltransferases have been widely reported to ameliorate several aspects of aging in different organisms [140]. In mammals, not only SIRT1, which is the nearest homolog to invertebrate Sir2, but also SIRT3 and SIRT6 contribute to healthy aging through beneficial effects on genomic stability, metabolic efficiency and nutrient sensing [141–148]. Another epigenetic modification during aging is global heterochromatin loss and redistribution, caused by decreased levels of chromatin remodeling factors such as heterochromatin protein 1α (HP1α), Polycomb proteins and the NuRD complex [149–151]. Finally, microarray comparative analysis of young versus old tissues from different species have highlighted aging-associated transcriptional signatures affecting not only mRNAs of inflammatory, mitochondrial and lysosomal degradation pathways, but also noncoding RNAs such as a class of miRNAs that target components of longevity networks [152, 153]. Because of their reversibility, epigenetic alterations have been extensively explored for the design of novel anti-aging therapies [154].

7.2.3.2 Mitotic Defects Associated with Aging Epigenetic Marks

Supporting the functional relevance of age-related epigenetically mediated chromatin alterations in mitosis, there is a notable connection between heterochromatin formation at repeated DNA domains and chromosomal stability. In particular, heterochromatin assembly at pericentromeric regions requires trimethylation of histones H3K9 and H4K20, as well as HP1α binding, and is important for chromosomal stability [155]. Mammalian telomeric repeats are also enriched for these chromatin modifications, indicating that chromosome ends are assembled into heterochromatin domains [156, 157].

Moreover, the changes in chromatin conformation and compaction during mitosis, namely transcriptional repression and chromosome condensation, depend largely on the deacetylation of lysine residues of H3 and H4 core histones [158]. Inhibition of histone deacetylation with trichostatin A shortly before mitosis impairs proper chromosome condensation resulting in poor sister chromatid resolution (chromatin bridges) [23]. Histone hyperacetylation also causes depletion of HP1 from the pericentromeric chromatin, which in turn leads to loosened centromeres that promote the formation of merotelic attachments and lagging chromosomes [23, 159] (Fig. 7.4a). In addition, displacement of heterochromatin proteins HP1α and HP1γ from chromatin induces premature chromatid separation concomitant with delocalization of cohesion proteins from the centromeres [160] (Fig. 7.4b).

Mitotic roles have been reported for several sirtuins thus suggesting that decreased sirtuin activity during aging might lead to age-associated aneuploidy. SIRT1 deficiency results in accumulation of cells in early mitotic stages due to incomplete chromosome condensation [147]. The defective chromosome condensation in sirt1$^{-/-}$ cells is due increased acetylation of H3K9 that impairs its trimethylation and, consequently, the recruitment of HP1α, which is required to establish a closed chromatin configuration. In addition, SIRT1 was found to associate with mitotic chromatin at prometaphase in order to mediate the chromosomal loading of

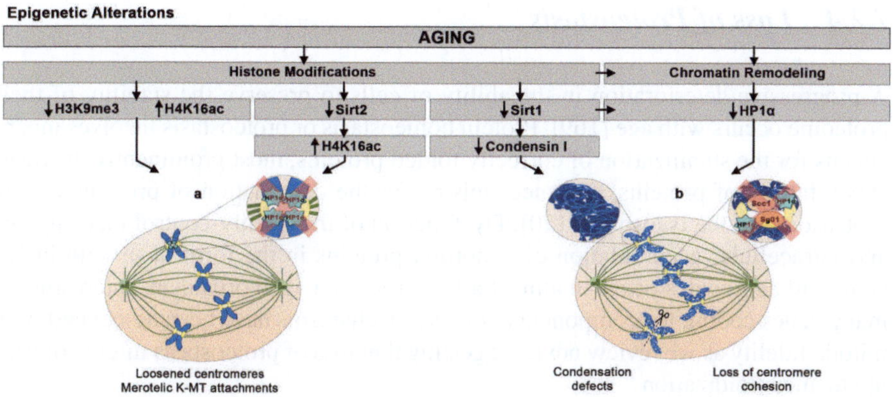

Fig. 7.4 Age-associated epigenetic alterations induce mitotic defects. Changes in histone modifications (for example, decreased H3K9 methylation and increased H4K16 acetylation) and decreased levels of sirtuins (Sirt1, Sirt2) and chromatin remodeling factors (such as HP1α), are epigenetic signatures of aged cells. (**a**) Increased H4K16 acetylation associated with decreased H3K9me3 or Sirt2 deacetylase activity leads to depletion of HP1α from pericentromeric heterochromatin, which causes centromere relaxation and formation of merotelic attachments (*inset*). (**b**) Reduced Sirt1 activity interferes with condensin I complex loading into chromatin leading to chromosome condensation defects (*inset*). Moreover, reduced Sirt1 activity also causes HP1α depletion from chromatin resulting in heterochromatin loss and premature cohesion loss due to defective recruitment of Scc1 and Sgo1 proteins (*inset*)

histone H1 and the condensin I complex needed for chromosome condensation [161] (Fig. 7.4b). SIRT2 has strong preference for acetylated H4K16 which enzymatic conversion into its deacetylated form at G2/M may be crucial for chromatin condensation at early mitosis [162] (Fig. 7.4a). Moreover, SIRT2 was shown to regulate the anaphase-promoting complex/cyclosome (APC/C)[20] activity through deacetylation of its coactivators, APC(CDH1) and CDC20 [163]. This leads to increased levels of mitotic regulators such as Aurora-A and -B in SIRT2 deficient cells that induce centrosome amplification, aneuploidy, and mitotic cell death. Microtubule polymerization increases in SIRT3 depleted cells suggesting that SIRT3 regulates spindle dynamics [164]. SIRT6 levels, similarly to SIRT1 and SIRT2, increase in mitosis, but SIRT6 partially co-localizes with the mitotic spindle instead of being associated with condensed chromosomes [165].

Finally, even though still elusive, some noncoding RNAs implicated in the senescence and aging processes in recent years, such as *geromiRs* [153] might impact in mitotic function. Some examples are: miR-1 that downregulates genes involved in DNA replication and mitosis [166]; miR-34 that suppresses SIRT1 activity [167]; and let-7b that targets Aurora B, causing increased rate of aneuploidy, polyploidy and multipolarity [168].

[20] Multiprotein complex with ubiquitin-ligase activity that is responsible for the ubiquitination of numerous key cell-cycle regulators.

7.2.4 Loss of Proteostasis

A progressive deterioration in the ability of cells to preserve the stability of their proteome occurs with age [169]. Protein homeostasis or proteostasis involves mechanisms for the stabilization of correctly folded proteins, most prominently the heat-shock family of proteins, and mechanisms for the degradation of proteins by the proteasome or the lysosome [170]. Dysfunction of the quality-control mechanisms and intracellular accumulation of abnormal proteins in the form of protein inclusions and aggregates occur in almost all tissues of an aged organism. Interestingly, many genes coding for components of these mechanisms have been associated with mitotic fidelity as we review next, suggesting that loss of proteostasis might contribute to aneuploidization.

7.2.4.1 The Protein Control Machinery and Aging

Chaperones or heat shock proteins (HSPs) are highly conserved molecules that act to assure that proteins acquire a stable folded conformation. Primary or secondary deficits in chaperone function have been reported for age-related diseases, the extent of which depends on the chaperone, the tissue and even the organism [171]. Upregulation of hsp70 in response to different stressors is decreased in fibroblasts aged in vitro and tissues from old organisms [172–174], due to the inability of the heat shock factor (HSF) transcriptional activator to bind the chaperone gene promoter [175, 176]. Conversely, extra copies of an hsp70 family member, as well as HSF over-expression, have been shown to increase lifespan [177–179].

If chaperone-driven folding attempts are unsuccessful, proteins are then delivered to the proteolytic machinery. The two main components of the ubiquitin-proteasome system (UPS), the ubiquitinization machinery and the proteasome core, also undergo age-dependent changes that lead to loss of proteostasis such as decreased levels of free ubiquitin [180], transcriptional repression of ubiquitin-conjugating enzymes or E3 ligases [181], defective expression of proteasomal subunits or its regulatory subunits [182, 183]. On the other hand, maintained UPS activity has been shown to promote lifespan extension in different model systems [184, 185]. Also the proteolytic activities of both macroautophagy and chaperone-mediated autophagy (CMA) have been described to decrease with age. In macroautophagy, a whole region of cytosol is sequestered inside double membrane vesicles (autophagosomes), which then fuse with lysosomes to degrade their cargo [186]. Age-associated malfunctioning of macroautophagy arises from impairment of autophagosome fusion with the lysosome [187], or inhibitory effect on lysosomal proteolysis [188]. In CMA, substrate proteins are selectively recognized by the chaperone hsc70, which then binds to the lysosome-associated receptor LAMP-2A, so that translocation of the substrate across the lysosomal membrane occurs [189]. Progressively lower levels of the CMA receptor at the lysosomal membrane were found with age likely due to changes in the lipid membrane composition [190].

Increasing evidence has shown that preventing the decline in autophagic activity slows down cellular aging and preserves organ function [191].

Being aberrant protein conformers key determinants of aging, the ability to control their inheritance is crucial for avoiding aging in specific cells [192]. Several reports have shown that yeast and mammalian cells use an intricate machinery to spatially sequester misfolded proteins in inclusions that interact with organelles and cytoskeleton to ensure the polarity of their inheritance after mitosis, as reviewed in [193]. However, the rejuvenation of daughter cells by asymmetric mitotic partitioning might gradually decline with advanced age as recently found in yeast [194]. Asymmetric cell division as a key determinant of aging and age-associated diseases is reviewed by Polymenis and Kennedy in the same volume.

7.2.4.2 Mitotic Defects Associated with Loss of Proteostasis

Molecular chaperones and protein control pathways are essential for the assembly and disassembly of macromolecular complexes. They are therefore expected to assist the proper and timely assembly of cytoskeletal structures during mitosis. Indeed, HSPs have been involved in microtubule dynamics and spindle assembly during mitosis. Hsp70 is phosphorylated by the mitotic kinase Plk1 and its centrosomal localization may interfere with spindle dynamics [195] (Fig. 7.5a). Hsp72, an inducible cytoplasmic isoform of the Hsp70 family, is required for assembly of a robust bipolar spindle capable of efficient chromosome congression [196]. Targeting of Hsp72 to the mitotic spindle is dependent on phosphorylation by the Nek6 kinase. Phosphorylated Hsp72 localizes to the spindle poles and sites of kinetochore-microtubule attachment and acts not only to stabilize K-fibers through the recruitment of the ch-TOG/TACC3 complex, but also to regulate spindle positioning through the attachment of astral microtubules to the cell cortex (Fig. 7.5a). Also, the molecular chaperone Hsp90 was found to interact with the Drosophila orthologue of ch-TOG (Msps) at centrosomes and spindle, as well as to regulate the efficient localization of cyclin B at those structures [197]. Disruption of Hsp90 function by mutations in the Drosophila gene or treatment of mammalian cells with the Hsp90 inhibitor geldanamycin, results in abnormal centrosome separation and maturation, aberrant spindles and impaired chromosome segregation [198]. Moreover, the Hsp90 cochaperone Sgt1, localizes to kinetochores and its phosphorylation by Plk1 enhances the association of the Hsp90-Sgt1 chaperone with the MIS12 complex at the kinetochores to promote stable microtubule attachment and chromosome alignment [199, 200]. The Hsp90-Sgt1 complex is also required for kinetochore assembly, as Hsp90 inhibition causes delocalization of several kinetochore proteins including CENP-I and CENP-H, leading to chromosome misalignment and aneuploidy [201] (Fig. 7.5a). Interestingly, limited binding and transactivating capacity of the chaperone transcriptional regulator HSF1 was found in mitotic cells, where the chromatin is tightly compacted, turning these cells vulnerable to proteotoxicity [202]. In dividing cells, when levels of aggregation-prone proteins exceed the capacity of the proteasome to degrade them, perinuclear aggresomes accumulate

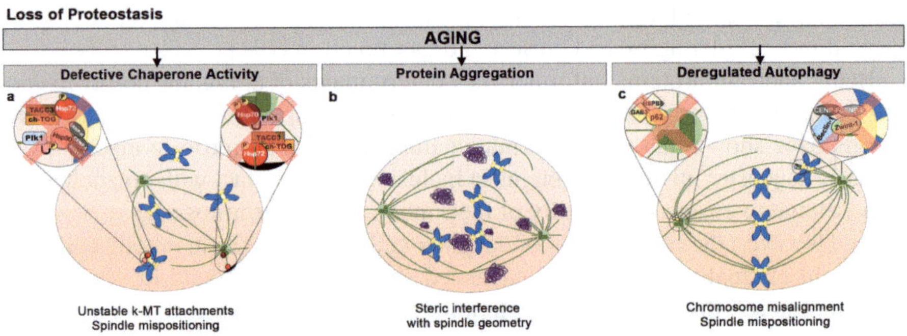

Fig. 7.5 Age-associated loss of proteostasis induces mitotic defects. During natural aging there is loss of proteostasis expressed by defective chaperone activity, protein aggregation and deregulated autophagy. (**a**) Defective chaperone activity compromises the assembly of many protein complexes required for spindle positioning and kinetochore-microtubule (k-MT) attachment stability. For instance, defective Hsp72 activity at spindle poles and kinetochores prevents the recruitment of the ch-TOG/TACC3 complex involved in the attachment of astral microtubules to the cell cortex and stabilization of k-MT fibers. Defective Hsp70 activity at centrosomes interferes with spindle dynamics. Defective Hsp90 at kinetochores causes delocalization of several kinetochore proteins (MIS12 complex, CENP-H, CENP-I) required for stable k-MT attachments. (**b**) Accumulation of protein aggregates interferes sterically with spindle geometry. (**c**) Deregulated autophagy compromises the mitotic function of p62 and Beclin-1 autophagic proteins leading to spindle mispositioning and chromosome misalignment. The p62/HSPB8/BAG3 complex acts during mitosis to facilitate the remodeling of actin-based structures that guide spindle orientation (*inset*). Beclin-1 depletion causes reduction of several outer kinetochore proteins (ZW10, CENP-E, CENP-F) required for proper chromosome alignment (*inset*)

and have a detrimental effect on mitosis by steric interference with chromosome alignment, centrosome positioning, and spindle formation [203] (Fig. 7.5b).

Even though it remains controversial whether autophagy operates during mitosis [204, 205], there is cumulative evidence suggesting that defective autophagy induces mitotic anomalies. For instance, the multichaperone complex HSPB8-BAG3 that senses damaged cytoskeletal proteins and orchestrates their seclusion by selective autophagy, acts during mitosis together with the p62/SQSTM1 autophagic receptor to facilitate the proper and timely remodeling of actin-based mitotic structures that guide spindle orientation and proper chromosome segregation [206] (Fig. 7.5c). p62/SQSTM1 has been also implicated in the concluding step of cytokinesis, suggesting that overload of the autophagic pathway might lead to impaired clearing of midbody rings and cytokinesis failure [207]. Monoallelic depletion of Beclin-1, a subunit of the PI3K-III core complex involved in autophagy, reportedly causes chromosomal disorders such as aneuploidy and double-minute chromosomes [208]. Chromosome segregation errors associated to Beclin-1 depletion were found to arise from severe congression defects associated with a reduction in several outer kinetochore components, including ZW10, CENP-E and CENP-F [209] (Fig. 7.5c).

7.3 Aneuploidy Induces Aging

7.3.1 Aneuploidy-Driven Aging Hallmarks

In recent years, systematic analyses of disomic yeast, trisomic mouse and human cells, all cells with an extra chromosome, have elucidated the impact of aneuploidy in cellular fitness [210]. Two types of phenotypes can arise from changes in chromosome number: (1) karyotype-specific phenotypes caused by changes in copy number of specific genes and (2) phenotypes shared by different aneuploidies which reflect a decrease in cellular fitness due to a cumulative effect of copy number changes of many genes [211]. These pan-aneuploidy phenotypes include a number of aneuploidy-associated stresses found in both yeast and mammalian cells [212] that include proliferation defects [213–215], a gene expression profile similar to the environmental stress response (ESR) [216–218], multiple forms of genomic instability [219, 220], and proteotoxicity [213, 216, 221–226]. Below we summarize major findings on the impact of aneuploidy in genomic stability and protein quality-control machinery in order to emphasize how these pan-aneuploidy phenotypes recapitulate the genomic instability and proteotoxicity hallmarks of aged cells, reinforcing the causal role of aneuploidy in aging (Fig. 7.6). Whether other primary causes of cellular aging such as telomere attrition and epigenetic alterations are present in the aneuploid cell models remains unknown. Interestingly, for the human trisomy 21 or Down syndrome (DS), telomere shortening and aging epigenetic alterations have been reported. DS patients age prematurely and present early onset of Alzheimer's disease (AD). Telomere shortening in T-lymphocytes has been proposed as a biomarker of clinical progression of AD for adults with DS [227], even though limited and conflicting data exist as to whether DS individuals have shorter telomere lengths before birth or an accelerated rate loss after birth [228]. Moreover, analysis of the quantitative DNA methylation-based biological marker of aging known as 'epigenetic clock' has shown that trisomy 21 significantly increases the age of blood and brain tissue on average by 6.6 years [229].

7.3.1.1 Genomic Instability

Delineating the molecular mechanisms underlying aneuploidy-driven genomic instability has remained challenging due to the difficulty of separating the effects of aneuploidy from those of other associated genetic alterations when using heterogeneous aneuploid cell populations. Recently established cell models with defined aneuploid karyotypes have facilitated the analysis of the immediate consequences of aneuploidy *per se* [213, 215, 218, 230]. Observations in budding and fission yeasts suggested that aneuploidy impairs the fidelity of chromosome segregation, and increases mutation and recombination rates [220, 231]. Isogenic aneuploid yeast strains with ploidies between 1N and 2N obtained through a random meiotic process were found to exhibit various levels of whole-chromosome instability, with cells with a chromosomal content closer to 1N being more stable than cells with ploidy between 1.5N and 2N [220]. Even though these results suggested inability of the mitotic system to scale continuously with an increasing number of chromosomes, the presence of specific aneuploid chromosomes also seemed to determine the rate of CIN [220]. Aneuploidy-induced chromosomal instability was also shown for human trisomies 7 and 13, which exhibit increased rate of anaphase laggings [230]. Two independent studies demonstrated that lagging chromosomes arising from merotelic attachments can in turn induce genomic instability, especially DNA damage, either due to breakage during cytokinesis [232] or due to formation of micronuclei and chromothripsis [120]. Nevertheless, untransformed damaged cells will end up activating the stress kinase p38 and the stress-induced transcription factor p53 causing cell cycle arrest [21, 232, 233].

Replication defects also seem to be a widespread consequence of the aneuploidy condition. Analysis of disomic yeast strains revealed the presence of increased levels of Rad52 foci, which formed during S phase due DSBs generated by defects in DNA replication initiation and elongation [219]. Also, a series of trisomic and tetrasomic human cells derived from near-diploid and chromosomally stable parental cell lines were recently found to exhibit anaphase UFBs and DNA damage associated with replication stress, which in turn induced genomic rearrangements at common fragile sites [234]. Importantly, reduced expression of the replicative helicase MCM2-7 was shown to account for the genomic instability phenotype [213, 217, 234] (Fig. 7.6a). Aneuploid human pluripotent stem cells (hPSCs) were also found to undergo replication stress, resulting in defective chromosome condensation and segregation [235]. However, in this study, downregulation of the transcription factor SRF and its actin cytoskeletal gene targets were identified as the molecular mechanism behind chromosomal defects. Even though aneuploidy-associated replication stress might act to promote tumorigenesis in immortalized cells, in primary cells with efficient checkpoint activities, it will most likely lead to cell death or senescence.

7.3.1.2 Proteotoxicity

Changes in gene copy number generally translate into a corresponding change in gene expression [213, 221, 226]. Thus, gains or losses of entire chromosomes lead to massive alterations in relative abundance of many proteins. This impact in proteome composition results in proteoxicity, a state in which the protein quality-control machinery of the cell (protein chaperones, ubiquitin proteasome system, autophagy) is overwhelmed causing protein misfolding [236]. As mentioned above, yeast, mouse and human aneuploid cells exhibit upregulated expression of stress response genes, which include the chaperones [217] and autophagic proteins [213, 224]. Furthermore, analysis of disomic budding yeast revealed increased sensitivity to drugs that interfere with protein folding, synthesis and degradation [218], increased propensity to form protein aggregates, and decreased ability to fold HSP90 protein clients [223]. In line with these findings, trisomic mouse embryonic

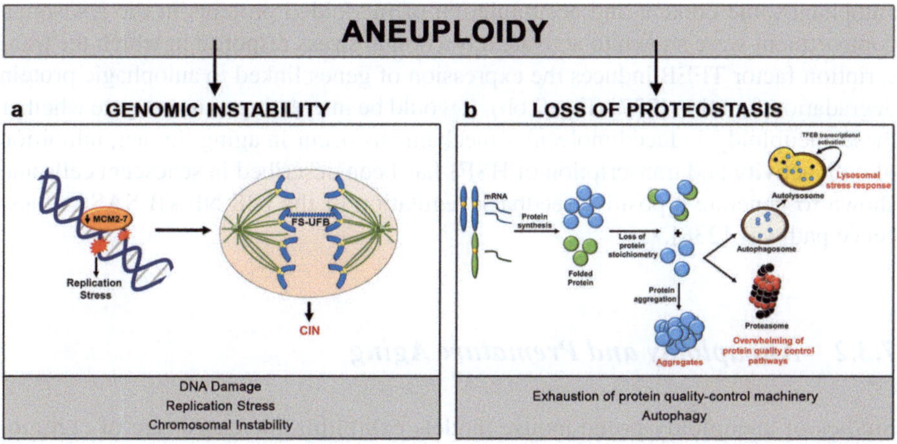

Fig. 7.6 Aneuploidy leads to aging hallmarks. Established cell models with defined aneuploid karyotypes exhibit a number of pan-aneuploidy phenotypes that include genomic instability and proteotoxicity. (**a**) DNA damage, replication stress and chromosomal instability (CIN) are types of genomic instability found in the aneuploidy cell models. Decreased levels of the replicative helicase MCM2-7 were recently shown to induce replication stress particularly at fragile sites (FS) leading to increased frequency of anaphase ultrafine bridges (FS-UBFs) that promote CIN. (**b**) Loss of protein stoichiometry due to karyotype imbalance, leads to accumulation of protein aggregates that overwhelm the capacity of the protein-control machinery comprised by the proteasomal and autophagic degradation pathways. An impaired clearance of misfolded/aggregated proteins in autolysosomes was recently shown to activate a lysosomal stress response that upregulates the expression of transcription factor TFEB and its targets with functional role in protein-control mechanisms

fibroblasts (MEFs) and chromosomally unstable aneuploid cancer cell lines were more sensitive to inhibition of the chaperone HSP90 than their euploid counterparts [225]. These observations clearly indicated impairment in protein quality control, either during the folding and/or disaggregation processes and/or at the level of protein degradation. Indeed, studies also uncovered mutations in the Ubp6 protein, a deubiquitinating enzyme that negatively regulates proteasome function, as conferring improved proliferation rate to a subset of disomic yeast strains, thereby implicating proteasome function in protein homeostasis in aneuploid cells [226]. In addition, effects on the autophagic protein degradation pathway were detected in mammalian aneuploid cells. Aneuploid human cell lines created by chromosome transfer and trisomic MEFs were shown to harbor significantly increased levels of LC3-II, an autophagosome-specific lipidated form of LC3, and SQSTM1/p62, the cytosolic receptor that targets ubiquitinated proteins to autophagy [213]. More recently, the molecular mechanisms behind these aneuploidy-associated phenotypes have been further elucidated. The impairment of protein folding capacity is due to defective HSF1-dependent activation of the heat shock response, as HSF1 overexpression can counteract the effects of aneuploidy in HSP90-dependent protein folding [222]. Regarding autophagy triggering, an impaired clearance of autophagosome content and accumulation of misfolded proteins in the lysosomal compartment were shown to activate a lysosomal stress response in which the transcription factor TFEB induces the expression of genes linked to autophagic protein degradation function [237] (Fig. 7.6b). It would be interesting to investigate whether these aneuploidy-induced molecular mechanisms occur in aging. In fact, inhibition of both activity and transcription of HSF1 has been described in senescent cells and shown to generate a positive feedback regulation of the p38-NF-κB-SASP senescence pathway [238].

7.3.2 Aneuploidy and Premature Aging

Studies of aneuploidy-prone mouse models exhibiting increased rate of chromosome mis-segregation uncovered a surprising link with the rate of aging and the development of age-related pathologies [239–242]. Mutant mice with low levels of the spindle assembly checkpoint protein BubR1 were found to develop progressive aneuploidy along with a variety of progeroid features, including short lifespan, growth retardation, sarcopenia, cataracts, loss of subcutaneous fat and impaired wound healing [240]. Reinforcing this link between BubR1 and aging, the majority of human patients suffering from Mosaic Variegated Aneuploidy (MVA) syndrome were found to have mutations in *BUBR1* that generate unstable gene products and cause progeroid features recapitulating those in BubR1 hypomorphic mice [243, 244]. Mice doubly haploinsufficient for the mitotic checkpoint genes *Bub3* and *Rae1* were another aneuploidy-prone mouse strain described to exhibit an accelerated aging phenotype [245]. However, progeroid features have not been found in many other aneuploidy mouse models. Possible explanations are the premature sacrifice of mice before they start developing aging phenotypes later in life and the superficial analysis for overt age-related degeneration that might miss restricted tissue-specific phenotypes [242]. Important to mention that, even though detrimental in most cells and organisms, in some tissues such as brain and liver, aneuploidy seems to be part of normal development [246]. One possibility is that neurons and hepatocytes somehow adapt to chromosome imbalances either by accumulating chromosome-specific gene products that provide them with a selective advantage or by regulating the expression of detrimental aneuploidy-induced targets [13]. However, very recently, genome-wide high-resolution analysis of chromosome copy number variations in mouse and human tissues by single cell sequencing, has shown that aneuploidy occurs much less frequently in normal brain and liver [16] and brain with Alzheimer disease [17] than previously reported. One potential explanation is that different cell types will follow different fates (senescence vs. apoptosis) in response to aneuploidy. Senescent endothelial cells have increased sensitivity to apoptosis in comparison to senescent fibroblasts [247]. Thus, aneuploid neurons might not be detected by single cell sequencing if cleared due to apoptosis.

7.4 Concluding Remarks and Future Perspectives

Aging and aneuploidy have profound impact on most cellular functions and their association has been reported by several independent studies. However, the molecular mechanisms by which aging induces aneuploidy and by which aneuploidy triggers aging remain largely unknown. Here, we presented a detailed revision on how primary hallmarks of aging have often been associated with mitotic defects, in the attempt of gathering mechanistic insights into age-associated aneuploidy. In addition, we reviewed the systemic impacts of aneuploidy in cellular physiology, such as genomic instability and proteoxicity, and brought our perspective on how these recapitulate aging hallmarks and could be the source of aneuploidy-induced aging (Fig. 7.7). Considering all the emergent data, we propose that there is a complex positive feedback regulation between aging and aneuploidy. Age-related aneuploidy

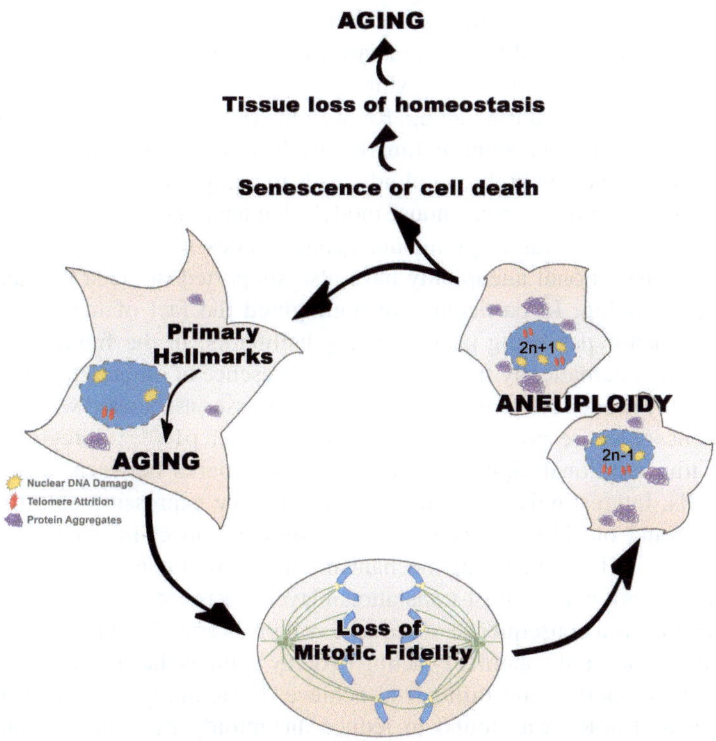

Fig. 7.7 Feedback positive regulation between aging and aneuploidy. Primary causes of cellular damage during natural aging, namely nuclear DNA damage, telomere attrition and protein aggregates, lead to loss of mitotic fidelity and generation of aneuploid daughter cells. Aneuploidy, in turn, induces cellular phenotypes that include primary aging hallmarks such as genomic instability and proteotoxicity. Telomere attrition in aneuploid cells is still elusive. Thus, age-related aneuploidy further enhances aging phenotypes that, in the presence of active cell cycle checkpoints, will likely lead to secondary aging hallmarks such as cellular senescence

arises from mitotic function decline induced by cellular damage, and in turn contributes to secondary aging hallmarks, such as cellular senescence. It is therefore reason to ask if *aneuploidy is an aging hallmark*. To be an aging hallmark, aneuploidy should meet the following criteria: (1) it should manifest during normal aging; (2) its experimental aggravation should accelerate aging; and (3) its experimental amelioration should delay aging [27]. We reason these criteria are indeed met. First, we compiled a number of early and recent studies indicating that aneuploidy increases with advancing age. However, accurate karyotyping would still be needed to unequivocally establish this correlation. Each karyotyping method presently used is limited as to which cytogenetic abnormalities can be detected and in which cell type (dividing or non-dividing) [248]. Therefore, upcoming studies should consider the use of different methods to measure aneuploidy, and definitely explore the emerging karyotyping platforms that combine single cell resolution with complete karyotyping. Moreover, it would be crucial to measure aneuploidy levels in different tissues during organismal natural aging. Considering the human lifespan and limited access to human tissues, study models such as mouse and zebrafish become essential for future analyses. Another imperative aspect to investigate is whether mitotic processes exhibit an age-related decline as to expect from the aneuploidy increase during aging. Reinforcing this possibility is the observation that many of the primary aging hallmarks result in mitotic phenotypes. Secondly, we gathered evidence on how aneuploidy leads to aging. In addition to the pioneering studies in aneuploidy-prone mouse models that have shown how experimentally induced aneuploidy leads to premature aging, studies in yeast, mice and human models of constitutional aneuploidy have also supported the idea that aneuploidy contributes to aging. In particular, we highlighted the fact of many aneuploidy-associated phenotypes being primary aging hallmarks. In the future, it would be interesting to systematically investigate for the presence of other aging hallmarks in the aneuploidy models, determine if the cellular stresses associated with aneuploidy engage senescence response pathways (for instance, $p16^{Ink4a}$ upregulation), and compare transcriptional signatures and epigenetic marks between aneuploid and elderly cells. Interestingly, previous analyses on gene expression data from aneuploid [217] and old [18, 19] cells revealed similar impact in genes functionally linked to mitosis. The underlying mechanisms are poorly understood and may occur at both transcriptional and post-translational levels, or mitotic gene expression may merely decline as a consequence of reduced cell proliferation with aging. Third and finally, amelioration of aneuploidy acting to delay aging is the criterion for hallmark most poorly supported and difficult to achieve. Intriguingly, sustained high-level expression of BubR1 was found to reduce aneuploidy by counteracting mitotic defects possibly associated with age-related decline [239]. Attenuated aneuploidy tightly correlated with reduced senescence and tissue deterioration. Therefore, this study provided a molecular entry point for modulation of aneuploidy as an opportunity to extend healthy lifespan. One question now is whether there are mitotic genes other than BubR1 whose levels can be modulated without any overt adverse effects to prevent age-related mitotic decline and aneuploidy. In addition, it will be interesting to determine whether mutations that suppress the adverse effects of aneuploidy [226] also delay aging and extend lifespan.

Why should modulation of aneuploidy be a preferential opportunity to delay aging? Experimental ameliorations of genomic instability, telomere erosion, epigenetic changes and proteotoxic stress have all been shown to successfully extend lifespan, suggesting these hallmarks add-up into cellular aging. Therefore, a favored hallmark to anti-aging therapies has been difficult to assign. If in one hand genomic instability has been emphasized as a major mechanism based on the observation that most premature aging phenotypes are caused by mutations in DNA repair genes, on the other hand proteotoxic stress has been highlighted in neuronal aging. Here we have shown that all primary hallmarks end up generating aneuploidy, providing a common feature to explore in anti-aging therapy. Even though the routes to aneuploidy seem to comprise many different mechanisms depending on the type of cellular damage, one should be aware that most of these mechanisms are likely occurring jointly during chronological aging and not separately as inferred from studies in models of aging disease caused by mutations in single genes. One way to address this would be to characterize the mitotic behavior of elderly cells under advanced light microscopy. This has been largely limited by the low proliferation indexes of old tissues and/or primary cell cultures. Also, it would be important to investigate if stress-signaling pathways activated by distinct types of damage might all converge downstream to inhibit mitotic proficiency. If this assumption is correct then (1) inhibition of stress response pathways should prevent mitotic decline and (2) overexpression of a mitotic gene downstream target of stress response should increase mitotic efficiency. Inactivation of a stress pathway in humans is an unfeasible approach as it would also eliminate critical tumor-suppressive pathways and induce cancer. However, one recent strategy based in drug-induced apoptosis was designed to selectively kill stressed cells expressing the $p16^{Ink4a}$ gene in the context of cellular senescence [249]. If $p16^{Ink4a}$-positive senescent cells come to be shown as being mainly aneuploid, then aneuploidy-selective antiproliferation compounds would be an attractive alternative to senescent cell clearance. Thus, it would be valuable to measure the extent of aneuploidy in the senescent cell population. Moreover, the aneuploidy levels induced by specific primary hallmarks could provide a means to understand the most relevant types of damage contributing to tissue-specific aging. Regarding one common downstream target of stress-signaling pathways, the Forkhead box (Fox) transcription factor family has emerged as an interesting candidate. The balance between rapid growth over maintenance of youth is largely regulated by the Fox class of transcription factors and the anaphase-promoting complex (APC) [250]. Specifically, FoxM1 and APC^{Cdc20} function together to maintain genomic stability by regulating separation of sister chromosomes and chromatin structure, while the FoxOs and APC^{Cdh1} regulate cellular repair and maintenance. The FoxO family has been reproducibly found to extend lifespan through reduced insulin-signaling in many model systems [250]. FoxM1 primarily drives the expression of G2/M specific genes [251] and seems to counter senescence [252]. Interestingly, artificial enrichment of FoxM1 improves liver regenerating capacity in older mice [253] and lung regeneration following injury [254], without being tumorigenic in those organs. Therefore, modulation of mitotic competence through a balanced/stoichiometric upregulation of several mitotic genes could act to prevent aneuploidy more efficiently than upregulation of specific genes such as BubR1. In the future, the crosstalk between stress response pathways and mitotic gene expression should be further investigated. Moreover, it would be interesting to investigate the impact of mitotic proficiency in cell fate decision in response to cellular stress. We support that aneuploid cells arising from age-associated mitotic decline in highly proliferating tissues and cell types are inherently more resistant to apoptosis and stay senescent than those that occasionally proliferate such as stem cells. Indeed, when chromosomal instability and aneuploidy were provoked in a tissue-specific manner, in mouse epidermis, epidermal hair follicle stem cells were rapidly depleted (likely through apoptosis), while the more committed transit amplifying cells tolerated the resulting aneuploidy quite well (even though we predict they become senescent) [255]. This could explain why aneuploid cells seem to accumulate in various somatic cell types in the ageing mouse, whereas aneuploidy in stem cell linages in the same mice remains rare [239]. In conclusion, we propose aneuploidy as a key aging hallmark and we argue for the beneficial effects of mitotic efficiency modulation at a molecular level as a strategic anti-aging therapy.

References

1. Jacobs PA, Court Brown WM (1966) Age and chromosomes. Nature 212:823–824
2. Nagaoka SI, Hassold TJ, Hunt PA (2012) Human aneuploidy: mechanisms and new insights into an age-old problem. Nat Rev Genet 13:493–504
3. Jacobs PA, Brunton M, Court Brown WM et al (1963) Change of human chromosome count distribution with age: evidence for a sex differences. Nature 197:1080–1081
4. Pierre RV, Hoagland HC (1972) Age-associated aneuploidy: loss of Y chromosome from human bone marrow cells with aging. Cancer 30:889–894
5. Stone JF, Sandberg AA (1995) Sex chromosome aneuploidy and aging. Mutat Res 338:107–113
6. Dumanski JP, Rasi C, Lönn M et al (2015) Mutagenesis. Smoking is associated with mosaic loss of chromosome Y. Science 347:81–83
7. Fitzgerald PH, McEwan CM (1977) Total aneuploidy and age-related sex chromosome aneuploidy in cultured lymphocytes of normal men and women. Hum Genet 39:329–337
8. Guttenbach M, Schakowski R, Schmid M (1994) Aneuploidy and ageing: sex chromosome exclusion into micronuclei. Hum Genet 94:295–298
9. Alejandro J, Payne S, Mukherjee AB, Thomas S (1996) Age-related aneuploidy analysis of human blood cells in vivo by fluorescence in situ hybridization (FISH). Mech Ageing Dev 90:145–156
10. Mukherjee AB, Thomas S (1997) A longitudinal study of human age-related chromosomal analysis in skin fibroblasts. Exp Cell Res 235:161–169
11. Nakagome Y, Abe T, Misawa S et al (1984) The "loss" of centromeres from chromosomes of aged women. Am J Hum Genet 36:398–404
12. Carere A, Antoccia A, Cimini D et al (1999) Analysis of chromosome loss and non-disjunction in cytokinesis-blocked lymphocytes of 24 male subjects. Mutagenesis 14:491–496

13. Faggioli F, Wang T, Vijg J, Montagna C (2012) Chromosome-specific accumulation of aneuploidy in the aging mouse brain. Hum Mol Genet 21:5246–5253
14. Thomas P, Fenech M (2008) Chromosome 17 and 21 aneuploidy in buccal cells is increased with ageing and in Alzheimer's disease. Mutagenesis 23:57–65
15. Yurov YB, Vorsanova SG, Liehr T et al (2014) X chromosome aneuploidy in the Alzheimer's disease brain. Mol Cytogenet 7(1):20
16. Knouse KA, Wu J, Whittaker CA, Amon A (2014) Single cell sequencing reveals low levels of aneuploidy across mammalian tissues. Proc Natl Acad Sci U S A 111:13409–13414
17. van den Bos H, Spierings DCJ, Taudt AS et al (2016) Single-cell whole genome sequencing reveals no evidence for common aneuploidy in normal and Alzheimer's disease neurons. Genome Biol 17:116
18. Geigl JB, Langer S, Barwisch S et al (2004) Analysis of gene expression patterns and chromosomal changes associated with aging. Cancer Res 64:8550–8557
19. Ly DH, Lockhart DJ, Lerner RA, Schultz PG (2000) Mitotic misregulation and human aging. Science 287:2486–2492
20. Nicholson JM, Cimini D (2011) How mitotic errors contribute to karyotypic diversity in cancer. Adv Cancer Res 112:43–75
21. Thompson SL, Bakhoum SF, Compton DA (2010) Mechanisms of chromosomal instability. Curr Biol 20:R285–R295
22. Loncarek J, Kisurina-Evgenieva O, Vinogradova T et al (2007) The centromere geometry essential for keeping mitosis error free is controlled by spindle forces. Nature 450:745–749
23. Cimini D, Mattiuzzo M, Torosantucci L, Degrassi F (2003) Histone hyperacetylation in mitosis prevents sister chromatid separation and produces chromosome segregation defects. Mol Biol Cell 14:3821–3833
24. Ganem NJ, Godinho SA, Pellman D (2009) A mechanism linking extra centrosomes to chromosomal instability. Nature 460:278–282
25. Silkworth WT, Nardi IK, Scholl LM, Cimini D (2009) Multipolar spindle pole coalescence is a major source of kinetochore mis-attachment and chromosome mis-segregation in cancer cells. PLoS One 4:e6564
26. Bakhoum SF, Compton DA (2012) Chromosomal instability and cancer: a complex relationship with therapeutic potential. J Clin Invest 122:1138–1143
27. López-Otín C, Blasco MA, Partridge L et al (2013) The hallmarks of aging. Cell 153:1194–1217
28. Hoeijmakers JHJ (2009) DNA damage, aging, and cancer. N Engl J Med 361:1475–1485
29. Erol A (2011) Genotoxic stress-mediated cell cycle activities for the decision of cellular fate. Cell Cycle 10:3239–3248
30. Worman HJ (2012) Nuclear lamins and laminopathies. J Pathol 226:316–325
31. Coppedè F, Migliore L (2010) DNA repair in premature aging disorders and neurodegeneration. Curr Aging Sci 3:3–19
32. Puzianowska-Kuznicka M, Kuznicki J (2005) Genetic alterations in accelerated ageing syndromes. Do they play a role in natural ageing? Int J Biochem Cell Biol 37:947–960
33. Schumacher B, Garinis GA, Hoeijmakers JHJ (2008) Age to survive: DNA damage and aging. Trends Genet 24:77–85
34. Vermeij WP, Hoeijmakers JHJ, Pothof J (2014) Aging: not all DNA damage is equal. Curr Opin Genet Dev 26:124–130
35. Campisi J (2005) Senescent cells, tumor suppression, and organismal aging: good citizens, bad neighbors. Cell 120:513–522. doi:10.1016/j.cell.2005.02.003
36. Coppé J-P, Desprez P-Y, Krtolica A, Campisi J (2010) The senescence-associated secretory phenotype: the dark side of tumor suppression. Annu Rev Pathol 5:99–118
37. Edifizi D, Schumacher B (2015) Genome instability in development and aging: insights from nucleotide excision repair in humans, mice, and worms. Biomol Ther 5:1855–1869
38. Gorbunova V, Seluanov A, Mao Z, Hine C (2007) Changes in DNA repair during aging. Nucleic Acids Res 35:7466–7474

39. Goukassian D, Gad F, Yaar M et al (2000) Mechanisms and implications of the age-associated decrease in DNA repair capacity. FASEB J 14:1325–1334
40. Jackson SP, Bartek J (2009) The DNA-damage response in human biology and disease. Nature 461:1071–1078
41. van Vugt MATM, Gardino AK, Linding R et al (2010) A mitotic phosphorylation feedback network connects Cdk1, Plk1, 53BP1, and Chk2 to inactivate the G(2)/M DNA damage checkpoint. PLoS Biol 8:e1000287
42. Benada J, Burdová K, Lidak T et al (2015) Polo-like kinase 1 inhibits DNA damage response during mitosis. Cell Cycle 14:219–231
43. Orthwein A, Fradet-Turcotte A, Noordermeer SM et al (2014) Mitosis inhibits DNA double-strand break repair to guard against telomere fusions. Science 344:189–193
44. Bakhoum SF, Kabeche L, Murnane JP et al (2014) DNA-damage response during mitosis induces whole-chromosome missegregation. Cancer Discov 4:1281–1289
45. Mehta PA, Harris RE, Davies SM et al (2010) Numerical chromosomal changes and risk of development of myelodysplastic syndrome – acute myeloid leukemia in patients with Fanconi anemia. Cancer Genet Cytogenet 203:180–186
46. Pulliam-Leath AC, Ciccone SL, Nalepa G et al (2010) Genetic disruption of both Fancc and Fancg in mice recapitulates the hematopoietic manifestations of Fanconi anemia. Blood 116:2915–2920
47. Naim V, Rosselli F (2009) The FANC pathway and BLM collaborate during mitosis to prevent micro-nucleation and chromosome abnormalities. Nat Cell Biol 11:761–768
48. Nalepa G, Enzor R, Sun Z et al (2013) Fanconi anemia signaling network regulates the spindle assembly checkpoint. J Clin Invest 123:3839–3847
49. Singh DK, Ahn B, Bohr VA (2009) Roles of RECQ helicases in recombination based DNA repair, genomic stability and aging. Biogerontology 10:235–252
50. Honma M, Tadokoro S, Sakamoto H et al (2002) Chromosomal instability in B-lymphoblasotoid cell lines from Werner and Bloom syndrome patients. Mutat Res 520:15–24
51. Kaloustian Der VM, McGill JJ, Vekemans M, Kopelman HR (1990) Clonal lines of aneuploid cells in Rothmund-Thomson syndrome. Am J Med Genet 37:336–339
52. Melaragno MI, Pagni D, Smith MA (1995) Cytogenetic aspects of Werner's syndrome lymphocyte cultures. Mech Ageing Dev 78:117–122
53. Mukherjee AB, Costello C (1998) Aneuploidy analysis in fibroblasts of human premature aging syndromes by FISH during in vitro cellular aging. Mech Ageing Dev 103:209–222
54. Salk D, Au K, Hoehn H, Martin GM (1981) Cytogenetics of Werner's syndrome cultured skin fibroblasts: variegated translocation mosaicism. Cytogenet Cell Genet 30:92–107
55. Chang S, Multani AS, Cabrera NG et al (2004) Essential role of limiting telomeres in the pathogenesis of Werner syndrome. Nat Genet 36:877–882
56. Chester N, Babbe H, Pinkas J et al (2006) Mutation of the murine Bloom's syndrome gene produces global genome destabilization. Mol Cell Biol 26:6713–6726
57. Mann MB, Hodges CA, Barnes E et al (2005) Defective sister-chromatid cohesion, aneuploidy and cancer predisposition in a mouse model of type II Rothmund-Thomson syndrome. Hum Mol Genet 14:813–825
58. Chan KL, Palmai-Pallag T, Ying S, Hickson ID (2009) Replication stress induces sister-chromatid bridging at fragile site loci in mitosis. Nat Cell Biol 11:753–760
59. Chung L, Onyango D, Guo Z et al (2015) The FEN1 E359K germline mutation disrupts the FEN1-WRN interaction and FEN1 GEN activity, causing aneuploidy-associated cancers. Oncogene 34:902–911
60. Larsen NB, Hickson ID (2013) RecQ helicases: conserved guardians of genomic integrity. Adv Exp Med Biol 767:161–184
61. Chan KL, North PS, Hickson ID (2007) BLM is required for faithful chromosome segregation and its localization defines a class of ultrafine anaphase bridges. EMBO J 26:3397–3409

62. Liu Y, Nielsen CF, Yao Q, Hickson ID (2014) The origins and processing of ultra fine anaphase DNA bridges. Curr Opin Genet Dev 26:1–5
63. Ke Y, Huh J-W, Warrington R et al (2011) PICH and BLM limit histone association with anaphase centromeric DNA threads and promote their resolution. EMBO J 30:3309–3321
64. Ito S, Tan LJ, Andoh D et al (2010) MMXD, a TFIIH-independent XPD-MMS19 protein complex involved in chromosome segregation. Mol Cell 39:632–640
65. Tan LJ, Saijo M, Kuraoka I et al (2012) Xeroderma pigmentosum group F protein binds to Eg5 and is required for proper mitosis: implications for XP-F and XFE. Genes Cells 17:173–185
66. Naim V, Wilhelm T, Debatisse M, Rosselli F (2013) ERCC1 and MUS81-EME1 promote sister chromatid separation by processing late replication intermediates at common fragile sites during mitosis. Nat Cell Biol 15:1008–1015
67. Dechat T, Pfleghaar K, Sengupta K et al (2008) Nuclear lamins: major factors in the structural organization and function of the nucleus and chromatin. Genes Dev 22:832–853
68. Gruenbaum Y, Foisner R (2015) Lamins: nuclear intermediate filament proteins with fundamental functions in nuclear mechanics and genome regulation. Annu Rev Biochem 84:131–164
69. Cau P, Navarro C, Harhouri K et al (2014) Nuclear matrix, nuclear envelope and premature aging syndromes in a translational research perspective. Semin Cell Dev Biol 29:125–147
70. Cadiñanos J, Varela I, López-Otín C, Freije JMP (2005) From immature lamin to premature aging: molecular pathways and therapeutic opportunities. Cell Cycle 4:1732–1735
71. Goldman RD, Shumaker DK, Erdos MR et al (2004) Accumulation of mutant lamin a causes progressive changes in nuclear architecture in Hutchinson-Gilford progeria syndrome. Proc Natl Acad Sci U S A 101:8963–8968
72. Yang SH, Chang SY, Ren S et al (2011) Absence of progeria-like disease phenotypes in knock-in mice expressing a non-farnesylated version of progerin. Hum Mol Genet 20:436–444
73. Mehta IS, Eskiw CH, Arican HD et al (2011) Farnesyltransferase inhibitor treatment restores chromosome territory positions and active chromosome dynamics in Hutchinson-Gilford progeria syndrome cells. Genome Biol 12:R74
74. Yang SH, Bergo MO, Toth JI et al (2005) Blocking protein farnesyltransferase improves nuclear blebbing in mouse fibroblasts with a targeted Hutchinson-Gilford progeria syndrome mutation. Proc Natl Acad Sci U S A 102:10291–10296
75. Moulson CL, Go G, Gardner JM et al (2005) Homozygous and compound heterozygous mutations in ZMPSTE24 cause the laminopathy restrictive dermopathy. J Invest Dermatol 125:913–919
76. Navarro CL, Cadiñanos J, De Sandre-Giovannoli A et al (2005) Loss of ZMPSTE24 (FACE-1) causes autosomal recessive restrictive dermopathy and accumulation of Lamin a precursors. Hum Mol Genet 14:1503–1513
77. Jamin A, Wiebe MS (2015) Barrier to Autointegration factor (BANF1): interwoven roles in nuclear structure, genome integrity, innate immunity, stress responses and progeria. Curr Opin Cell Biol 34:61–68
78. Scaffidi P, Misteli T (2005) Reversal of the cellular phenotype in the premature aging disease Hutchinson-Gilford progeria syndrome. Nat Med 11:440–445
79. McClintock D, Ratner D, Lokuge M et al (2007) The mutant form of lamin a that causes Hutchinson-Gilford progeria is a biomarker of cellular aging in human skin. PLoS One 2:e1269
80. Scaffidi P, Misteli T (2006) Lamin A-dependent nuclear defects in human aging. Science 312:1059–1063
81. Freund A, Laberge R-M, Demaria M, Campisi J (2012) Lamin B1 loss is a senescence-associated biomarker. Mol Biol Cell 23:2066–2075
82. Shimi T, Butin-Israeli V, Adam SA et al (2011) The role of nuclear lamin B1 in cell proliferation and senescence. Genes Dev 25:2579–2593

83. Cao K, Capell BC, Erdos MR et al (2007) A lamin a protein isoform overexpressed in Hutchinson-Gilford progeria syndrome interferes with mitosis in progeria and normal cells. Proc Natl Acad Sci U S A 104:4949–4954
84. Dechat T, Shimi T, Adam SA et al (2007) Alterations in mitosis and cell cycle progression caused by a mutant lamin a known to accelerate human aging. Proc Natl Acad Sci U S A 104:4955–4960
85. Pratt CH, Curtain M, Donahue LR, Shopland LS (2011) Mitotic defects lead to pervasive aneuploidy and accompany loss of RB1 activity in mouse Lmna Dhe dermal fibroblasts. PLoS One 6:e18065
86. Gruber J, Lampe T, Osborn M, Weber K (2005) RNAi of FACE1 protease results in growth inhibition of human cells expressing lamin a: implications for Hutchinson-Gilford progeria syndrome. J Cell Sci 118:689–696
87. Liu B, Wang J, Chan KM et al (2005) Genomic instability in laminopathy-based premature aging. Nat Med 11:780–785
88. Qi R, Xu N, Wang G et al (2015) The lamin-A/C-LAP2α-BAF1 protein complex regulates mitotic spindle assembly and positioning. J Cell Sci 128:2830–2841
89. Schweizer N, Weiss M, Maiato H (2014) The dynamic spindle matrix. Curr Opin Cell Biol 28:1–7
90. Ma L, Tsai M-Y, Wang S et al (2009) Requirement for Nudel and dynein for assembly of the lamin B spindle matrix. Nat Cell Biol 11:247–256
91. Zheng Y (2010) A membranous spindle matrix orchestrates cell division. Nat Rev Mol Cell Biol 11:529–535
92. Blackburn EH, Epel ES, Lin J (2015) Human telomere biology: a contributory and interactive factor in aging, disease risks, and protection. Science 350:1193–1198
93. de Lange T (2005) Telomere-related genome instability in cancer. Cold Spring Harb Symp Quant Biol 70:197–204
94. Palm W, de Lange T (2008) How shelterin protects mammalian telomeres. Annu Rev Genet 42:301–334
95. Fumagalli M, Rossiello F, Clerici M et al (2012) Telomeric DNA damage is irreparable and causes persistent DNA-damage-response activation. Nat Cell Biol 14:355–365
96. Hewitt G, Jurk D, Marques FDM et al (2012) Telomeres are favoured targets of a persistent DNA damage response in ageing and stress-induced senescence. Nat Commun 3:708
97. Greider CW, Blackburn EH (1985) Identification of a specific telomere terminal transferase activity in Tetrahymena extracts. Cell 43:405–413
98. Hayflick L, Moorhead PS (1961) The serial cultivation of human diploid cell strains. Exp Cell Res 25:585–621
99. Olovnikov AM (1996) Telomeres, telomerase, and aging: origin of the theory. Exp Gerontol 31:443–448
100. Bodnar AG, Ouellette M, Frolkis M et al (1998) Extension of life-span by introduction of telomerase into normal human cells. Science 279:349–352
101. Batista LFZ (2014) Telomere biology in stem cells and reprogramming. Prog Mol Biol Transl Sci 125:67–88
102. Simm A, Campisi J (2014) Stress and aging. Exp Gerontol 59:1–2
103. Blasco MA, Lee HW, Hande MP et al (1997) Telomere shortening and tumor formation by mouse cells lacking telomerase RNA. Cell 91:25–34
104. Armanios M, Blackburn EH (2012) The telomere syndromes. Nat Rev Genet 13:693–704
105. Blasco MA (2005) Telomeres and human disease: ageing, cancer and beyond. Nat Rev Genet 6:611–622
106. Jaskelioff M, Muller FL, Paik J-H et al (2011) Telomerase reactivation reverses tissue degeneration in aged telomerase-deficient mice. Nature 469:102–106
107. Lee D-H, Acharya SS, Kwon M et al (2014) Dephosphorylation enables the recruitment of 53BP1 to double-strand DNA breaks. Mol Cell 54:512–525

108. Cesare AJ (2014) Mitosis, double strand break repair, and telomeres: a view from the end: how telomeres and the DNA damage response cooperate during mitosis to maintain genome stability. BioEssays 36:1054–1061
109. Cesare AJ, Hayashi MT, Crabbe L, Karlseder J (2013) The telomere deprotection response is functionally distinct from the genomic DNA damage response. Mol Cell 51:141–155
110. Kaul Z, Cesare AJ, Huschtscha LI et al (2012) Five dysfunctional telomeres predict onset of senescence in human cells. EMBO Rep 13:52–59
111. d'Adda di Fagagna F, Reaper PM, Clay-Farrace L et al (2003) A DNA damage checkpoint response in telomere-initiated senescence. Nature 426:194–198
112. Takai H, Smogorzewska A, de Lange T (2003) DNA damage foci at dysfunctional telomeres. Curr Biol 13:1549–1556
113. van Steensel B, Smogorzewska A, de Lange T (1998) TRF2 protects human telomeres from end-to-end fusions. Cell 92:401–413
114. Soler D, Genescà A, Arnedo G et al (2005) Telomere dysfunction drives chromosomal instability in human mammary epithelial cells. Genes Chromosomes Cancer 44:339–350
115. Tusell L, Pampalona J, Soler D et al (2010) Different outcomes of telomere-dependent anaphase bridges. Biochem Soc Trans 38:1698–1703
116. Tusell L, Soler D, Agostini M et al (2008) The number of dysfunctional telomeres in a cell: one amplifies; more than one translocate. Cytogenet Genome Res 122:315–325
117. Pampalona J, Soler D, Genescà A, Tusell L (2010) Whole chromosome loss is promoted by telomere dysfunction in primary cells. Genes Chromosomes Cancer 49:368–378
118. Pampalona J, Roscioli E, Silkworth WT et al (2016) Chromosome bridges maintain kinetochore-microtubule attachment throughout mitosis and rarely break during anaphase. PLoS One 11:e0147420
119. Zhang C-Z, Spektor A, Cornils H et al (2015) Chromothripsis from DNA damage in micronuclei. Nature 522:179–184
120. Crasta K, Ganem NJ, Dagher R et al (2012) DNA breaks and chromosome pulverization from errors in mitosis. Nature 482:53–58
121. Maciejowski J, Li Y, Bosco N et al (2015) Chromothripsis and Kataegis induced by telomere crisis. Cell 163:1641–1654
122. Roberts SA, Sterling J, Thompson C et al (2012) Clustered mutations in yeast and in human cancers can arise from damaged long single-strand DNA regions. Mol Cell 46:424–435
123. Dynek JN, Smith S (2004) Resolution of sister telomere association is required for progression through mitosis. Science 304:97–100
124. Hsiao SJ, Smith S (2008) Tankyrase function at telomeres, spindle poles, and beyond. Biochimie 90:83–92
125. Canudas S, Smith S (2009) Differential regulation of telomere and centromere cohesion by the Scc3 homologues SA1 and SA2, respectively, in human cells. J Cell Biol 187:165–173
126. Canudas S, Houghtaling BR, Kim JY et al (2007) Protein requirements for sister telomere association in human cells. EMBO J 26:4867–4878
127. Smith S, Giriat I, Schmitt A, de Lange T (1998) Tankyrase, a poly(ADP-ribose) polymerase at human telomeres. Science 282:1484–1487
128. Bisht KK, Daniloski Z, Smith S (2013) SA1 binds directly to DNA through its unique AT-hook to promote sister chromatid cohesion at telomeres. J Cell Sci 126:3493–3503
129. Bisht KK, Dudognon C, Chang WG et al (2012) GDP-mannose-4,6-dehydratase is a cytosolic partner of tankyrase 1 that inhibits its poly(ADP-ribose) polymerase activity. Mol Cell Biol 32:3044–3053
130. Kim MK, Smith S (2014) Persistent telomere cohesion triggers a prolonged anaphase. Mol Biol Cell 25:30–40
131. Yalon M, Gal S, Segev Y et al (2004) Sister chromatid separation at human telomeric regions. J Cell Sci 117:1961–1970
132. Han S, Brunet A (2012) Histone methylation makes its mark on longevity. Trends Cell Biol 22:42–49

133. Johnson AA, Akman K, Calimport SRG et al (2012) The role of DNA methylation in aging, rejuvenation, and age-related disease. Rejuvenation Res 15:483–494
134. Maegawa S, Hinkal G, Kim HS et al (2010) Widespread and tissue specific age-related DNA methylation changes in mice. Genome Res 20:332–340
135. Hannum G, Guinney J, Zhao L et al (2013) Genome-wide methylation profiles reveal quantitative views of human aging rates. Mol Cell 49:359–367
136. Horvath S (2013) DNA methylation age of human tissues and cell types. Genome Biol 14:R115
137. Weidner CI, Wagner W (2014) The epigenetic tracks of aging. Biol Chem 395:1307–1314
138. Fraga MF, Esteller M (2007) Epigenetics and aging: the targets and the marks. Trends Genet 23:413–418
139. Jin C, Li J, Green CD et al (2011) Histone demethylase UTX-1 regulates C. elegans life span by targeting the insulin/IGF-1 signaling pathway. Cell Metab 14:161–172
140. Guarente L (2011) Sirtuins, aging, and metabolism. Cold Spring Harb Symp Quant Biol 76:81–90
141. Brown K, Xie S, Qiu X et al (2013) SIRT3 reverses aging-associated degeneration. Cell Rep 3:319–327
142. Kanfi Y, Peshti V, Gil R et al (2010) SIRT6 protects against pathological damage caused by diet-induced obesity. Aging Cell 9:162–173
143. Kawahara TLA, Michishita E, Adler AS et al (2009) SIRT6 links histone H3 lysine 9 deacetylation to NF-kappaB-dependent gene expression and organismal life span. Cell 136:62–74
144. Nogueiras R, Habegger KM, Chaudhary N et al (2012) Sirtuin 1 and sirtuin 3: physiological modulators of metabolism. Physiol Rev 92:1479–1514
145. Oberdoerffer P, Michan S, McVay M et al (2008) SIRT1 redistribution on chromatin promotes genomic stability but alters gene expression during aging. Cell 135:907–918
146. Someya S, Yu W, Hallows WC et al (2010) Sirt3 mediates reduction of oxidative damage and prevention of age-related hearing loss under caloric restriction. Cell 143:802–812
147. Wang R-H, Sengupta K, Li C et al (2008) Impaired DNA damage response, genome instability, and tumorigenesis in SIRT1 mutant mice. Cancer Cell 14:312–323
148. Zhong L, D'Urso A, Toiber D et al (2010) The histone deacetylase Sirt6 regulates glucose homeostasis via Hif1alpha. Cell 140:280–293
149. Pegoraro G, Misteli T (2009) The central role of chromatin maintenance in aging. Aging (Albany NY) 1:1017–1022
150. Shumaker DK, Dechat T, Kohlmaier A et al (2006) Mutant nuclear lamin a leads to progressive alterations of epigenetic control in premature aging. Proc Natl Acad Sci U S A 103:8703–8708
151. Tsurumi A, Li WX (2012) Global heterochromatin loss: a unifying theory of aging? Epigenetics 7:680–688
152. de Magalhães JP, Budovsky A, Lehmann G et al (2009) The human ageing genomic resources: online databases and tools for biogerontologists. Aging Cell 8:65–72
153. Ugalde AP, Español Y, López-Otín C (2011) Micromanaging aging with miRNAs: new messages from the nuclear envelope. Nucleus 2:549–555
154. Rando TA, Chang HY (2012) Aging, rejuvenation, and epigenetic reprogramming: resetting the aging clock. Cell 148:46–57
155. Guppy BJ, McManus KJ (2015) Mitotic accumulation of dimethylated lysine 79 of histone H3 is important for maintaining genome integrity during mitosis in human cells. Genetics 199:423–433
156. Blasco MA (2007) The epigenetic regulation of mammalian telomeres. Nat Rev Genet 8:299–309
157. Gonzalo S, Jaco I, Fraga MF et al (2006) DNA methyltransferases control telomere length and telomere recombination in mammalian cells. Nat Cell Biol 8:416–424
158. Cheung P, Tanner KG, Cheung WL et al (2000) Synergistic coupling of histone H3 phosphorylation and acetylation in response to epidermal growth factor stimulation. Mol Cell 5:905–915

159. Inoue A, Hyle J, Lechner MS, Lahti JM (2008) Perturbation of HP1 localization and chromatin binding ability causes defects in sister-chromatid cohesion. Mutat Res 657:48–55
160. Shimura M, Toyoda Y, Iijima K et al (2011) Epigenetic displacement of HP1 from heterochromatin by HIV-1 Vpr causes premature sister chromatid separation. J Cell Biol 194:721–735
161. Fatoba ST, Okorokov AL (2011) Human SIRT1 associates with mitotic chromatin and contributes to chromosomal condensation. Cell Cycle 10:2317–2322
162. Vaquero A, Scher MB, Lee DH et al (2006) SirT2 is a histone deacetylase with preference for histone H4 Lys 16 during mitosis. Genes Dev 20:1256–1261
163. Kim H-S, Vassilopoulos A, Wang R-H et al (2011) SIRT2 maintains genome integrity and suppresses tumorigenesis through regulating APC/C activity. Cancer Cell 20:487–499
164. Choi B-S, Park JE, Jang C-Y (2014) Sirt3 controls chromosome alignment by regulating spindle dynamics during mitosis. Biochem Biophys Res Commun 444:662–669
165. Ardestani PM, Liang F (2014) Sub-cellular localization, expression and functions of Sirt6 during the cell cycle in HeLa cells. Nucleus 3:442–451
166. Hudson RS, Yi M, Esposito D et al (2012) MicroRNA-1 is a candidate tumor suppressor and prognostic marker in human prostate cancer. Nucleic Acids Res 40:3689–3703
167. Yamakuchi M, Ferlito M, Lowenstein CJ (2008) miR-34a repression of SIRT1 regulates apoptosis. Proc Natl Acad Sci U S A 105:13421–13426
168. Mäki-Jouppila JHE, Pruikkonen S, Tambe MB et al (2015) MicroRNA let-7b regulates genomic balance by targeting Aurora B kinase. Mol Oncol 9:1056–1070
169. Morimoto RI, Cuervo AM (2009) Protein homeostasis and aging: taking care of proteins from the cradle to the grave. J Gerontol A Biol Sci Med Sci 64:167–170
170. Koga H, Kaushik S, Cuervo AM (2011) Protein homeostasis and aging: the importance of exquisite quality control. Ageing Res Rev 10:205–215
171. Swindell WR, Masternak MM, Kopchick JJ et al (2009) Endocrine regulation of heat shock protein mRNA levels in long-lived dwarf mice. Mech Ageing Dev 130:393–400
172. Fargnoli J, Kunisada T, Fornace AJ et al (1990) Decreased expression of heat shock protein 70 mRNA and protein after heat treatment in cells of aged rats. Proc Natl Acad Sci U S A 87:846–850
173. Hall DM, Xu L, Drake VJ et al (2000) Aging reduces adaptive capacity and stress protein expression in the liver after heat stress. J Appl Physiol 89:749–759
174. Pahlavani MA, Harris MD, Moore SA et al (1995) The expression of heat shock protein 70 decreases with age in lymphocytes from rats and rhesus monkeys. Exp Cell Res 218:310–318
175. Heydari AR, You S, Takahashi R et al (2000) Age-related alterations in the activation of heat shock transcription factor 1 in rat hepatocytes. Exp Cell Res 256:83–93
176. Locke M, Tanguay RM (1996) Diminished heat shock response in the aged myocardium. Cell Stress Chaperones 1:251–260
177. Hsu A-L, Murphy CT, Kenyon C (2003) Regulation of aging and age-related disease by DAF-16 and heat-shock factor. Science 300:1142–1145
178. Morley JF, Morimoto RI (2004) Regulation of longevity in Caenorhabditis elegans by heat shock factor and molecular chaperones. Mol Biol Cell 15:657–664
179. Walker GA, Lithgow GJ (2003) Lifespan extension in C. elegans by a molecular chaperone dependent upon insulin-like signals. Aging Cell 2:131–139
180. Jahngen JH, Lipman RD, Eisenhauer DA et al (1990) Aging and cellular maturation cause changes in ubiquitin-eye lens protein conjugates. Arch Biochem Biophys 276:32–37
181. Min J-N, Whaley RA, Sharpless NE et al (2008) CHIP deficiency decreases longevity, with accelerated aging phenotypes accompanied by Altered protein quality control. Mol Cell Biol 28:4018–4025
182. Ferrington DA, Husom AD, Thompson LV (2005) Altered proteasome structure, function, and oxidation in aged muscle. FASEB J 19:644–646
183. Keller JN, Huang FF, Markesbery WR (2000) Decreased levels of proteasome activity and proteasome expression in aging spinal cord. Neuroscience 98:149–156

184. Li W, Gao B, Lee S-M et al (2007) RLE-1, an E3 ubiquitin ligase, regulates C. elegans aging by catalyzing DAF-16 polyubiquitination. Dev Cell 12:235–246
185. Yun C, Stanhill A, Yang Y et al (2008) Proteasomal adaptation to environmental stress links resistance to proteotoxicity with longevity in Caenorhabditis elegans. Proc Natl Acad Sci U S A 105:7094–7099
186. Mizushima N, Levine B, Cuervo AM, Klionsky DJ (2008) Autophagy fights disease through cellular self-digestion. Nature 451:1069–1075
187. Terman A (1995) The effect of age on formation and elimination of autophagic vacuoles in mouse hepatocytes. Gerontology 41(Suppl 2):319–326
188. Terman A, Brunk UT (1998) Ceroid/lipofuscin formation in cultured human fibroblasts: the role of oxidative stress and lysosomal proteolysis. Mech Ageing Dev 104:277–291
189. Dice JF (2007) Chaperone-mediated autophagy. Autophagy 3:295–299
190. Kiffin R, Kaushik S, Zeng M et al (2007) Altered dynamics of the lysosomal receptor for chaperone-mediated autophagy with age. J Cell Sci 120:782–791
191. Zhang C, Cuervo AM (2008) Restoration of chaperone-mediated autophagy in aging liver improves cellular maintenance and hepatic function. Nat Med 14:959–965
192. Lindner AB, Madden R, Demarez A et al (2008) Asymmetric segregation of protein aggregates is associated with cellular aging and rejuvenation. Proc Natl Acad Sci U S A 105: 3076–3081
193. Amen T, Kaganovich D (2015) Dynamic droplets: the role of cytoplasmic inclusions in stress, function, and disease. Cell Mol Life Sci 72:401–415
194. Zhou C, Slaughter BD, Unruh JR et al (2014) Organelle-based aggregation and retention of damaged proteins in asymmetrically dividing cells. Cell 159:530–542
195. Chen Y-J, Lai K-C, Kuo H-H et al (2014) HSP70 colocalizes with PLK1 at the centrosome and disturbs spindle dynamics in cells arrested in mitosis by arsenic trioxide. Arch Toxicol 88:1711–1723
196. O'Regan L, Sampson J, Richards MW et al (2015) Hsp72 is targeted to the mitotic spindle by Nek6 to promote K-fiber assembly and mitotic progression. J Cell Biol 209:349–358
197. Basto R, Gergely F, Draviam VM et al (2007) Hsp90 is required to localise cyclin B and Msps/ch-TOG to the mitotic spindle in Drosophila and humans. J Cell Sci 120:1278–1287
198. Lange BM, Bachi A, Wilm M, González C (2000) Hsp90 is a core centrosomal component and is required at different stages of the centrosome cycle in Drosophila and vertebrates. EMBO J 19:1252–1262
199. Davies AE, Kaplan KB (2010) Hsp90-Sgt1 and Skp1 target human Mis12 complexes to ensure efficient formation of kinetochore-microtubule binding sites. J Cell Biol 189: 261–274
200. Liu XS, Song B, Tang J et al (2012) Plk1 phosphorylates Sgt1 at the kinetochores to promote timely kinetochore-microtubule attachment. Mol Cell Biol 32:4053–4067
201. Niikura Y, Ohta S, Vandenbeldt KJ et al (2006) 17-AAG, an Hsp90 inhibitor, causes kinetochore defects: a novel mechanism by which 17-AAG inhibits cell proliferation. Oncogene 25:4133–4146
202. Vihervaara A, Sergelius C, Vasara J et al (2013) Transcriptional response to stress in the dynamic chromatin environment of cycling and mitotic cells. Proc Natl Acad Sci U S A 110:E3388–E3397
203. Lu M, Boschetti C, Tunnacliffe A (2015) Long term Aggresome accumulation leads to DNA damage, p53-dependent cell cycle arrest, and steric interference in mitosis. J Biol Chem 290:27986–28000
204. Eskelinen E-L, Prescott AR, Cooper J et al (2002) Inhibition of autophagy in mitotic animal cells. Traffic 3:878–893
205. Liu L, Xie R, Nguyen S et al (2009) Robust autophagy/mitophagy persists during mitosis. Cell Cycle 8:1616–1620
206. Fuchs M, Luthold C, Guilbert SM et al (2015) A role for the chaperone complex BAG3-HSPB8 in actin dynamics, spindle orientation and proper chromosome segregation during mitosis. PLoS Genet 11:e1005582

207. Pohl C (2009) Dual control of cytokinesis by the ubiquitin and autophagy pathways. Autophagy 5:561–562
208. Mathew R, Kongara S, Beaudoin B et al (2007) Autophagy suppresses tumor progression by limiting chromosomal instability. Genes Dev 21:1367–1381
209. Frémont S, Gérard A, Galloux M et al (2013) Beclin-1 is required for chromosome congression and proper outer kinetochore assembly. EMBO Rep 14:364–372
210. Gordon DJ, Resio B, Pellman D (2012) Causes and consequences of aneuploidy in cancer. Nat Rev Genet 13:189–203
211. Tang Y-C, Amon A (2013) Gene copy-number alterations: a cost-benefit analysis. Cell 152:394–405
212. Santaguida S, Amon A (2015) Short- and long-term effects of chromosome mis-segregation and aneuploidy. Nat Rev Mol Cell Biol 16:473–485
213. Stingele S, Stoehr G, Peplowska K et al (2012) Global analysis of genome, transcriptome and proteome reveals the response to aneuploidy in human cells. Mol Syst Biol 8:608
214. Thorburn RR, Gonzalez C, Brar GA et al (2013) Aneuploid yeast strains exhibit defects in cell growth and passage through START. Mol Biol Cell 24:1274–1289
215. Williams BR, Prabhu VR, Hunter KE et al (2008) Aneuploidy affects proliferation and spontaneous immortalization in mammalian cells. Science 322:703–709
216. Sheltzer JM, Amon A (2011) The aneuploidy paradox: costs and benefits of an incorrect karyotype. Trends Genet 27:446–453
217. Sheltzer JM, Torres EM, Dunham MJ, Amon A (2012) Transcriptional consequences of aneuploidy. Proc Natl Acad Sci U S A 109:12644–12649
218. Torres EM, Sokolsky T, Tucker CM et al (2007) Effects of aneuploidy on cellular physiology and cell division in haploid yeast. Science 317:916–924
219. Blank HM, Sheltzer JM, Meehl CM, Amon A (2015) Mitotic entry in the presence of DNA damage is a widespread property of aneuploidy in yeast. Mol Biol Cell 26:1440–1451
220. Zhu J, Pavelka N, Bradford WD et al (2012) Karyotypic determinants of chromosome instability in Aneuploid budding yeast. PLoS Genet 8:e1002719
221. Dephoure N, Hwang S, O'Sullivan C et al (2014) Quantitative proteomic analysis reveals posttranslational responses to aneuploidy in yeast. eLife 3:e03023
222. Donnelly N, Passerini V, Dürrbaum M et al (2014) HSF1 deficiency and impaired HSP90-dependent protein folding are hallmarks of aneuploid human cells. EMBO J 33:2374–2387
223. Oromendia AB, Dodgson SE, Amon A (2012) Aneuploidy causes proteotoxic stress in yeast. Genes Dev 26:2696–2708
224. Stingele S, Stoehr G, Storchová Z (2013) Activation of autophagy in cells with abnormal karyotype. Autophagy 9:246–248
225. Tang Y-C, Williams BR, Siegel JJ, Amon A (2011) Identification of aneuploidy-selective antiproliferation compounds. Cell 144:499–512
226. Torres EM, Dephoure N, Panneerselvam A et al (2010) Identification of aneuploidy-tolerating mutations. Cell 143:71–83
227. Jenkins EC, Ye L, Krinsky-McHale SJ et al (2016) Telomere longitudinal shortening as a biomarker for dementia status of adults with down syndrome. Am J Med Genet B Neuropsychiatr Genet 171:169–174
228. Wenger SL, Hansroth J, Shackelford AL (2014) Decreased telomere length in metaphase and interphase cells from newborns with trisomy 21. Gene 542:87
229. Horvath S, Garagnani P, Bacalini MG et al (2015) Accelerated epigenetic aging in down syndrome. Aging Cell 14:491–495
230. Nicholson JM, Macedo JC, Mattingly AJ et al (2015) Chromosome mis-segregation and cytokinesis failure in trisomic human cells. eLife 4:e05068
231. Sheltzer JM, Blank HM, Pfau SJ et al (2011) Aneuploidy drives genomic instability in yeast. Science 333:1026–1030
232. Janssen A, van der Burg M, Szuhai K et al (2011) Chromosome segregation errors as a cause of DNA damage and structural chromosome aberrations. Science 333:1895–1898

233. Lee K, Kenny AE, Rieder CL (2010) P38 mitogen-activated protein kinase activity is required during mitosis for timely satisfaction of the mitotic checkpoint but not for the fidelity of chromosome segregation. Mol Biol Cell 21:2150–2160
234. Passerini V, Ozeri-Galai E, de Pagter MS et al (2016) The presence of extra chromosomes leads to genomic instability. Nat Commun 7:10754
235. Lamm N, Ben-David U, Golan-Lev T et al (2016) Genomic instability in human pluripotent stem cells arises from replicative stress and chromosome condensation defects. Cell Stem Cell 18:253–261
236. Oromendia AB, Amon A (2014) Aneuploidy: implications for protein homeostasis and disease. Dis Model Mech 7:15–20
237. Santaguida S, Vasile E, White E, Amon A (2015) Aneuploidy-induced cellular stresses limit autophagic degradation. Genes Dev 29:2010–2021
238. Kim G, Meriin AB, Gabai VL et al (2012) The heat shock transcription factor Hsf1 is downregulated in DNA damage-associated senescence, contributing to the maintenance of senescence phenotype. Aging Cell 11:617–627
239. Baker DJ, Dawlaty MM, Wijshake T et al (2013) Increased expression of BubR1 protects against aneuploidy and cancer and extends healthy lifespan. Nat Cell Biol 15:96–102
240. Baker DJ, Jeganathan KB, Cameron JD et al (2004) BubR1 insufficiency causes early onset of aging-associated phenotypes and infertility in mice. Nat Genet 36:744–749
241. Malureanu LA, Wijshake T, Baker DJ et al (2012) Reduced life- and Healthspan in mice carrying a mono-allelic BubR1 MVA mutation. PLoS Genet 8:e1003138
242. Ricke RM, van Deursen JM (2013) Aneuploidy in health, disease, and aging. J Cell Biol 201:11–21
243. Hanks S, Coleman K, Reid S et al (2004) Constitutional aneuploidy and cancer predisposition caused by biallelic mutations in BUB1B. Nat Genet 36:1159–1161
244. Suijkerbuijk SJE, van Osch MHJ, Bos FL et al (2010) Molecular causes for BUBR1 dysfunction in the human cancer predisposition syndrome mosaic variegated aneuploidy. Cancer Res 70:4891–4900
245. Baker DJ, Jeganathan KB, Malureanu L et al (2006) Early aging–associated phenotypes in Bub3/Rae1 haploinsufficient mice. J Cell Biol 172:529–540
246. Faggioli F, Vijg J, Montagna C (2011) Chromosomal aneuploidy in the aging brain. Mech Ageing Dev 132:429–436
247. Childs BG, Baker DJ, Kirkland JL et al (2014) Senescence and apoptosis: dueling or complementary cell fates? EMBO Rep 15:1139–1153
248. Bakker B, van den Bos H, Lansdorp PM, Foijer F (2015) How to count chromosomes in a cell: an overview of current and novel technologies. BioEssays 37:570–577
249. Baker DJ, Childs BG, Durik M et al (2016) Naturally occurring p16(Ink4a)-positive cells shorten healthy lifespan. Nature 530:184–189
250. Postnikoff SDL, Harkness TAA (2012) Mechanistic insights into aging, cell-cycle progression, and stress response. Front Physiol 3:183
251. Laoukili J, Kooistra MRH, Brás A et al (2005) FoxM1 is required for execution of the mitotic programme and chromosome stability. Nat Cell Biol 7:126–136
252. Laoukili J, Stahl M, Medema RH (2007) FoxM1: at the crossroads of ageing and cancer. Biochim Biophys Acta 1775:92–102
253. Wang X, Quail E, Hung NJ et al (2001) Increased levels of forkhead box M1B transcription factor in transgenic mouse hepatocytes prevent age-related proliferation defects in regenerating liver. Proc Natl Acad Sci U S A 98:11468–11473
254. Kalinichenko VV, Gusarova GA, Tan Y et al (2003) Ubiquitous expression of the forkhead box M1B transgene accelerates proliferation of distinct pulmonary cell types following lung injury. J Biol Chem 278:37888–37894
255. Foijer F, DiTommaso T, Donati G et al (2013) Spindle checkpoint deficiency is tolerated by murine epidermal cells but not hair follicle stem cells. Proc Natl Acad Sci U S A 110: 2928–2933

Chapter 8
Unbalanced Growth, Senescence and Aging

Michael Polymenis and Brian K. Kennedy

Abstract Usually, cells balance their growth with their division. Coordinating growth inputs with cell division ensures the proper timing of division when sufficient cell material is available and affects the overall rate of cell proliferation. At a very fundamental level, cellular replicative lifespan—defined as the number of times a cell can divide, is a manifestation of cell cycle control. Hence, control of mitotic cell divisions, especially when the commitment is made to a new round of cell division, is intimately linked to replicative aging of cells. In this chapter, we review our current understanding, and its shortcomings, of how unbalanced growth and division, can dramatically influence the proliferative potential of cells, often leading to cellular and organismal aging phenotypes. The interplay between growth and division also underpins cellular senescence (i.e., inability to divide) and quiescence, when cells exit the cell cycle but still retain their ability to divide.

Keywords Cell senescence • Quiescence • Aging • Cell growth • Protein translation • mTOR signaling • Asymmetric division • Hypertrophy and Cdk

8.1 Introduction

The two cells generated at the end of the cell cycle usually inherit sufficient amounts of essential constituents, ensuring their survival. Moreover, the composition of proliferating cells varies very little from generation to generation, implying that proliferating cells somehow balance their growth (increase in biomass) with their division. Since different levels of nutrients and growth factors sustain different rates of cell proliferation, cells have elaborate mechanisms to sense nutrient and growth signals, adjusting their metabolic and proliferative activity accordingly. A detailed

M. Polymenis (✉)
Department of Biochemistry and Biophysics, Texas A&M University,
2128 TAMU, College Station, TX 77845, USA
e-mail: polymenis@tamu.edu

B.K. Kennedy (✉)
The Buck Institute for Research on Aging, 8001 Redwood Blvd, Novato, CA 94945, USA
e-mail: bkennedy@buckinstitute.org

M. Gotta, P. Meraldi (eds.), *Cell Division Machinery and Disease*, Advances in Experimental Medicine and Biology 1002, DOI 10.1007/978-3-319-57127-0_8

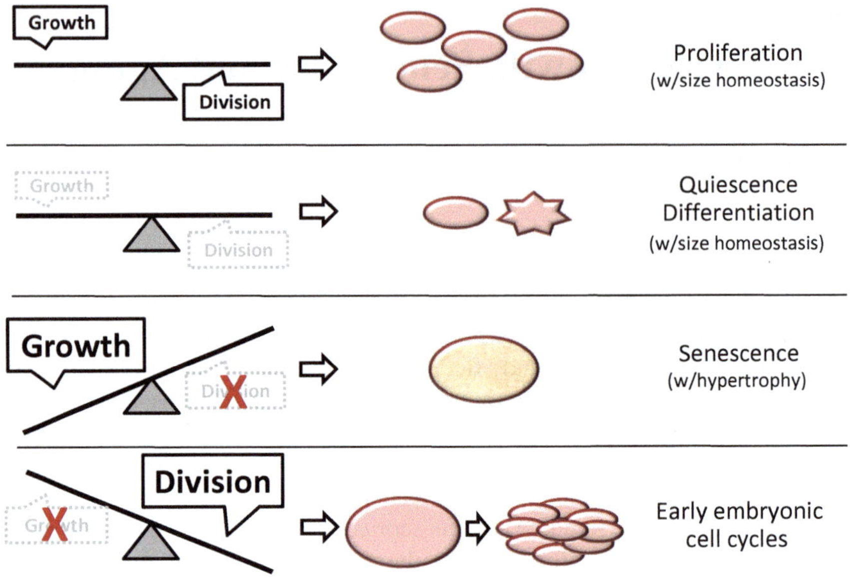

Fig. 8.1 Schematic representation of all possible outcomes when growth and division are balanced or unbalanced

mechanistic understanding of this coupling between growth and division has remained elusive. Nonetheless, properly coupling growth with division is thought to determine the rate at which cells proliferate [1–6].

Nutrient and growth factor limitations do not delay all cell cycle transitions uniformly. Instead, transit through some cell cycle phases is delayed disproportionately. Overwhelmingly, poor growth conditions prolong the G1 phase of the cell cycle, preceding initiation of DNA synthesis, while transit through the remaining cell cycle phases is not delayed significantly [1, 2, 7–12]. In yeast, the point of commitment to a new round of cell division is called START [2] and in animal cells the Restriction Point [6]. Once cells pass through these points in late G1 phase, they will initiate and complete their division even if they encounter growth limitations [1, 2, 6, 12]. Hence, although there may be some nutrient and growth factor inputs in later stages of the cell cycle [13, 14], it is in G1 that cells delay committing to a new round of cell division in the face of weak growth prospects.

The consequences of uncoupling growth from division are profound and accompanied by changes in the size of cells (see Fig. 8.1 for a schematic). In this Chapter, we describe possible outcomes when growth and division are not balanced. We also discuss models that envision imbalances between growth and division as a critical component of senescence and aging mechanisms [15, 16]. Our discussion will include examples from animal models systems and unicellular organisms, especially the budding yeast *S. cerevisiae*.

8.2 Growth and Division: A Tight Balancing Act

Intuitively, it makes sense that for a cell to successfully divide and give rise to two viable cells, it must produce enough macromolecules, membranes, and organelles for the two cells that will arise at the end of cytokinesis. Since these cellular components are determinants of the cell's volume, it is not surprising that cell size has often been used as an "umbrella" metric for cell growth [3, 5, 14, 17, 18]. Lately, there is renewed interest in the development of methodologies that report accurately and precisely on the size of animal cells [5, 19–21]. In addition, asymmetric segregation of cellular constituents between two products of a mitotic cell division is also tightly regulated and will be discussed below.

Although cell size is a very useful "growth" metric, the events most closely associated with cell proliferation are anabolic processes that yield the macromolecules necessary to build new cells. Among those macromolecules, proteins are often considered the most important component of growth, and for good reason. The protein fraction of dry mass is ≈55% in *E. coli* [22] and ≈40–45% in the budding yeast *S. cerevisiae* [22, 23]. The protein content of mammalian cells varies in different tissues. For commonly used cell lines, such as mouse fibroblasts (NIH3T3 cells) or human HeLa cells, the protein molecules per unit volume is roughly the same as in budding yeast cells (1-2E + 06 proteins/fL; [22]). Also, a significant fraction of the proteome (>20%) is dedicated to making ribosomal proteins and translation factors that will, in turn, promote the synthesis of more proteins [24].

Making ribosomal components and assembling them into functional ribosomes involves a broad repertoire of cellular constituents and processes [25–27]. In budding yeast, the cytoplasmic ribosome contains 78 ribosomal proteins encoded by the RP regulon of 138 genes. Note that 59 of the 78 yeast ribosomal proteins are encoded by pairs of very similar paralogs [28, 29]. The ribosomal proteins together with the four rRNAs (5S, 5.8S, 18S, and 25S) make up the ribosome. The rRNA genes are encoded by rDNA tandem repeats, whose number is dynamic (usually ≈100–200) and varies with growth conditions. Greater than 200 protein assembly and accessory factors are needed at many stages to put a functional ribosome together. Their expression is thought to be regulated coordinately, through the ribosome biogenesis (Ribi) regulon. In the Ribi regulon, one also finds the various tRNA synthetases, rRNA processing and modifying enzymes, and translation factors, which collectively control translational capacity [30, 31]. Most of the cell's transcriptional activity is devoted to building and maintaining the translational machinery. Of all the RNA in the cell, 80% is rRNA, 15% is tRNA, and 5% is mRNA, and a large fraction of mRNA is devoted to ribosome synthesis [25, 32].

Transcription of RP genes alone is responsible for approximately 50% of all RNA PolII-mediated transcription initiation events. The energetic cost of making the translation machinery is astounding, consuming as much as ≈90% of the total energy of fast-proliferating yeast cells [25]. Estimates of the ribosome content of cells give an even more impressive view of the centrality of ribosome biogenesis in governing the growth of cells. From super-resolution, single-molecule imaging techniques, it seems that *E. coli* cells contain 30,000–50,000 ribosomes per fL [33]. Analogous quantitative measurements are lacking in eukaryotes, but prior estimates in yeast put the number at about 200,000 ribosomes per cell [25]. On average then, during one cell cycle lasting ≈100 min, a yeast cell must produce ≈2000 ribosomes per minute. Based on these metrics of the cellular economy, one can easily see why for decades protein synthesis has been viewed as *the* fundamental measure of cell growth in considerations of balancing growth with cell division [34].

Building and maintaining the ability to synthesize proteins is such a costly process that would be expected to influence if, and when, cells commit to a new round of cell division. The earliest evidence for specific effects on the cell cycle due to translational control was the isolation of budding conditional yeast *cdc* (cell division cycle) mutants in what turned out to be translation factors [2]. Hypomorphic mutations in translation initiation factors impair the capacity of cells to initiate a new round of cell division [12, 35–40]. Moreover, signaling pathways that control initiation of division, such as the Target of Rapamycin (TOR) pathway, may do so, at least in part, by regulating translation initiation. Loss of TOR function causes G1 arrest in mammalian cells [41, 42] and yeast [43, 44]. Conversely, overexpression of translation initiation factor eIF4E in mammals is oncogenic [45], and the translational output of TOR signaling is critical for cancer initiation [46]. Moreover, inhibiting translation elongation with cycloheximide also prolongs the G1 phase of the cell cycle [12, 47]. In budding yeast, cycloheximide reduces the newborn cell size [12, 47] and the rate at which cells increase in size [48]. It also increases the critical size threshold for START [47, 48]. Together, these results support the notion that a critical rate of protein synthesis is required for G1 transit and completion of START in budding yeast [49] and animal cells [50, 51].

If ribosome biogenesis and protein synthesis are such integral parts of cell growth, propelling cells to divide, how do cells control ribosome biogenesis? In all eukaryotes, the principal regulator of catabolic processes leading to energy production is protein kinase A (PKA), while the analogous "master" regulator of anabolic, biosynthetic processes is the TOR kinase (the TORC1 complex). As we will describe in subsequent sections, these two pathways have overlapping roles as determinants of cellular and organismal replicative potential and aging. In rapidly proliferating cells, however, for maximal growth and rates of cell division, both PKA and TORC1 are active, and they are both needed to activate ribosome biogenesis fully by derepressing the Ribi regulon [44, 52, 53].

In yeast, replicative aging is defined by the number of buds that can be produced by one mother cell, indicative of the number of times it can progress through the cell cycle and undergo mitosis [54]. An increase in cell size (i.e., cellular hypertrophy) has been linked to replicative aging. The cellular hypertrophy model of aging was formulated to account for the yeast replicative aging, invoking the existence of a maximal cell size beyond which cells could not maintain division [55, 56]. According to the cellular hypertrophy model, the large cell size of old yeast mother cells is incompatible with some cellular function that is necessary for cell division. Every time mother cells divide, they increase in size until they reach the terminal, large size, at which point they will enter a state of proliferative arrest. The hypertrophy model predicts that small cells would be able to divide more times before reaching the terminal size, resulting in longer replicative lifespan. On the other hand, large cells will reach the terminal size after fewer divisions, having a shorter replicative lifespan. Overall, the model predicted that changes in replicative lifespan and cell size ought to be proportional. However, based on genome-wide measurements of cell size, we recently reported that the mean cell size of long-lived yeast mutants was not significantly different from the size of mutants that were not long-lived [57]. This finding is incompatible with the key prediction of the hypertrophy model that long-lived mutants would have a small overall cell size, allowing these cells to divide more times until they reach the terminal size and enter senescence. Therefore, based on these experiments in yeast, it seems that the cellular enlargement is not a primary determinant of aging.

The factors linked to cell growth and protein translation definitely affect aging, however. Reduced mTOR signaling leads to lifespan extension is yeast [58, 59], worms [60], flies [61], mice [62, 63], and initial studies suggest possibly humans [64, 65]. The mechanisms that underlie lifespan extension remain to be fully determined, partly due to the complexity of these signaling pathways, but evidence exists both for altered protein synthesis and enhanced protein turnover through autophagy [66]. The former comes from findings that reduced expression of several translation initiation factors and ribosomal components also lead to lifespan extension in a range of organisms [67]. It is clear at least for replicative lifespan in yeast (the number of times one cell can divide to produce daughter cells), however, that globally and uniformly inhibiting protein synthesis in insufficient to slow aging since cycloheximide is unable to mediate this effect [68]. This finding indicates that translational changes to specific mRNAs are likely conferring, at least in part, longevity phenotypes in mTORC1 and translation factor mutants.

Interestingly, one downstream factor linked to lifespan extension is *GCN4*, a transcription factor regulated itself by translation and dependent on the presence of small upstream open reading frames in its mRNA [68, 69]. Inhibition of mTORC1, reduced 60S ribosomal subunit levels and calorie restriction all lead to enhanced *GCN4* translation [70–74], which in yeast induces expression of stress and nutrient response pathways [69]. Loss of *GCN4* at least partially abrogates lifespan extension by these interventions. The mammalian ortholog of *GCN4*, *ATF4*, is also induced in cells and mice from a range of interventions conferring long lifespan [75, 76], arguing for conservation of this pathway.

Reduced mTORC1 signaling also leads to enhanced autophagy, which has been linked to lifespan extension as well in a range of conditions [77]. In non-vertebrates, for instance, inhibition of autophagy is sufficient to block lifespan extension by mutants that affect translation [78–81]. Moreover, at least in flies and mice, induction of autophagy through overexpression of autophagy components is reported to lead to lifespan, and sometimes healthspan, extension [82–85].

The mTORC1 complex regulates transcription as well and this phenomenon has been studied extensively in yeast. For chronological aging, defined as survival in a post-replicative state, reduced mTORC1 signaling enhances longevity [59, 86], likely through mechanisms leading to enhanced transcription of a stress response transcription factor network driven by Msn2/4 and Gis1 [54]. Interestingly, reduced PKA signaling promotes chronological lifespan extension through overlapping mechanisms [87, 88].

Protein kinase A signaling has been connected to aging in multiple organisms [89]. In yeast, under maximal growth conditions in rich media, where both PKA and mTOR collaborate to drive protein synthesis, mutations leading to reduced PKA activity promote replicative and chronological lifespan extension [90, 91]. Although not studied as extensively in multi-cellular organisms, this phenomenon may also be conserved as mice lacking the protein kinase subunit RIIβ are long-lived [92, 93], as well as those lacking *ADCY5*, encoding type 5-adenylyl cyclase (AC5) that converts ATP into cAMP in turn activating PKA. These mice are stress resistant and experience a 30% increase in median lifespan [94, 95]. In addition, genetic variants leading to reduced production of the adenylyl cyclase-activating β2-adrenergic receptor, are prevalent in men from a Chinese centenarian population [95].

In conclusion, in balanced growth and division, the increased ribosome biogenesis and protein synthesis is coupled to cell division, maintaining the overall cellular homeostasis and macromolecular composition. These processes also have robust effects on aging, although the links are far from straightforward.

8.3 Asymmetric Segregation During Cell Division

When cells divide, their cellular constituents have to be divided between the two offspring and studies have started to address mechanisms underlying this partitioning. In mammalian cells, for instance primary fibroblasts in culture, division is symmetric and while partitioning may occur, it is hard to distinguish morphologically. Interestingly the culture of primary fibroblasts senescences at a similar number of population doublings, suggesting that with respect to cellular aging damaged molecules may not be partitioned specifically to one cell after division.

Yeast, being a single-celled organism, has to maintain continuous division in the colony in the face of the challenges of aging. By virtue of their division by budding, which produces a larger mother cell and a smaller bud that are easily distinguishable, yeast offers a great opportunity to detect differential segregation of cellular materials. While the mother cell ages, the daughter remains youthful, suggesting that damaged cellular constituents driving aging may remain in the mother cell [96]. Some components of the daughter cell, such as the cell wall are largely the result of new synthesis during division, representing one method of segregating old material to mothers. However, many cytoplasmic factors partition and several aging factors are reported to remain in mothers. For instance, extrachromosomal rDNA circles (ERCs), small episomes containing rDNA repeats that drive aging possibly by competing for replication factors with chromosomal origins [97], are heavily partitioned toward mothers [98]. The mechanism likely relates to the closed mitosis of yeast, which maintain a nuclear structure. Originally nuclear pore association was proposed as a mechanism by which ERCs were retained in the mother, with the assertion that nuclear pores have very restricted access to daughters [99]. Later reports called that into question [100], and suggested that ERCs may simply not diffuse efficiency through the bud neck [101–103]. Thus geometry drives asymmetry.

Damaged aggregated proteins are also retained in the mother cells, likely through restriction of access to daughters and also by active transport of damaged molecules from daughters to back to mothers [96]. The former process again has been reported to involved both active retention and limits to passive diffusion to daughters, while the latter likely involves the actin cytoskeleton and requires Sir2, a protein deacetylase linked to aging [104]. The mechanisms underlying these processes remain to be fully elaborated.

Mitochondria are reported to undergo asymmetric inheritance in yeast, with fitter mitochondria finding their way to daughters [105]. More work is required to access whether and how partition occurs in this and other organelles. In fact, cellular processes to maintain asymmetry may be broader that we currently appreciate as recent single cell based screens have identified hundreds of asymmetrically partitioned proteins during budding. One screen identified 74 proteins partitioned to mothers and 60 to daughters [106]. Interestingly, strains lacking genes for the mother-specific proteins are more likely to have an enhanced lifespan. Whether it comes to individual proteins, damaged protein aggregates, extrachromosomal rDNA circles or organelles, evidence suggests that asymmetry breaks down with the age of the mother and this is consistent with observations that daughters from old mothers do not enjoy a full replicative lifespan.

Of course, asymmetry in cell division has massive impacts during development and cell differentiation throughout the mammalian organism. A classic example is an adult stem cell that divides to produce another stem cell and a cell committed to a differentiation pathway. Asymmetry of cellular constituents plays a role in defining cell fate in this context and it is highly likely that damaged molecules are partitioned to the more committed cell [107]. Clearly cell- autonomous and –non autonomous mechanisms are in play and it will be intriguing to determine to what extent the more elaborated mechanisms in yeast are conserved with the cell autonomous mechanisms.

8.4 Quiescence: Not Dividing, but Keep on Ticking

Cells can enter a quiescent state, in response to a range of signals, in which they do not divide, but maintain a metabolically active state and can resume the capacity to divide, when conditions permit. When cells adopt a differentiated state they exit the cell cycle, sometimes permanently [108]. In all eukaryotes, cyclin-dependent kinase (Cdk) protein complexes are at the core of the cell division machinery [109]. Initiation of cell division requires an increase in Cdk activity. Later cell cycle transitions also need high Cdk activity, while a drop in Cdk activity triggers exit from mitosis. Cdks are Ser/Thr protein kinases, similar in structure to most kinases [110]. However, all Cdks are active only when they bind other activating proteins, such as cyclins. Cdk activity is further regulated by phosphorylation or binding of additional protein subunits. These layers of control can raise or lower overall Cdk activity, depending on the phosphorylated Cdk residue, or the interacting protein, in each case.

Changes in Cdk activity underlie transitions from resting to proliferative cellular states in health and disease. Indeed, high Cdk activity contributes to most proliferative disorders, including cancer cell development [111, 112]. On the other hand, low Cdk activity is associated with terminal differentiation [113], and accompanies poor organ regeneration, for example, in hepatic [114], cardiac [115], neuronal [116], or appendage tissues [117].

It is clear that in quiescent cells there is a strong albeit potentially reversible block in cell division. Maintaining the *potential* to divide, however, is a key feature that distinguishes quiescent from senescent cells. This concept was put to the test almost two decades ago, in a particularly lucid experiment. Microinjection of preformed active Cdk protein complexes was sufficient to initiate cell division in quiescent human fibroblasts, in the absence of growth factors [118].

In quiescent cells, the block in cell division is also accompanied by a profound reprogramming of cellular metabolism. The cells remain metabolically active, enabling them to stay alive (e.g., quiescent yeast cells) or perform the functions prescribed by their terminally differentiated state. Interestingly, balanced downregulation of the master "growth" signaling pathways we described above, the PKA and the TOR pathways, is observed in quiescent yeast cells [119], and this is important for chronological lifespan extension [119], which is the period of time a cell can remain viable in a non-proliferative state.

More recently, it was reported that quiescent cells have a massively re-organized chromatin structure [120]. In yeast cells entering quiescence, the conserved lysine deacetylase Rpd3p establishes a repressive transcriptional state, reducing by $\approx$30-fold steady-state mRNA levels [120]. Cells lacking Rpd3p also have a 2–3 fold reduction in their mean chronological lifespan [120]. The replicative lifespan of these cells, however, is not affected [121]. This is not surprising since there is no significant overlap of gene deletions that extend lifespan in both the chronological and replicative lifespan assays [122], at least under the assay conditions tested.

Interestingly, however, there are connections between the two types of yeast aging, as chronologically aged cells have reduced replicative lifespan when returned to the cell cycle [123–125]. This is clearly linked to metabolic state, as dietary restriction during the replicative phase of this experiment results in suppression of the short lifespan [126]. Quiescent cells certainly accumulate damage, but once a quiescent cell reenters the cell cycle, this damage may stay with the mother cell [127]. In that scenario, the proliferative capacity and fitness of the *population* as a whole would be maintained. While this nice model needs further testing, what is clear is that growth and division are still balanced in the quiescent state, and homeostasis is maintained (see Fig. 8.1, the second case from top).

In quiescence, the down-regulation of TOR and PKA leads to significantly reduced ribosome biogenesis and overall protein synthesis [128]. Cell growth and metabolic activity is generally low in quiescent cells [129]. But because this is happening in the context of cell cycle arrest [129, 130], the general properties and macromolecular composition of quiescent cells remain stable and they are easily recognized [16, 128]. Overall, quiescence likely represents a physiological extreme in the normal range of balancing growth with division, a case where both growth and division are coordinately downregulated.

8.5 Senescence: Growing Desperately, with No Possibility of Ever Dividing Again

It has become clearer in recent years that cell cycle arrest can come in different flavors, especially in the context of unabated cell growth. If a cell continues to make proteins and other macromolecules at a high rate in the face of a cell cycle block, then there are only a few possible outcomes. (1) The cell must find ways to get rid of the large excess (e.g., lysosomal degradation, secretion). (2) The cell must somehow accommodate the extra macromolecular amounts within its boundaries, inevitably leading to increased cell size. In fact, the above are typical properties of senescent cells [128, 131] (described in more detail below) and exemplify a clear case of unbalanced growth and division (see Fig. 8.1, the third case from top). The strong growth of senescent cells (often the result of oncogenic stimulation), is not balanced with cell division. Instead, it persists in the face of stable cell cycle blocks.

An important component in the cell cycle arrest of quiescent and senescent cells is the accumulation of Cdk inhibitor molecules. The kinds of Cdk inhibitors employed in each case, however, are different. The cell cycle arrest of quiescent or fully differentiated cells is usually imposed by members of the $p27^{KIP1}$ family of Cdk inhibitors, which inhibit multiple cyclin/Cdk complexes by interacting with both the cyclin and the Cdk. On the other hand, in senescent cells there is a buildup of $p16^{INK4}$ Cdk inhibitors, which bind to monomeric Cdk4/6 and reduce cyclin binding affinity [128, 131].

Likewise, while both in quiescent and senescent cells there is an accumulation of tumor suppressors that broadly inhibit transcription associated with entry into the cell cycle, the molecular players are different in each case. Quiescence is associated with the pRB-like proteins p107 and p130, which interact with the transcription factor E2F during G1 phase to inhibit G1/S transcription and commitment to division. Instead, senescent cells have high levels of pRB, and there is also a buildup of p53, a regulator of multiple processes (e.g., DNA damage response) that impinge on the cell cycle [128, 131]. Hence, the molecular effectors of the cell cycle arrest are different. Furthermore, the exit point of the cell cycle may be different in quiescent vs. senescent cells. Quiescent cells uniformly exit the cell cycle before initiation of DNA replication in G1 phase [128, 131]. G1 arrest is also common in senescence. Surprisingly, however, in several cases senescent cells appear to have a permanent G2 phase block in later stages of the cell cycle [128, 132–135].

The unbalanced growth and division observed in senescence is associated with a variety of phenotypes typical of extremely stressed cells. The exact signatures are still a matter of debate [128]. In addition to the cell cycle markers we mentioned above, other traits associated with senescence often include: short or dysfunctional telomeres, lysosomal stress and expression of β-galactosidase, DNA damage response, stress granule formation, hyper-secretory functions, formation of heterochromatic foci, and the senescence-associated secretory phenotype (SASP) [128, 131, 136]. Overall, senescent cells have been aptly compared to automobiles that simultaneously attempt to accelerate (i.e., hyperactive growth pathways) and stop (i.e., strong, permanent cell cycle block), putting the cell on its way to a highly stressed, irreversibly aged state [15, 16, 137].

The phenomenon of cell senescence was discovered more than 50 years ago and it was almost immediately hypothesized to be associated with organismal aging [138]. While it has been clearly established that cell senescence serves as a tumor suppressive mechanism [131], support for the aging theory has waxed and waned over the years. Currently, it is buoyed by a series of recent studies linking cell senescence to aging in mice.

A principle argument against a role for cell senescence in aging has been that even in old individuals, only a small fraction of cells within a tissue appear to be senescent. How could a phenomenon affecting only a small percentage of cells seriously impair an entire tissue? This question has been potentially resolved with the discovery and characterization of the SASP, wherein senescent cells secrete a novel panel of factors in part comprised of inflammatory cytokines that can have potent paracrine and endocrine effects on non-senescent cells [139, 140]. Moreover, a better understanding has emerged regarding the events that can drive cellular senescence. These now include a wide range of cellular stresses [131], which are associated with chronic diseases of aging, suggesting that aging events may drive cell senescence that in turn promote increased aging.

Senescent cells do accumulate with aging and the Cdk inhibitor, $p16^{INK4}$, has been proposed as a biomarker of aging [128]. Indeed, in selected T cell populations, $p16^{INK4}$ levels do show a statistically significant predictive value for human age. In addition, panels of inflammatory cytokines have been proposed as aging biomarkers and these may be at least in part related to the SASP. Several recent studies have reported mechanistic insights into SASP induction in senescent cells. Several pathways appear to be involved, including those related to cell growth, such as mTOR, and cell proliferation, such as p53. Rapamycin suppresses aspects of the SASP, but must be delivered continuously to have this effect [141]. This is in contrast to organismal aging, where a transient three-month exposure to rapamycin in middle age is sufficient to extend the lifespan of mice [142, 143].

To test the role of senescence in aging, two different strategies were employed to conditionally ablate senescent cells, both related to the specificity of $p16^{INK4}$ expression in this cellular condition. Findings in these mice appear promising as ablation of senescent cells is linked to partial suppression of pathology in a mouse progeria model, the BubR1 mice [144], and can extend the lifespan and some healthspan parameters in normal mice [145].

The connection between BubR1 and progeria is interesting in its own right as the gene encodes a component of the mitotic spindle assembly checkpoint, which prevents cells from initiating anaphase if one or more kinetochores are not attached to the mitotic spindle [146]. Mice hypomorphic for BubR1 rapidly develop aging features, including kyphosis, cachexia, and cataracts [147, 148]. They also have a severely reduced lifespan. With age, BubR1 expression declines in a number of tissues, suggesting that reduced expression of the protein late in life may contribute to normal aging [147]. Moreover, overexpression of BubR1 delays aspects of aging [149]. A potential unifying model is that reduced BubR1 expression leads to mitotic defects, driving cell senescence and that the senescent cells drive aging phenotypes through the SASP or other mechanisms [144]. Ablation of senescent cells improves a range of healthspan parameters.

The promise of research in senescence has led to drug discovery approaches designed to specifically kill senescent cells. Several candidates have already emerged, and these compounds have shown efficacy in preclinical models of chronic disease states [150–153]. While the clinical work remains to be done, the last 10 years have seen cell senescence emerge as one pathway likely to modulate organismal aging and many new pathways of therapeutics for age-associated diseases.

8.6 Division Without Growth

In the classic experiments by Hartwell and colleagues, it was established that in most cases growth controls cell division and not the other way around [1, 2, 12]. Stopping cell growth will also stop cell division, but stopping cell division does not usually stop cell growth (as displayed in senescent cells, see discussion in the previous section). From these principles, it follows then that cell division in the absence of growth is untenable, at least when the mass of the daughter cells is reduced below a threshold necessary to sustain their viability. This is precisely what happens during the early embryonic cell cycles after fertilization until the mid-blastula transition ([154]; see Fig. 8.1, last case). At the mid-blastula transition, before the re-establishment of the normal somatic cell cycles, the block in cell division is imposed by Cdk inhibitors. In mutants lacking these inhibitors, cells usually undergo just one extra division [154–156]. Overall, these early embryonic cell cycles do not necessarily violate the fundamental need to balance growth with division. They just reflect the fact that growth needs have been satisfied during oogenesis.

8.7 Outlook

In yeast, it is implicit that aging, both replicative and chronological, must be linked to critical cell cycle decisions. Balancing cell growth with division to maintain cellular homeostasis is a critical component of this process, whether cells are dividing or in a non-proliferative state. The key pathways that coordinate cell growth signals are intimately linked to aging in yeast, and considerable evidence suggests that they have conserved effects on aging in multicellular organisms. Therefore, continued efforts to understand yeast aging in the context of cell growth and division are likely to continue to inform about human aging. A major challenge now in yeast is to understand aging at the systems level, taking a holistic approach to integrate the contributions of different aging mechanisms and pathways in order to model the aging condition. This approach involves combining large-scale studies, including transcriptomics and epistasis network analysis, with directed studies with the goal of establishing as complete as possible a picture of single cell aging that can set the stage for similar studies in multicellular organisms.

In the multi-cellular context, a major challenge has been to understand the links between aging at the level of the organism and (causal?) changes to cells in the aging body. In that context, cell senescence has emerged as a major candidate driver of aging processes. Major insights in this arena have led to the identification of candidate therapeutics to kill senescent cells as means of offsetting or treating age-related chronic diseases. The next few years will help define the merits of this new therapeutic route based on aging studies.

More broadly, aging is linked to several pathways involved in cell growth and specifically in protein synthesis and turnover. It is clear for instance that reduced mTORC1 signaling leads to lifespan extension, but further work needs to be done to identify whether aging benefits come from altered protein translation, increased turnover of damaged macromolecules, suppression of the SASP, or some other mechanism. Moreover, it is important to identify in what tissues reduced mTOR signaling, and other pathways such as PKA, promotes longevity. With dramatic increases in the aging population and new insights from research on aging and longevity, the promise is there for major new advances that could refocus medical care toward interventions that slow aging and keep people healthy longer. Understanding links between cell growth, division and aging are integral to achieving this goal.

References

1. Johnston GC, Pringle JR, Hartwell LH (1977) Coordination of growth with cell division in the yeast *Saccharomyces cerevisiae*. Exp Cell Res 105(1):79–98
2. Pringle JR, Hartwell LH (1981) The *Saccharomyces cerevisiae* cell cycle. In: The molecular and cellular biology of the yeast saccharomyces, vol 1. Cold Spring Harbor Laboratory Press, New York, pp 97–142
3. Turner JJ, Ewald JC, Skotheim JM (2012) Cell size control in yeast. Curr Biol 22(9):R350–R359. doi:10.1016/j.cub.2012.02.041
4. Jorgensen P, Tyers M (2004) How cells coordinate growth and division. Curr Biol 14(23):R1014–R1027. doi:10.1016/j.cub.2004.11.027
5. Ginzberg MB, Kafri R, Kirschner M (2015) Cell biology. On being the right (cell) size. Science 348(6236):1245075. doi:10.1126/science.1245075
6. Blagosklonny MV, Pardee AB (2002) The restriction point of the cell cycle. Cell Cycle 1(2):103–110

7. Carter BL, Jagadish MN (1978) Control of cell division in the yeast *Saccharomyces cerevisiae* cultured at different growth rates. Exp Cell Res 112(2):373–383
8. Jagadish MN, Carter BL (1977) Genetic control of cell division in yeast cultured at different growth rates. Nature 269(5624):145–147
9. Jagadish M, Carter B (1978) Effects of temperature and nutritional conditions on the mitotic cell cycle of *Saccharomyces cerevisiae*. J Cell Sci 31(1):71–78
10. Johnston GC, Singer RA, McFarlane S (1977) Growth and cell division during nitrogen starvation of the yeast *Saccharomyces cerevisiae*. J Bacteriol 132(2):723–730
11. Lillie SH, Pringle JR (1980) Reserve carbohydrate metabolism in *Saccharomyces cerevisiae*: responses to nutrient limitation. J Bacteriol 143(3):1384–1394
12. Hartwell LH, Unger MW (1977) Unequal division in *Saccharomyces cerevisiae* and its implications for the control of cell division. J Cell Biol 75(2 Pt 1):422–435
13. Messier V, Zenklusen D, Michnick SW (2013) A nutrient-responsive pathway that determines M phase timing through control of B-cyclin mRNA stability. Cell 153(5):1080–1093
14. Wood E, Nurse P (2015) Sizing up to divide: mitotic cell-size control in fission yeast. Annu Rev Cell Dev Biol 31:11–29. doi:10.1146/annurev-cellbio-100814-125601
15. Blagosklonny MV (2012) Answering the ultimate question "what is the proximal cause of aging?". Aging 4(12):861–877
16. Blagosklonny MV (2014) Geroconversion: irreversible step to cellular senescence. Cell Cycle 13(23):3628–3635. doi:10.4161/15384101.2014.985507
17. Schmoller KM, Skotheim JM (2015) The biosynthetic basis of cell size control. Trends Cell Biol 25(12):793–802. doi:10.1016/j.tcb.2015.10.006
18. Johnson A, Skotheim JM (2013) Start and the restriction point. Curr Opin Cell Biol. doi:10.1016/j.ceb.2013.07.010
19. Son S, Tzur A, Weng Y, Jorgensen P, Kim J, Kirschner MW, Manalis SR (2012) Direct observation of mammalian cell growth and size regulation. Nat Methods 9(9):910–912
20. Son S, Kang JH, Oh S, Kirschner MW, Mitchison TJ, Manalis S (2015) Resonant microchannel volume and mass measurements show that suspended cells swell during mitosis. J Cell Biol 211(4):757–763. doi:10.1083/jcb.201505058
21. Godin M, Delgado FF, Son S, Grover WH, Bryan AK, Tzur A, Jorgensen P, Payer K, Grossman AD, Kirschner MW (2010) Using buoyant mass to measure the growth of single cells. Nat Methods 7(5):387–390
22. Milo R (2013) What is the total number of protein molecules per cell volume? A call to rethink some published values. Bioessays 35(12):1050–1055. doi:10.1002/bies.201300066
23. Lange HC, Heijnen JJ (2001) Statistical reconciliation of the elemental and molecular biomass composition of *Saccharomyces cerevisiae*. Biotechnol Bioeng 75(3):334–344
24. Liebermeister W, Noor E, Flamholz A, Davidi D, Bernhardt J, Milo R (2014) Visual account of protein investment in cellular functions. Proc Natl Acad Sci U S A 111(23):8488–8493. doi:10.1073/pnas.1314810111
25. Warner JR (1999) The economics of ribosome biosynthesis in yeast. Trends Biochem Sci 24(11):437–440
26. Fatica A, Tollervey D (2002) Making ribosomes. Curr Opin Cell Biol 14(3):313–318
27. Venema J, Tollervey D (1999) Ribosome synthesis in *Saccharomyces cerevisiae*. Annu Rev Genet 33(1):261–311
28. Wapinski I, Pfiffner J, French C, Socha A, Thompson DA, Regev A (2010) Gene duplication and the evolution of ribosomal protein gene regulation in yeast. Proc Natl Acad Sci 107(12):5505–5510
29. Planta RJ, Mager WH (1998) The list of cytoplasmic ribosomal proteins of *Saccharomyces cerevisiae*. Yeast 14(5):471–477. doi:10.1002/(sici)1097-0061(19980330)14:5<471::aid-yea241>3.0.co;2-u
30. Wade CH, Umbarger MA, McAlear MA (2006) The budding yeast rRNA and ribosome biosynthesis (RRB) regulon contains over 200 genes. Yeast 23(4):293–306. doi:10.1002/yea.1353

31. Jorgensen P, Nishikawa JL, Breitkreutz BJ, Tyers M (2002) Systematic identification of pathways that couple cell growth and division in yeast. Science 297(5580):395–400. doi:10.1126/science.1070850
32. Velculescu VE, Zhang L, Zhou W, Vogelstein J, Basrai MA, Bassett DE, Hieter P, Vogelstein B, Kinzler KW (1997) Characterization of the yeast transcriptome. Cell 88(2):243–251
33. Bakshi S, Siryaporn A, Goulian M, Weisshaar JC (2012) Superresolution imaging of ribosomes and RNA polymerase in live *Escherichia coli* cells. Mol Microbiol 85(1):21–38. doi:10.1111/j.1365-2958.2012.08081.x
34. Mitchison JM (1971) Cell growth and protein synthesis. In: The biology of the cell cycle. Cambridge University Press, New York, p 129
35. Brenner C, Nakayama N, Goebl M, Tanaka K, Toh-e A, Matsumoto K (1988) CDC33 encodes mRNA cap-binding protein eIF-4E of *Saccharomyces cerevisiae*. Mol Cell Biol 8(8):3556–3559
36. Hanic-Joyce PJ, Johnston GC, Singer RA (1987) Regulated arrest of cell proliferation mediated by yeast prt1 mutations. Exp Cell Res 172(1):134–145
37. Bedard DP, Johnston GC, Singer RA (1981) New mutations in the yeast *Saccharomyces cerevisiae* affecting completion of "start". Curr Genet 4(3):205–214. doi:10.1007/BF00420500
38. Unger MW, Hartwell LH (1976) Control of cell division in *Saccharomyces cerevisiae* by methionyl-tRNA. Proc Natl Acad Sci U S A 73(5):1664–1668
39. Yu L, Pena Castillo L, Mnaimneh S, Hughes TR, Brown GW (2006) A survey of essential gene function in the yeast cell division cycle. Mol Biol Cell 17(11):4736–4747. doi:10.1091/mbc.E06-04-0368
40. Polymenis M, Aramayo R (2015) Translate to divide: control of the cell cycle by protein synthesis. Microbial Cell 2(4):94–104
41. Wicker LS, Boltz RC, Matt V, Nichols EA, Peterson LB, Sigal NH (1990) Suppression of B cell activation by cyclosporin a, FK506 and rapamycin. Eur J Immunol 20(10):2277–2283
42. Zoncu R, Efeyan A, Sabatini DM (2010) mTOR: from growth signal integration to cancer, diabetes and ageing. Nat Rev Mol Cell Biol 12(1):21–35
43. Heitman J, Movva NR, Hall MN (1991) Targets for cell cycle arrest by the immunosuppressant rapamycin in yeast. Science 253(5022):905–909
44. Loewith R, Hall MN (2011) Target of rapamycin (TOR) in nutrient signaling and growth control. Genetics 189(4):1177–1201. doi:10.1534/genetics.111.133363
45. Lazaris-Karatzas A, Montine KS, Sonenberg N (1990) Malignant transformation by a eukaryotic initiation factor subunit that binds to mRNA 5′ cap. Nature 345(6275):544–547. doi:10.1038/345544a0
46. Hsieh AC, Liu Y, Edlind MP, Ingolia NT, Janes MR, Sher A, Shi EY, Stumpf CR, Christensen C, Bonham MJ, Wang S, Ren P, Martin M, Jessen K, Feldman ME, Weissman JS, Shokat KM, Rommel C, Ruggero D (2012) The translational landscape of mTOR signalling steers cancer initiation and metastasis. Nature 485(7396):55–61. doi:10.1038/nature10912
47. Popolo L, Vanoni M, Alberghina L (1982) Control of the yeast cell cycle by protein synthesis. Exp Cell Res 142(1):69–78
48. Jorgensen P, Rupes I, Sharom JR, Schneper L, Broach JR, Tyers M (2004) A dynamic transcriptional network communicates growth potential to ribosome synthesis and critical cell size. Genes Dev 18(20):2491–2505. doi:10.1101/gad.1228804
49. Moore SA (1988) Kinetic evidence for a critical rate of protein synthesis in the *Saccharomyces cerevisiae* yeast cell cycle. J Biol Chem 263(20):9674–9681
50. Rossow PW, Riddle VG, Pardee AB (1979) Synthesis of labile, serum-dependent protein in early G1 controls animal cell growth. Proc Natl Acad Sci U S A 76(9):4446–4450
51. Brooks R (1977) Continuous protein synthesis is required to maintain the probability of entry into S phase. Cell 12(1):311–317
52. Lippman SI, Broach JR (2009) Protein kinase a and TORC1 activate genes for ribosomal biogenesis by inactivating repressors encoded by Dot6 and its homolog Tod6. Proc Natl Acad Sci 106(47):19928–19933. doi:10.1073/pnas.0907027106

53. Broach JR (2012) Nutritional control of growth and development in yeast. Genetics 192(1):73–105. doi:10.1534/genetics.111.135731
54. Longo VD, Shadel GS, Kaeberlein M, Kennedy B (2012) Replicative and chronological aging in *Saccharomyces cerevisiae*. Cell Metab 16(1):18–31. doi:10.1016/j.cmet.2012.06.002
55. Bilinski T, Zadrag-Tecza R, Bartosz G (2012) Hypertrophy hypothesis as an alternative explanation of the phenomenon of replicative aging of yeast. FEMS Yeast Res 12(1):97–101. doi:10.1111/j.1567-1364.2011.00759.x
56. Zadrag R, Kwolek-Mirek M, Bartosz G, Bilinski T (2006) Relationship between the replicative age and cell volume in *Saccharomyces cerevisiae*. Acta Biochim Pol 53(4):747–751
57. He C, Tsuchiyama SK, Nguyen QT, Plyusnina EN, Terrill SR, Sahibzada S, Patel B, Faulkner AR, Shaposhnikov MV, Tian R, Tsuchiya M, Kaeberlein M, Moskalev AA, Kennedy BK, Polymenis M (2014) Enhanced longevity by ibuprofen, conserved in multiple species, occurs in yeast through inhibition of tryptophan import. PLoS Genet 10(12):e1004860. doi:10.1371/journal.pgen.1004860
58. Kaeberlein M, Powers RW 3rd, Steffen KK, Westman EA, Hu D, Dang N, Kerr EO, Kirkland KT, Fields S, Kennedy BK (2005) Regulation of yeast replicative life span by TOR and Sch9 in response to nutrients. Science 310(5751):1193–1196. doi:10.1126/science.1115535
59. Powers RW 3rd, Kaeberlein M, Caldwell SD, Kennedy BK, Fields S (2006) Extension of chronological life span in yeast by decreased TOR pathway signaling. Genes Dev 20(2):174–184
60. Vellai T, Takacs-Vellai K, Zhang Y, Kovacs AL, Orosz L, Muller F (2003) Genetics: influence of TOR kinase on lifespan in *C. elegans*. Nature 426:620
61. Kapahi P, Zid BM, Harper T, Koslover D, Sapin V, Benzer S (2004) Regulation of lifespan in *Drosophila* by modulation of genes in the TOR signaling pathway. Curr Biol 14:885–890
62. Harrison DE, Strong R, Sharp ZD, Nelson JF, Astle CM, Flurkey K, Nadon NL, Wilkinson JE, Frenkel K, Carter CS, Pahor M, Javors MA, Fernandez E, Miller RA (2009) Rapamycin fed late in life extends lifespan in genetically heterogeneous mice. Nature 460(7253):392–395. doi:10.1038/nature08221
63. Chen C, Liu Y, Zheng P (2009) mTOR regulation and therapeutic rejuvenation of aging hematopoietic stem cells. Sci Signal 2(98):ra75. doi:10.1126/scisignal.2000559
64. Mannick JB, Del Giudice G, Lattanzi M, Valiante NM, Praestgaard J, Huang B, Lonetto MA, Maecker HT, Kovarik J, Carson S, Glass DJ, Klickstein LB (2014) mTOR inhibition improves immune function in the elderly. Sci Transl Med 6(268):268ra179. doi:10.1126/scitranslmed.3009892
65. Passtoors WM, Beekman M, Deelen J, van der Breggen R, Maier AB, Guigas B, Derhovanessian E, van Heemst D, de Craen AJ, Gunn DA, Pawelec G, Slagboom PE (2013) Gene expression analysis of mTOR pathway: association with human longevity. Aging Cell 12(1):24–31. doi:10.1111/acel.12015
66. Kennedy BK, Lamming DW (2016) The mechanistic target of rapamycin: the grand conductor of metabolism and aging. Cell Metab 23(6):990–1003. doi:10.1016/j.cmet.2016.05.009
67. Steffen KK, Dillin A (2016) A ribosomal perspective on proteostasis and aging. Cell Metab 23(6):1004–1012. doi:10.1016/j.cmet.2016.05.013
68. Steffen KK, MacKay VL, Kerr EO, Tsuchiya M, Hu D, Fox LA, Dang N, Johnston ED, Oakes JA, Tchao BN, Pak DN, Fields S, Kennedy BK, Kaeberlein M (2008) Yeast life span extension by depletion of 60s ribosomal subunits is mediated by Gcn4. Cell 133(2):292–302. doi:10.1016/j.cell.2008.02.037
69. Hinnebusch AG (2005) Translational regulation of GCN4 and the general amino acid control of yeast. Annu Rev Microbiol 59:407–450
70. Foiani M, Cigan AM, Paddon CJ, Harashima S, Hinnebusch AG (1991) GCD2, a translational repressor of the GCN4 gene, has a general function in the initiation of protein synthesis in *Saccharomyces cerevisiae*. Mol Cell Biol 11(6):3203–3216
71. Cherkasova VA, Hinnebusch AG (2003) Translational control by TOR and TAP42 through dephosphorylation of eIF2alpha kinase GCN2. Genes Dev 17(7):859–872. doi:10.1101/gad.1069003

72. Kubota H, Obata T, Ota K, Sasaki T, Ito T (2003) Rapamycin-induced translational derepression of GCN4 mRNA involves a novel mechanism for activation of the eIF2 alpha kinase GCN2. J Biol Chem 278(23):20457–20460
73. Valenzuela L, Aranda C, Gonzalez A (2001) TOR modulates GCN4-dependent expression of genes turned on by nitrogen limitation. J Bacteriol 183(7):2331–2334. doi:10.1128/JB.183.7.2331-2334.2001
74. Martin-Marcos P, Hinnebusch AG, Tamame M (2007) Ribosomal protein L33 is required for ribosome biogenesis, subunit joining and repression of GCN4 translation. Mol Cell Biol. 27(17):5968–5985
75. Li W, Li X, Miller RA (2014) ATF4 activity: a common feature shared by many kinds of slow-aging mice. Aging Cell 13(6):1012–1018. doi:10.1111/acel.12264
76. Li W, Miller RA (2014) Elevated ATF4 function in fibroblasts and liver of slow-aging mutant mice. J Gerontol A Biol Sci Med Sci. doi:10.1093/gerona/glu040
77. Lapierre LR, Kumsta C, Sandri M, Ballabio A, Hansen M (2015) Transcriptional and epigenetic regulation of autophagy in aging. Autophagy 11(6):867–880. doi:10.1080/15548627.2015.1034410
78. Alvers AL, Wood MS, Hu D, Kaywell AC, Dunn WA Jr, Aris JP (2009) Autophagy is required for extension of yeast chronological life span by rapamycin. Autophagy 5(6):847–849
79. Toth ML, Sigmond T, Borsos E, Barna J, Erdelyi P, Takacs-Vellai K, Orosz L, Kovacs AL, Csikos G, Sass M, Vellai T (2008) Longevity pathways converge on autophagy genes to regulate life span in *Caenorhabditis elegans*. Autophagy 4(3):330–338
80. Hansen M, Chandra A, Mitic LL, Onken B, Driscoll M, Kenyon C (2008) A role for autophagy in the extension of lifespan by dietary restriction in *C. elegans*. PLoS Genet 4(2):e24. doi:10.1371/journal.pgen.0040024
81. Bjedov I, Toivonen JM, Kerr F, Slack C, Jacobson J, Foley A, Partridge L (2010) Mechanisms of life span extension by rapamycin in the fruit fly *Drosophila melanogaster*. Cell Metab 11(1):35–46. doi:10.1016/j.cmet.2009.11.010
82. Pyo JO, Yoo SM, Ahn HH, Nah J, Hong SH, Kam TI, Jung S, Jung YK (2013) Overexpression of Atg5 in mice activates autophagy and extends lifespan. Nat Commun 4:2300. doi:10.1038/ncomms3300
83. Simonsen A, Cumming RC, Brech A, Isakson P, Schubert DR, Finley KD (2008) Promoting basal levels of autophagy in the nervous system enhances longevity and oxidant resistance in adult Drosophila. Autophagy 4(2):176–184
84. Bai H, Kang P, Hernandez AM, Tatar M (2013) Activin signaling targeted by insulin/dFOXO regulates aging and muscle proteostasis in Drosophila. PLoS Genet 9(11):e1003941. doi:10.1371/journal.pgen.1003941
85. Ulgherait M, Rana A, Rera M, Graniel J, Walker DW (2014) AMPK modulates tissue and organismal aging in a non-cell-autonomous manner. Cell Rep 8(6):1767–1780. doi:10.1016/j.celrep.2014.08.006
86. Fabrizio P, Pozza F, Pletcher SD, Gendron CM, Longo VD (2001) Regulation of longevity and stress resistance by Sch9 in yeast. Science 292:288–290
87. Longo VD (1999) Mutations in signal transduction proteins increase stress resistance and longevity in yeast, nematodes, fruit flies, and mammalian neuronal cells. Neurobiol Aging 20(5):479–486
88. Fabrizio P, Liou LL, Moy VN, Diaspro A, SelverstoneValentine J, Gralla EB, Longo VD (2003) SOD2 functions downstream of Sch9 to extend longevity in yeast. Genetics 163(1):35–46
89. Enns LC, Ladiges W (2010) Protein kinase A signaling as an anti-aging target. Ageing Res Rev 9(3):269–272. doi:10.1016/j.arr.2010.02.004
90. Lin SJ, Kaeberlein M, Andalis AA, Sturtz LA, Defossez PA, Culotta VC, Fink GR, Guarente L (2002) Calorie restriction extends *Saccharomyces cerevisiae* lifespan by increasing respiration. Nature 418(6895):344–348

91. Gendron CM, Minois N, Longo VD, Pletcher SD, Vaupel JW (2003) Biodemographic trajectories of age-specific reproliferation from stationary phase in the yeast *Saccharomyces cerevisiae* seem multiphasic. Mech Ageing Dev 124:1059–1063
92. Enns LC, Morton JF, Mangalindan RS, McKnight GS, Schwartz MW, Kaeberlein MR, Kennedy BK, Rabinovitch PS, Ladiges WC (2009) Attenuation of age-related metabolic dysfunction in mice with a targeted disruption of the Cbeta subunit of protein kinase a. J Gerontol A Biol Sci Med Sci 64(12):1221–1231. doi:10.1093/gerona/glp133
93. Enns LC, Morton JF, Treuting PR, Emond MJ, Wolf NS, McKnight GS, Rabinovitch PS, Ladiges WC (2009) Disruption of protein kinase a in mice enhances healthy aging. PLoS One 4(6):e5963. doi:10.1371/journal.pone.0005963
94. Yan L, Vatner DE, O'Connor JP, Ivessa A, Ge H, Chen W, Hirotani S, Ishikawa Y, Sadoshima J, Vatner SF (2007) Type 5 adenylyl cyclase disruption increases longevity and protects against stress. Cell 130(2):247–258. doi:10.1016/j.cell.2007.05.038
95. Vatner SF, Park M, Yan L, Lee GJ, Lai L, Iwatsubo K, Ishikawa Y, Pessin J, Vatner DE (2013) Adenylyl cyclase type 5 in cardiac disease, metabolism, and aging. Am J Phys Heart Circ Phys 305(1):H1–H8. doi:10.1152/ajpheart.00080.2013
96. Higuchi-Sanabria R, Pernice WM, Vevea JD, Alessi Wolken DM, Boldogh IR, Pon LA (2014) Role of asymmetric cell division in lifespan control in *Saccharomyces cerevisiae*. FEMS Yeast Res 14(8):1133–1146. doi:10.1111/1567-1364.12216
97. Kwan EX, Foss EJ, Tsuchiyama S, Alvino GM, Kruglyak L, Kaeberlein M, Raghuraman MK, Brewer BJ, Kennedy BK, Bedalov A (2013) A natural polymorphism in rDNA replication origins links origin activation with calorie restriction and lifespan. PLoS Genet 9(3):e1003329. doi:10.1371/journal.pgen.1003329
98. Sinclair DA, Guarente L (1997) Extrachromosomal rDNA circles--a cause of aging in yeast. Cell 91(7):1033–1042
99. Shcheprova Z, Baldi S, Frei SB, Gonnet G, Barral Y (2008) A mechanism for asymmetric segregation of age during yeast budding. Nature 454(7205):728–734. doi:10.1038/nature07212
100. Khmelinskii A, Keller PJ, Lorenz H, Schiebel E, Knop M (2010) Segregation of yeast nuclear pores. Nature 466(7305):E1. doi:10.1038/nature09255
101. Kennedy BK, McCormick MA (2011) Asymmetric segregation: the shape of things to come? Curr Biol 21(4):R149–R151. doi:10.1016/j.cub.2011.01.018
102. Khmelinskii A, Meurer M, Knop M, Schiebel E (2011) Artificial tethering to nuclear pores promotes partitioning of extrachromosomal DNA during yeast asymmetric cell division. Curr Biol 21(1):R17–R18. doi:10.1016/j.cub.2010.11.034
103. Gehlen LR, Nagai S, Shimada K, Meister P, Taddei A, Gasser SM (2011) Nuclear geometry and rapid mitosis ensure asymmetric episome segregation in yeast. Curr Biol 21(1):25–33. doi:10.1016/j.cub.2010.12.016
104. Erjavec N, Nystrom T (2007) Sir2p-dependent protein segregation gives rise to a superior reactive oxygen species management in the progeny of *Saccharomyces cerevisiae*. Proc Natl Acad Sci U S A 104(26):10877–10881. doi:10.1073/pnas.0701634104
105. McFaline-Figueroa JR, Vevea J, Swayne TC, Zhou C, Liu C, Leung G, Boldogh IR, Pon LA (2011) Mitochondrial quality control during inheritance is associated with lifespan and mother-daughter age asymmetry in budding yeast. Aging Cell 10(5):885–895. doi:10.1111/j.1474-9726.2011.00731.x
106. Yang J, McCormick MA, Zheng J, Xie Z, Tsuchiya M, Tsuchiyama S, El-Samad H, Ouyang Q, Kaeberlein M, Kennedy BK, Li H (2015) Systematic analysis of asymmetric partitioning of yeast proteome between mother and daughter cells reveals "aging factors" and mechanism of lifespan asymmetry. Proc Natl Acad Sci U S A 112(38):11977–11982. doi:10.1073/pnas.1506054112
107. Chen C, Fingerhut JM, Yamashita YM (2016) The ins(ide) and outs(ide) of asymmetric stem cell division. Curr Opin Cell Biol 43:1–6. doi:10.1016/j.ceb.2016.06.001

108. Raff M (2006) The mystery of intracellular developmental programmes and timers. Biochem Soc Trans 34(Pt 5):663–670. doi:10.1042/BST0340663
109. Malumbres M (2014) Cyclin-dependent kinases. Genome Biol 15(6):122
110. Pavletich NP (1999) Mechanisms of cyclin-dependent kinase regulation: structures of Cdks, their cyclin activators, and Cip and INK4 inhibitors. J Mol Biol 287(5):821–828. doi:10.1006/jmbi.1999.2640
111. Fisher RP (2010) Coming full circle: cyclin-dependent kinases as anti-cancer drug targets. Subcell Biochem 50:1–15. doi:10.1007/978-90-481-3471-7_1
112. Asghar U, Witkiewicz AK, Turner NC, Knudsen ES (2015) The history and future of targeting cyclin-dependent kinases in cancer therapy. Nat Rev Drug Discov 14(2):130–146. doi:10.1038/nrd4504
113. Chandler H, Peters G (2013) Stressing the cell cycle in senescence and aging. Curr Opin Cell Biol 25(6):765–771. doi:10.1016/j.ceb.2013.07.005
114. Menjo M, Ikeda K, Nakanishi M (1998) Regulation of G1 cyclin-dependent kinases in liver regeneration. J Gastroenterol Hepatol 13(Suppl):S100–S105
115. Tamamori-Adachi M, Takagi H, Hashimoto K, Goto K, Hidaka T, Koshimizu U, Yamada K, Goto I, Maejima Y, Isobe M, Nakayama KI, Inomata N, Kitajima S (2008) Cardiomyocyte proliferation and protection against post-myocardial infarction heart failure by cyclin D1 and Skp2 ubiquitin ligase. Cardiovasc Res 80(2):181–190. doi:10.1093/cvr/cvn183
116. Han IS, Seo TB, Kim KH, Yoon JH, Yoon SJ, Namgung U (2007) Cdc2-mediated Schwann cell migration during peripheral nerve regeneration. J Cell Sci 120(Pt 2):246–255. doi:10.1242/jcs.03322
117. Bedelbaeva K, Snyder A, Gourevitch D, Clark L, Zhang XM, Leferovich J, Cheverud JM, Lieberman P, Heber-Katz E (2010) Lack of p21 expression links cell cycle control and appendage regeneration in mice. Proc Natl Acad Sci U S A 107(13):5845–5850. doi:10.1073/pnas.1000830107
118. Connell-Crowley L, Elledge SJ, Harper JW (1998) G1 cyclin-dependent kinases are sufficient to initiate DNA synthesis in quiescent human fibroblasts. Curr Biol 8:65–68
119. Wei M, Fabrizio P, Hu J, Ge H, Cheng C, Li L, Longo VD (2008) Life span extension by calorie restriction depends on Rim15 and transcription factors downstream of Ras/PKA, tor, and Sch9. PLoS Genet 4(1):e13. doi:10.1371/journal.pgen.0040013
120. McKnight JN, Boerma JW, Breeden LL, Tsukiyama T (2015) Global promoter targeting of a conserved lysine deacetylase for transcriptional shutoff during quiescence entry. Mol Cell 59(5):732–743. doi:10.1016/j.molcel.2015.07.014
121. McCormick MA, Delaney JR, Tsuchiya M, Tsuchiyama S, Shemorry A, Sim S, Chou AC, Ahmed U, Carr D, Murakami CJ, Schleit J, Sutphin GL, Wasko BM, Bennett CF, Wang AM, Olsen B, Beyer RP, Bammler TK, Prunkard D, Johnson SC, Pennypacker JK, An E, Anies A, Castanza AS, Choi E, Dang N, Enerio S, Fletcher M, Fox L, Goswami S, Higgins SA, Holmberg MA, Hu D, Hui J, Jelic M, Jeong KS, Johnston E, Kerr EO, Kim J, Kim D, Kirkland K, Klum S, Kotireddy S, Liao E, Lim M, Lin MS, Lo WC, Lockshon D, Miller HA, Moller RM, Muller B, Oakes J, Pak DN, Peng ZJ, Pham KM, Pollard TG, Pradeep P, Pruett D, Rai D, Robison B, Rodriguez AA, Ros B, Sage M, Singh MK, Smith ED, Snead K, Solanky A, Spector BL, Steffen KK, Tchao BN, Ting MK, Vander Wende H, Wang D, Welton KL, Westman EA, Brem RB, Liu XG, Suh Y, Zhou Z, Kaeberlein M, Kennedy BK (2015) A comprehensive analysis of replicative lifespan in 4,698 single-gene deletion strains uncovers conserved mechanisms of aging. Cell Metab. doi:10.1016/j.cmet.2015.09.008
122. Burtner CR, Murakami CJ, Olsen B, Kennedy BK, Kaeberlein M (2011) A genomic analysis of chronological longevity factors in budding yeast. Cell Cycle 10(9):1385–1396
123. Ashrafi K, Sinclair D, Gordon JI, Guarente L (1999) Passage through stationary phase advances replicative aging in *Saccharomyces cerevisiae*. Proc Natl Acad Sci U S A 96(16):9100–9105
124. Piper PW, Harris NL, MacLean M (2006) Preadaptation to efficient respiratory maintenance is essential both for maximal longevity and the retention of replicative potential in chronologically ageing yeast. Mech Ageing Dev 127(9):733–740. doi:10.1016/j.mad.2006.05.004

125. Polymenis M, Kennedy BK (2012) Chronological and replicative lifespan in yeast: do they meet in the middle? Cell Cycle 11(19):3531–3532. doi:10.4161/cc.22041
126. Delaney JR, Murakami C, Chou A, Carr D, Schleit J, Sutphin GL, An EH, Castanza AS, Fletcher M, Goswami S, Higgins S, Holmberg M, Hui J, Jelic M, Jeong KS, Kim JR, Klum S, Liao E, Lin MS, Lo W, Miller H, Moller R, Peng ZJ, Pollard T, Pradeep P, Pruett D, Rai D, Ros V, Schuster A, Singh M, Spector BL, Wende HV, Wang AM, Wasko BM, Olsen B, Kaeberlein M (2013) Dietary restriction and mitochondrial function link replicative and chronological aging in *Saccharomyces cerevisiae*. Exp Gerontol 48(10):1006–1013. doi:10.1016/j.exger.2012.12.001
127. Murakami C, Delaney JR, Chou A, Carr D, Schleit J, Sutphin GL, An EH, Castanza AS, Fletcher M, Goswami S, Higgins S, Holmberg M, Hui J, Jelic M, Jeong KS, Kim JR, Klum S, Liao E, Lin MS, Lo W, Miller H, Moller R, Peng ZJ, Pollard T, Pradeep P, Pruett D, Rai D, Ros V, Schuster A, Singh M, Spector BL, Vander Wende H, Wang AM, Wasko BM, Olsen B, Kaeberlein M (2012) pH neutralization protects against reduction in replicative lifespan following chronological aging in yeast. Cell Cycle 11(16):3087–3096. doi:10.4161/cc.21465
128. Sharpless NE, Sherr CJ (2015) Forging a signature of in vivo senescence. Nat Rev Cancer 15(7):397–408. doi:10.1038/nrc3960
129. Pardee AB (1989) G1 events and regulation of cell proliferation. Science 246(4930): 603–608
130. Sherr CJ, Roberts JM (1995) Inhibitors of mammalian G1 cyclin-dependent kinases. Genes Dev 9(10):1149–1163
131. Campisi J (2013) Aging, cellular senescence, and cancer. Annu Rev Physiol 75:685–705. doi:10.1146/annurev-physiol-030212-183653
132. Mao Z, Ke Z, Gorbunova V, Seluanov A (2012) Replicatively senescent cells are arrested in G1 and G2 phases. Aging 4(6):431–435
133. Baus F, Gire V, Fisher D, Piette J, Dulic V (2003) Permanent cell cycle exit in G2 phase after DNA damage in normal human fibroblasts. EMBO J 22(15):3992–4002. doi:10.1093/emboj/cdg387
134. Gire V, Dulic V (2015) Senescence from G2 arrest, revisited. Cell Cycle 14(3):297–304. doi:10.1080/15384101.2014.1000134
135. Sherwood SW, Rush D, Ellsworth JL, Schimke RT (1988) Defining cellular senescence in IMR-90 cells: a flow cytometric analysis. Proc Natl Acad Sci U S A 85(23):9086–9090
136. Rodier F, Campisi J (2011) Four faces of cellular senescence. J Cell Biol 192(4):547–556. doi:10.1083/jcb.201009094
137. Blagosklonny MV (2012) Cell cycle arrest is not yet senescence, which is not just cell cycle arrest: terminology for TOR-driven aging. Aging 4(3):159–165
138. Hayflick L (1965) The limited in vitro lifetime of human diploid cell strains. Exp Cell Res 37:614–636
139. Coppe JP, Patil CK, Rodier F, Sun Y, Munoz DP, Goldstein J, Nelson PS, Desprez PY, Campisi J (2008) Senescence-associated secretory phenotypes reveal cell-nonautonomous functions of oncogenic RAS and the p53 tumor suppressor. PLoS Biol 6(12):2853–2868. doi:10.1371/journal.pbio.0060301
140. Kuilman T, Michaloglou C, Vredeveld LC, Douma S, van Doorn R, Desmet CJ, Aarden LA, Mooi WJ, Peeper DS (2008) Oncogene-induced senescence relayed by an interleukin-dependent inflammatory network. Cell 133(6):1019–1031. doi:10.1016/j.cell.2008.03.039
141. Laberge RM, Sun Y, Orjalo AV, Patil CK, Freund A, Zhou L, Curran SC, Davalos AR, Wilson-Edell KA, Liu S, Limbad C, Demaria M, Li P, Hubbard GB, Ikeno Y, Javors M, Desprez PY, Benz CC, Kapahi P, Nelson PS, Campisi J (2015) MTOR regulates the pro-tumorigenic senescence-associated secretory phenotype by promoting IL1A translation. Nat Cell Biol 17(8):1049–1061. doi:10.1038/ncb3195
142. Bitto A, Ito TK, Pineda VV, LeTexier NJ, Huang HZ, Sutlief E, Tung H, Vizzini N, Chen B, Smith K, Meza D, Yajima M, Beyer RP, Kerr KF, Davis DJ, Gillespie CH, Snyder JM, Treuting PM, Kaeberlein M (2016) Transient rapamycin treatment can increase lifespan and healthspan in middle-aged mice Elife 5. doi:10.7554/eLife.16351

143. Chen C, Liu Y, Liu Y, Zheng P (2009) mTOR regulation and therapeutic rejuvenation of aging hematopoietic stem cells. Science Signal 2(98):ra75. doi:10.1126/scisignal.2000559
144. Baker DJ, Wijshake T, Tchkonia T, LeBrasseur NK, Childs BG, van de Sluis B, Kirkland JL, van Deursen JM (2011) Clearance of p16Ink4a-positive senescent cells delays ageing-associated disorders. Nature 479(7372):232–236. doi:10.1038/nature10600
145. Baker DJ, Childs BG, Durik M, Wijers ME, Sieben CJ, Zhong J, Saltness RA, Jeganathan KB, Verzosa GC, Pezeshki A, Khazaie K, Miller JD, van Deursen JM (2016) Naturally occurring p16(Ink4a)-positive cells shorten healthy lifespan. Nature 530(7589):184–189. doi:10.1038/nature16932
146. Kapanidou M, Lee S, Bolanos-Garcia VM (2015) BubR1 kinase: protection against aneuploidy and premature aging. Trends Mol Med 21(6):364–372. doi:10.1016/j.molmed.2015.04.003
147. Baker DJ, Jeganathan KB, Cameron JD, Thompson M, Juneja S, Kopecka A, Kumar R, Jenkins RB, de Groen PC, Roche P, van Deursen JM (2004) BubR1 insufficiency causes early onset of aging-associated phenotypes and infertility in mice. Nat Genet 36(7):744–749. doi:10.1038/ng1382
148. Matsumoto T, Baker DJ, d'Uscio LV, Mozammel G, Katusic ZS, van Deursen JM (2007) Aging-associated vascular phenotype in mutant mice with low levels of BubR1. Stroke 38(3):1050–1056. doi:10.1161/01.str.0000257967.86132.01
149. Baker DJ, Dawlaty MM, Wijshake T, Jeganathan KB, Malureanu L, van Ree JH, Crespo-Diaz R, Reyes S, Seaburg L, Shapiro V, Behfar A, Terzic A, van de Sluis B, van Deursen JM (2013) Increased expression of BubR1 protects against aneuploidy and cancer and extends healthy lifespan. Nat Cell Biol 15(1):96–102. doi:10.1038/ncb2643
150. Roos CM, Zhang B, Palmer AK, Ogrodnik MB, Pirtskhalava T, Thalji NM, Hagler M, Jurk D, Smith LA, Casaclang-Verzosa G, Zhu Y, Schafer MJ, Tchkonia T, Kirkland JL, Miller JD (2016) Chronic senolytic treatment alleviates established vasomotor dysfunction in aged or atherosclerotic mice. Aging Cell doi:10.1111/acel.12458
151. Zhu Y, Tchkonia T, Fuhrmann-Stroissnigg H, Dai HM, Ling YY, Stout MB, Pirtskhalava T, Giorgadze N, Johnson KO, Giles CB, Wren JD, Niedernhofer LJ, Robbins PD, Kirkland JL (2016) Identification of a novel senolytic agent, navitoclax, targeting the Bcl-2 family of anti-apoptotic factors. Aging Cell 15(3):428–435. doi:10.1111/acel.12445
152. Zhu Y, Tchkonia T, Pirtskhalava T, Gower AC, Ding H, Giorgadze N, Palmer AK, Ikeno Y, Hubbard GB, Lenburg M, O'Hara SP, LaRusso NF, Miller JD, Roos CM, Verzosa GC, LeBrasseur NK, Wren JD, Farr JN, Khosla S, Stout MB, McGowan SJ, Fuhrmann-Stroissnigg H, Gurkar AU, Zhao J, Colangelo D, Dorronsoro A, Ling YY, Barghouthy AS, Navarro DC, Sano T, Robbins PD, Niedernhofer LJ, Kirkland JL (2015) The Achilles' heel of senescent cells: from transcriptome to senolytic drugs. Aging Cell 14(4):644–658. doi:10.1111/acel.12344
153. Chang J, Wang Y, Shao L, Laberge RM, Demaria M, Campisi J, Janakiraman K, Sharpless NE, Ding S, Feng W, Luo Y, Wang X, Aykin-Burns N, Krager K, Ponnappan U, Hauer-Jensen M, Meng A, Zhou D (2016) Clearance of senescent cells by ABT263 rejuvenates aged hematopoietic stem cells in mice. Nat Med 22(1):78–83. doi:10.1038/nm.4010
154. Farrell JA, O'Farrell PH (2014) From egg to gastrula: how the cell cycle is remodeled during the Drosophila mid-blastula transition. Annu Rev Genet 48:269–294. doi:10.1146/annurev-genet-111212-133531
155. de Nooij JC, Letendre MA, Hariharan IK (1996) A cyclin-dependent kinase inhibitor, Dacapo, is necessary for timely exit from the cell cycle during Drosophila embryogenesis. Cell 87(7):1237–1247
156. Lane ME, Sauer K, Wallace K, Jan YN, Lehner CF, Vaessin H (1996) Dacapo, a cyclin-dependent kinase inhibitor, stops cell proliferation during Drosophila development. Cell 87(7):1225–1235

Chapter 9
The Spindle Orientation Machinery Beyond Mitosis: When Cell Specialization Demands Polarization

Abigail L.D. Tadenev and Basile Tarchini

Abstract Mitosis is a process requiring strict spatial organization of cellular components. In particular, the orientation of the mitotic spindle with respect to the tissue defines the division plane. In turn, the orientation of cell division can regulate tissue morphology or the fate of daughter cells. While we have learned much about the mechanisms of mitotic spindle orientation, recent studies suggest that the proteins implicated can also play important roles in post-mitotic cells. Interestingly, post-mitotic protein function often involves polarizing the cell cytoskeleton during differentiation, mirroring its ability to orient the mitotic spindle during division. This review focuses on alternative functions of the spindle orientation machinery after division, when the cell undergoes a specialization process associated with differentiation or mature function, and discusses diseases associated to those alternative functions.

Keywords Cell polarity • Oriented cell division • Mitotic spindle • Cytoskeleton polarization • Post-mitotic cell morphogenesis • Inscuteable • LGN (leu-gly-asn) / Gpsm2 (G-protein signalling modulator 2) • Guanine nucleotide binding protein (G protein) • Alpha inhibiting

The control of cell proliferation, cell fate and cell organization in a tissue are major biological requirements at every stage of life. In the last 20 years, the regulation of cell division orientation has emerged as a prominent level of control in this context. On the one hand, the plane along which cells divide impacts tissue structure. The positioning of the two daughter cells is largely determined by the cleavage plane during cytokinesis, itself instructed by the orientation of the microtubule-based mitotic spindle. In an epithelium, for example, divisions along the apico-basal axis increase tissue thickness, while orthogonal divisions increase epithelial surface. Failure to properly regulate this process results in altered epithelial morphogenesis [1, 2],

A.L.D. Tadenev • B. Tarchini (✉)
The Jackson Laboratory, Bar Harbor, ME 04609, USA
e-mail: basile.tarchini@jax.org

M. Gotta, P. Meraldi (eds.), *Cell Division Machinery and Disease*, Advances in Experimental Medicine and Biology 1002, DOI 10.1007/978-3-319-57127-0_9

and was proposed to contribute to diseases such as polycystic kidney disease, microcephaly or lissencephaly [1–3]. On the other hand, the orientation of cell division can directly determine cell fate, at least in invertebrates. Since cellular components are not necessarily evenly distributed throughout the mitotic cell, their inheritance can be variably biased depending on the orientation of the mitotic spindle, influencing daughter cells' behavior. For example, stem cells can self-renew while generating a daughter cell with more restricted fate, and this asymmetric outcome involves asymmetric inheritance of RNA or protein fate determinants, membrane domains or organelles. In many cases, cell fate and tissue architecture are hard to separate however, as when failure to maintain planar division gives rise to delaminated cells that have lost their epithelial characteristics and become mesenchymal-like, which can promote proliferation and possibly tumor development [4, 5]. Although the importance of spindle orientation in strictly driving binary cell fate decision in vertebrates is still debated, it is clear that spindle misorientation can alter the total proliferative potential and cell type composition of a tissue.

The importance of proper control over the division plane has motivated a large-scale effort to identify and functionally characterize the molecular constituents of the machinery that orients the mitotic spindle. This prolific field of research has made tremendous strides in the recent past, and has been extensively reviewed already [1, 2, 6–11]. By contrast, our goal here is to spotlight the emergent, less well-known examples where key proteins regulating spindle orientation were found to take on different roles in post-mitotic cells. Analogous to their role during cell division, they largely appear to influence cytoskeleton polarization, and participate in specialized subcellular processes associated to cell differentiation or mature cell function. While such examples are still relatively scarce, their growing significance is underscored by relevance to disease like hearing loss and drug-seeking behavior.

9.1 The Core Machinery Behind Oriented Divisions

To provide context and draw parallels with their post-mitotic functions discussed further below, we will begin with a brief overview of the central players regulating mitotic spindle orientation. Generally speaking, these proteins become enriched at specific regions of the cell cortex in prometaphase dividing cells, guided by canonical markers of cell polarity. These regulators then locally recruit partner proteins that capture and pull on astral microtubules, the microtubules that emanate from each centrosome but do not participate in chromosome segregation. In essence, proteins of the core machinery are cortical landmarks used as reference to ensure that the mitotic spindle becomes aligned with the polarity of the cell, and that the resulting daughter cells are situated correctly within the tissue. The orientation machinery is strikingly conserved across tissues and organisms, and has been studied in a wide variety of model systems, including the first divisions of the *C. elegans* zygote [12–14], neuroblast lineages in the fly (see below), the murine embryonic epidermis [15–18], and neuroepithelial cells in the vertebrate central nervous system [19–22], to name only a few. Of note, however, there is much variation in the way the core spindle orientation proteins operate among different systems, a topic outside the scope of this chapter.

The initial discovery and much of the pioneering work addressing mitotic spindle orientation has been done in *C. elegans* and *Drosophila* [23]. In *Drosophila* embryonic and larval neuroblasts, which have become a choice model of self-renewing asymmetric stem cell division, the Par complex localized apically in the neuroectoderm is carried over when the neuroblast delaminates basally [24, 25]. This complex composed of Par-3, Par-6, and the atypical kinase aPKC is known as a master regulator of apico-basal polarity [26]. Par3 recruits the adapter protein Inscuteable (Insc; mInsc in mammals) to the apical cell cortex [27–29], and mInsc in turn binds to the TPR repeats of Partner of Inscuteable (Pins; LGN, mPins or Gpsm2 in mammals) [30–32] (Fig. 9.1a). Pins/LGN is further stabilized at the cortex through interaction of its GoLoco domains with GDP-bound $G\alpha_i$ ($G\alpha_{i}$-$_{GDP}$) anchored at the membrane via myristoylation [33]. As a result, Insc-Pins/LGN-$G\alpha_i$ colocalize in a crescent at the cell cortex during prophase and metaphase. This core spindle orientation complex then recruits the large coiled-coil protein Mud (NuMA in vertebrates) [34–38]. The transition is proposed to occur through a switch mechanism whereby Mud/NuMA replaces Insc, as both proteins compete for the TPR motifs in Pins/LGN and cannot bind simultaneously [39–41]. Mud/NuMA provides a link to the astral microtubules since it directly binds the Dynein-Dynactin motor complex [42]. Overall, the spindle becomes anchored to the cell cortex in a polarized manner, and pulling forces align the mitotic spindle to ensure apico-basal divisions where the apical daughter retains neuroblast identity and the basal daughter inherits basally located fate determinants, adopting a more restricted fate.

Biochemically, LGN and other GoLoco-containing proteins act as G protein dissociation inhibitors (GDI), effectively competing with $G\beta\gamma$ and preventing guanine nucleotide exchange by stabilizing $G\alpha_{i\text{-GDP}}$ [43–45]. In principle, this activity is known to uncouple trimeric G proteins from GPCRs at the membrane and reduce signaling, while potentially also prolonging stimulation of $G\beta\gamma$-dependent effectors. Interestingly however, there is only limited evidence that $G\alpha_i$ proteins relay GPCR signaling during spindle orientation [46]. While cell-autonomous guanine exchange factors (GEFs) have been implicated [47–51], it is generally accepted that LGN-$G\alpha_{i\text{-GDP}}$ is the active signaling complex acting on the spindle.

9.2 Roles of the Core mInsc-LGN-$G\alpha_i$ Complex Beyond Spindle Recruitment

A number of studies recently proposed that mInsc-LGN-$G\alpha_i$ proteins locally regulate cytoskeleton rearrangement in specialized cells, a fundamental role falling in line with their better-known ability to recruit the mitotic spindle during division.

We discuss below interesting novel findings where this protein complex is involved in such diverse post-mitotic processes as neuronal synaptic function,

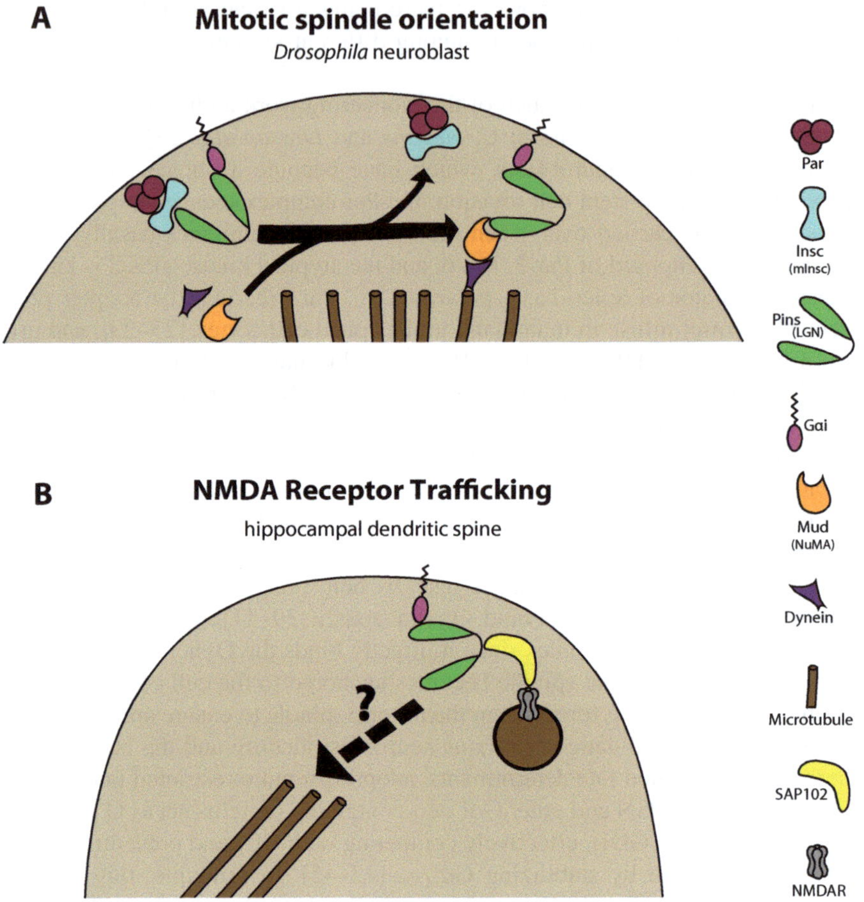

Fig. 9.1 Comparison of LGN's roles in mitotic spindle orientation and NMDA receptor trafficking. (**a**) In the *Drosophila* neuroblast, LGN is recruited to the membrane by the Par complex, $G\alpha_i$, and Insc. NuMA then displaces Insc from LGN, and NuMA's association with dynein recruits astral microtubules to the cortex. (**b**) In hippocampal dendritic spines, SAP102 binds LGN and NMDA receptors. By analogy with (**a**), LGN could provide a link to microtubules in order to help locally deliver NMDAR vesicles to the cell surface. See text for additional details

chemotaxis, and the generation of intrinsic cytoskeleton asymmetry in developing hair cells, a cellular patterning event crucial for sensory perception in the inner ear. Both the mitotic and post-mitotic actions of this complex are schematized in Fig. 9.2.

9.2.1 Modulating Neuronal Function

Components of the spindle orientation machinery have been shown to regulate the function of neuronal synapses. The NMDA receptor (NMDAR) is a glutamate receptor that is critical for proper neural development, learning and memory, affect,

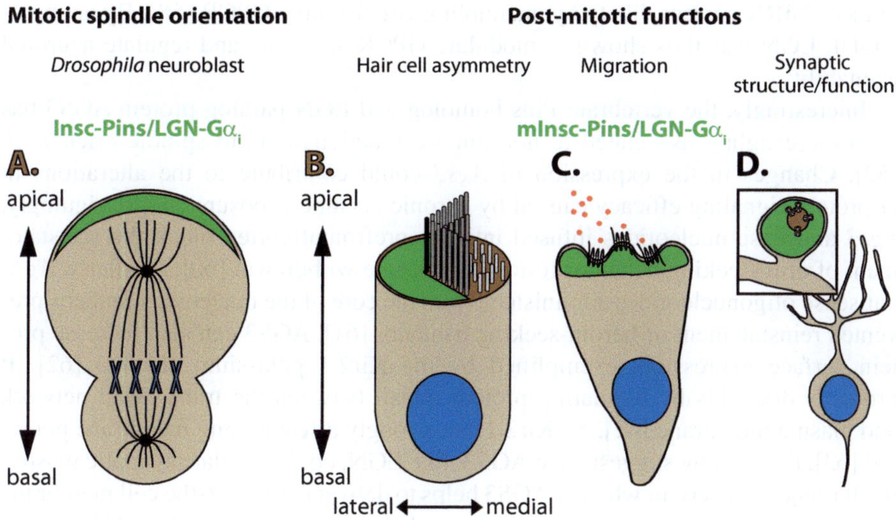

Fig. 9.2 The roles and localization of the Insc-Pins/LGN-$G\alpha_i$ complex in polarized cell activities in dividing and post-mitotic cells. (**a**) In *Drosophila* neuroblasts, Insc-Pins-$G\alpha_i$ (*green*) colocalize at the apical cell cortex and help orient the mitotic spindle along the apico-basal axis. (**b–d**) Functions of mInsc-LGN-$G\alpha_i$ in post-mitotic cells. (**b**) mInsc-LGN-$G\alpha_i$ localize to the "bare zone", a lateral subset of the apical membrane devoid of microvilli in inner ear hair cells. mInsc-LGN-$G\alpha_i$ were proposed to help define the lateral edge of the stereocilia bundle. Stereocilia and microvilli are depicted in *dark and light grey*, respectively, and the primary cilium, or kinocilium, is shown in *black*. (**c**) mInsc-LGN-$G\alpha_i$ are found at the leading edge of chemotaxing neutrophils, where they signal downstream of GPCRs to stabilize actin-based pseudopods. (**d**) Within the dendritic spines of neurons, Insc-Pins/LGN-$G\alpha_i$ interact with NMDA receptors, potentially influencing their delivery to the plasma membrane and influencing synaptic function

and cognition [52]. In experiments designed to elucidate the regulation of glutamate receptor trafficking, LGN was found to bind SAP102 (Dlg3) [53], a member of the MAGUK protein family important for scaffolding proteins at neuronal synapses [54]. Overexpression of LGN in cultured hippocampal neurons leads to changes in both number and morphology of dendritic spines [53]. LGN and SAP102 also bind NMDA receptor subunits and $G\alpha_{i\text{-}GDP}$, forming an NMDAR-SAP102-LGN-$G\alpha_i$ complex, which was proposed to be important for proper NMDAR trafficking [53]. Similar to its role in recruiting astral microtubules to the cell cortex during mitosis, in this model LGN could regulate receptor trafficking by acting as a bridge between microtubules and receptor-containing vesicles [55] (Fig. 9.1b). Collectively, these results suggest that the LGN-$G\alpha_i$ complex acts in multiple ways to influence synaptic signaling, as both spine morphology and NMDAR dynamics are mediators of synaptic plasticity [56, 57]. In hippocampal neurons, LGN also modulates current through the G protein-activated inwardly rectifying potassium channel (GIRK) [58]. In this study, the authors suggest that, under basal conditions, LGN enhances GIRK current by binding and stabilizing $G\alpha_{i\text{-}GDP}$, enhancing the activity of $G\beta\gamma$, which then activates GIRK. Following GPCR stimulation, however, LGN actually reduces GIRK current, likely by uncoupling Gα_i from the GPCR [58]. By acting as a GDI, LGN was thus shown to modulate GPCR signaling and regulate neuronal excitability.

Interestingly, the vertebrate Pins homolog and LGN paralog protein AGS3 has been more tightly associated to non-mitotic functions than to spindle orientation [59]. Changes in the expression of *Ags3* could contribute to the alterations in G-protein signaling efficacy caused by chronic cocaine exposure and, intriguingly, *Ags3* antisense nucleotides infused into the prefrontal cortex block the reinstatement of drug-seeking behavior following cocaine withdrawal [60]. Similarly, *Ags3* antisense oligonucleotides administered into the core of the nucleus accumbens prevented reinstatement of heroin-seeking behavior [61]. AGS3 can also increase protein surface expression, exemplified by the Kir2.1 potassium channel [62]. It probably does this by regulating protein transit between the trans-Golgi network and plasma membrane [62]. As Kir2.1 can strongly affect resting membrane potential [63], this finding suggests that AGS3, like LGN, could regulate synaptic plasticity. It remains uncertain whether AGS3 helps to deliver cargoes to the cell membrane by coupling to the cytoskeleton, as suggested above for LGN and the NMDAR.

9.2.2 Regulating Cellular Movement

Interestingly, mInsc can drive polarized responses in post-mitotic cells downstream of G-protein coupled receptor (GPCR) signaling. Neutrophils must chemotax toward the source of chemoattractants in order to help mediate immune responses. This directed motility is achieved by polymerization of filamentous actin at the leading edge of the cell and contraction at the opposite end of the cell mediated by myosin II [64]. Neutrophils express GPCRs that are locally activated by chemoattractants and, via coupling specifically to the Gαi family of heterotrimeric G proteins [65] at the leading edge, activation results in the generation of G$\alpha_{i\text{-GTP}}$ and free G$\beta\gamma$, which play separate but complementary roles in directed migration. Much work has focused on the role of free G$\beta\gamma$, which promotes motility via activation of molecules including PI3K [66]. Recently, it has also been suggested that G$\beta\gamma$-free G$\alpha_{i\text{-GDP}}$ produced by hydrolysis of G$\alpha_{i\text{-GTP}}$ plays an important role in maintaining appropriate directionality during chemotaxis [67]. Strikingly, this pathway uses many of the proteins involved in orienting the mitotic spindle: G$\alpha_{i\text{-GDP}}$ probably generated downstream of GPCR activation by chemoattractants recruits LGN/AGS3, which recruits mInsc and subsequently the Par complex to the leading edge [67] (Fig. 9.2c). Depletion of mInsc affects only directionality during chemotaxis, and not overall motility [67], suggesting that mInsc does not affect G$\beta\gamma$ function. It remains unclear, however, how G$\alpha_{i\text{-GDP}}$-LGN/AGS3-mInsc-Par stabilize the directionality of migrating neutrophils.

LGN can also control changes in cellular shape. Recent work suggests that LGN regulates sprouting angiogenesis, perhaps via destabilization of cell-cell and cell-matrix adhesions downstream of altered microtubule dynamics in endothelial cells [68].

9.2.3 Regulating Hair Cell Morphogenesis in the Inner Ear

We and others have discovered a surprising new role for the mInsc-LGN-Gα_i complex during early hair cell differentiation in the inner ear [69–71]. Here, these proteins are involved in organizing the apical membrane of hair cells, the highly specialized cells ensuring the detection of sounds, acceleration, and gravity.

Hair cells are crowned with a bundle of apical protrusions, termed stereocilia, that respond to mechanical deflection by modulating electric currents in the cell. Stereocilia derive from microvilli that initially cover the apical membrane and, under largely unknown influence, grow in girth and length. The stereocilia bundle is characterized by a strong radial asymmetry along the epithelial plane in each cell. Asymmetry is manifested notably by the V-or arched shape of the bundle, and the staircase-like organization of stereocilia, which align into rows harboring graded heights (Fig. 9.3a). Cytoskeleton polarization is also manifested at the tissue level. In the cochlea, hair cells are organized in four rows (Fig. 9.3b), and all cells adopt a strikingly uniform planar orientation of their bundle. This occurs by the planar cell polarity (PCP) pathway, which is generally responsible for the coordinated orientation of cells along the epithelial plane [72]. Cell-intrinsic and tissue level polarization are essential for sensory function, and notably account for direction-sensitivity to stimuli: hair cells only respond to bundle deflections toward or away from the tallest stereocilia row, while orthogonal deflections have no effect [73].

As in dividing progenitors, mInsc-LGN-Gα_i colocalize as a protein complex in early post-mitotic hair cells [69–71] (Fig. 9.3a–c). This complex is asymmetrically enriched in the plane, forming a lateral crescent at the apical membrane. mInsc-LGN-Gα_i both label and are required to generate a patch of membrane devoid of microvilli, which we termed the "bare zone" [70]. As the hair cell develops, this region expands and closely abuts the lateral edge of the forming bundle, which hosts the tallest stereocilia. Disrupting the protein complex reduces or eliminates the bare zone, leading to severe stereocilia placement defects. It thus appears that mInsc-LGN-Gα_i act by defining an exclusion zone for microvilli as a strategy to define the contour of the forming bundle.

The influence of mInsc-LGN-Gα_i is not limited to regulating the placement of actin-based stereocilia. Early during differentiation, the hair cells' one true cilium, the kinocilium, moves from the cell center to the periphery (Fig. 9.3c). Although the underlying mechanism remains obscure, the eccentric shift is required for bundle morphogenesis [74], and its normal lateral direction depends on tissue-level planar cell polarity (PCP) [75]. Since mInsc-LGN-Gα_i recruit astral microtubules during mitosis, it is tempting to speculate that these proteins could pull on microtubules connected to the basal body nucleating the kinocilium to trigger the shift. Accordingly, one study proposed that the shift depends on Gα_i signaling based on results in organotypic culture [69], although off-center kinocilium shifts were still observed when Gα_i inactivation was achieved in vivo [70]. Later during hair cell differentiation, LGN and Gα_i also play an important role to ensure the precise localization of the kinocilium in the center of the arched stereocilia bundle

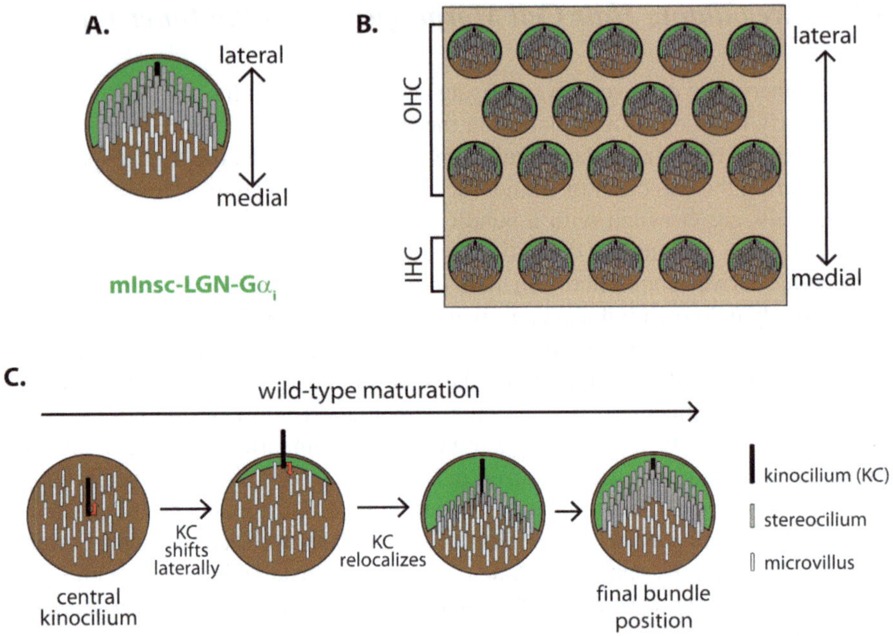

Fig. 9.3 Hair cell organization in the mammalian cochlea. (**a**) At the single cell level, each hair cell is highly asymmetric along the planar axis. mInsc-LGN-Gα_i (*green*) localize to a lateral crescent at the apical surface and mark a region devoid of microvilli (the "bare zone"). The mechanosensitive stereocilia bundle grows in a chevron pattern in the central region of the apical membrane. Stereocilia and microvilli are depicted in *dark* and *light grey*, respectively, and the primary cilium, or kinocilium, is shown in *black*. (**b**) At the tissue level, cochlear hair cells are organized in four rows (OHC: outer hair cells; IHC: inner hair cells). Hair cells are uniformly oriented, with the chevron shape of the bundle, the tallest stereocilia row and the mInsc-LGN- Gα_i crescent facing the lateral edge. (**c**) During early hair cell differentiation, the kinocilium (KC) or primary cilium is first observed at the center of the cell, amid a full covering of microvilli. The kinocilium then shifts laterally as mInsc-LGN-Gα_i become detectable at the lateral edge. mInsc-LGN-Gα_i expand medially, creating the microvilli-free bare zone, and the kinocilium relocalizes more centrally. At the same time, select microvilli grow into stereocilia that become precisely aligned and adopt graded heights to form the mature bundle

(Fig. 9.3c). Together, these results suggest that the mInsc-LGN-Gα_i complex is required to spatially coordinate apical membrane domains with both the microtubule- and actin-based cytoskeleton.

As mInsc-LGN-Gα_i work at the single cell level, their activity must somehow become coordinated with the PCP pathway to ensure that all hair cells orient their asymmetric bundle in the same planar direction. Interestingly, inactivating Gα_i signaling results not only in bundle defects in single hair cells, but also in hair cell misorientation [70]. This suggests the intriguing possibility that Gα_i signaling could link cell-intrinsic morphogenesis with PCP signaling initiated at apical junctions by cell–cell interactions.

The emerging molecular function of mInsc-LGN-Gα_i in hair cells is of particular interest since LGN mutations were recently shown to underlie congenital hereditary hearing loss in multiple human families [71, 76–80]. Loss-of-function mutations in *LGN* (*GPSM2*) were originally identified in patients classified as having nonsyndromic hearing loss [79, 80]. Mutations in *LGN* were subsequently identified in patients with Chudley-McCullough syndrome [76–78], a condition first described in 1997 [81] where profound congenital hearing loss coincides with partial agenesis of the corpus callosum, grey matter heterotopia, and often hydrocephaly [82]. Interestingly, the authors then expanded their analyses to the first reported *LGN* pedigrees and identified subclinical brain malformations consistent with Chudley-McCullough syndrome [78]. As mice expressing a truncated LGN protein are profoundly deaf [71], it is now tempting to speculate that hearing loss stems from defective apical cytoskeleton polarization in hair cells during development. In contrast, brain malformation could result from defects in mitotic spindle orientation, as described in the spinal cord and the cortex in model animals [19, 20]. If true, it could seem curious that mutations in a core mitotic spindle protein would have hearing loss as their most severe clinical presentation. However, hair cells are highly specialized, and many proteins that generate or compose their unique stereocilia bundle appear essential for this task in particular, resulting in non-syndromic hearing loss when defective (for review, see [83]). In contrast, given the importance of keeping cell proliferation and tissue architecture in check in all tissues, mitotic spindle orientation must be particularly robust mechanistically.

To date, no association to disease has been made for *mInsc*. Given that there are no clear paralogs of *mInsc*, mutations could be incompatible with life. However, *mInsc* knockout mice are viable and display no gross phenotypes [21, 67, 70], which does not support this idea. Rather, since *mInsc* mutation mildly affects hair cell morphology compared to disruption of *Lgn* or Gα_i [70], mutations may not lead to clinically noticeable phenotypes. It remains possible, however, that more subtle issues exist, such as reduced immune response due to defective neutrophil chemotaxis [67]. Gα_i proteins are involved in a multitude of signaling functions, making any particular connection between mutation and defects in cytoskeleton polarity challenging.

9.3 Further Evidence: Examples of Partner Proteins with Post-Mitotic Functions

9.3.1 Canoe/Afadin

Some proteins with well-established roles in mitotic spindle orientation in *Drosophila* were first studied in a post-mitotic context in vertebrates prior to being implicated in mitosis. For example, the *Drosophila* protein Canoe helps mediate spindle orientation [84] by binding Pins and helping recruit Mud (NuMA homolog), thus providing a link between Pins and microtubules [85]. A role for Canoe's mammalian homolog, Afadin/AF-6, in orienting the mitotic spindle has only recently been demonstrated. Studies in human cell lines suggest that Afadin is important for recruiting LGN to the cortex and providing a bridge to F-actin [86, 87]. Post-mitotically, Afadin is directly involved in the formation and/or maintenance of cellular junctions, including adherens junctions, tight junctions [88], and neuronal synapses [89]. Afadin is also important in remodeling the architecture of dendritic spines downstream of NMDA receptor activity [90]. Reminiscent of the role of mInsc in neutrophil chemotaxis, Afadin specifically regulates the directionality but not the overall motility of NIH3T3 cells [91]. In addition, Canoe can affect axon pathfinding by regulating Slit/Robo signaling at the *Drosophila* CNS midline [92].

9.3.2 Myosin VI

Myosin VI may be more accurately categorized as an "effector" rather than a "regulator" of spindle orientation. In *Drosophila* neuroblasts, Myosin VI targets the protein Miranda [93] and cell fate determinants Prospero, Brat, and Numb to the basal portion of the cell [9]. Myosin VI has not been associated to mitotic spindle orientation in vertebrates, but has interesting post-mitotic functions. In spite of being widely expressed in animal tissues [94] and the sole characterized minus end-directed myosin [95], Myosin VI predominantly causes deafness when absent [96], an interesting parallel to the case of LGN described above. Following up on this discovery, human deafness has also been linked to mutations in *MYO6* [97, 98]. In *Myo6* mutant mouse cochlear hair cells, stereocilia fuse together into giant stereocilia [99]. In addition, Myosin VI is also required at the basal end of hair cells to generate the ribbon synapses, a subtype of synapse specialized for fast, sustained, and graded neurotransmitter release, which transmit sound information to ganglion neurons [100]. Furthermore, like LGN and Afadin, Myosin VI is involved in neuronal synaptic function. Myosin VI is enriched at the postsynaptic density, and *Myo6* mutant hippocampal neurons have fewer dendritic spines and synapses and impaired internalization of AMPA receptors [101]. Strikingly, like mInsc and Afadin, Myosin VI was also proposed to regulate the directionality of cell migration without affecting overall motility by regulating transport of epidermal growth factor receptor to the leading edge [102]. Accordingly, Myosin VI is found at the leading edge of growth factor-stimulated fibroblasts [103] and is important for motility of *Drosophila* border cells [104].

9.3.3 Additional Candidates

The recurring patterns of protein function discussed above suggest that future work will uncover more links between the spindle orientation machinery and polarized responses in post-mitotic cells. For instance, the $G\alpha_i$ guanine nucleotide exchange factor Ric8 not only helps orient the mitotic spindle [47–51, 105], but it is also implicated in *Dictyostelium* chemotaxis by amplifying Gα_i signal initiated downstream of chemoattractant receptor signaling [106]. Additionally, huntingtin appears to regulate protein transport in mitotic and non-mitotic contexts. It mediates cortical localization of dynein-dynactin-LGN-NuMA in dividing cells, thus helping to orient the spindle [107, 108]. Huntingtin also regulates apical localization of Par3-aPKC during mouse mammary epithelial morphogenesis [109] and microtubule-based transport in neurons [110–112].

9.4 Summary

In conclusion, proteins that orient the mitotic spindle are emerging as also playing a variety of essential roles in post-mitotic cells. Examples detailed above represent relatively disparate systems and processes, suggesting they could be the tip of the iceberg. In these alternate contexts, mInsc-LGN-Gα_i and partners appear to use their ability to mark and organize subcellular domains for a wide variety of processes. They generally act by scaffolding partner proteins together and/or by regulating the cytoskeleton. We thus anticipate that several additional processes relying on mInsc-LGN-Gα_i will be uncovered in the future when their role is progressively studied in new post-mitotic contexts. In addition, new or known partners of mInsc-LGN-Gα_i in the spindle orientation machinery will be obvious candidates to pursue in these novel contexts. Finally, the large body of knowledge gathered over the years by studying spindle orientation will be invaluable to accelerate the understanding of normal biological processes and disease mechanisms where spindle proteins play a post-mitotic role.

References

1. Noatynska A, Gotta M, Meraldi P (2012) Mitotic spindle (DIS)orientation and DISease: cause or consequence? J Cell Biol 199(7):1025–1035
2. Bergstralh DT, St Johnston D (2014) Spindle orientation: what if it goes wrong? Semin Cell Dev Biol 34:140–145
3. Fischer E, Legue E, Doyen A, Nato F, Nicolas JF, Torres V, Yaniv M, Pontoglio M (2006) Defective planar cell polarity in polycystic kidney disease. Nat Genet 38(1):21–23
4. Nakajima Y, Meyer EJ, Kroesen A, McKinney SA, Gibson MC (2013) Epithelial junctions maintain tissue architecture by directing planar spindle orientation. Nature 500(7462): 359–362
5. Pease JC, Tirnauer JS (2011) Mitotic spindle misorientation in cancer—out of alignment and into the fire. J Cell Sci 124(Pt 7):1007–1016
6. Gillies TE, Cabernard C (2011) Cell division orientation in animals. Curr Biol 21(15):R599–R609
7. Kulukian A, Fuchs E (2013) Spindle orientation and epidermal morphogenesis. Philos Trans R Soc Lond Ser B Biol Sci 368(1629):20130016

8. Lancaster MA, Knoblich JA (2012) Spindle orientation in mammalian cerebral cortical development. Curr Opin Neurobiol 22(5):737–746
9. Lu MS, Johnston CA (2013) Molecular pathways regulating mitotic spindle orientation in animal cells. Development 140(9):1843–1856
10. Morin X, Bellaiche Y (2011) Mitotic spindle orientation in asymmetric and symmetric cell divisions during animal development. Dev Cell 21(1):102–119
11. Poulson ND, Lechler T (2012) Asymmetric cell divisions in the epidermis. Int Rev Cell Mol Biol 295:199–232
12. Nguyen-Ngoc T, Afshar K, Gonczy P (2007) Coupling of cortical dynein and G alpha proteins mediates spindle positioning in *Caenorhabditis elegans*. Nat Cell Biol 9(11):1294–1302
13. Couwenbergs C, Labbe JC, Goulding M, Marty T, Bowerman B, Gotta M (2007) Heterotrimeric G protein signaling functions with dynein to promote spindle positioning in *C. elegans*. J Cell Biol 179(1):15–22
14. Park DH, Rose LS (2008) Dynamic localization of LIN-5 and GPR-1/2 to cortical force generation domains during spindle positioning. Dev Biol 315(1):42–54
15. Lechler T, Fuchs E (2005) Asymmetric cell divisions promote stratification and differentiation of mammalian skin. Nature 437(7056):275–280
16. Poulson ND, Lechler T (2010) Robust control of mitotic spindle orientation in the developing epidermis. J Cell Biol 191(5):915–922
17. Williams SE, Beronja S, Pasolli HA, Fuchs E (2011) Asymmetric cell divisions promote Notch-dependent epidermal differentiation. Nature 470(7334):353–358
18. Williams SE, Ratliff LA, Postiglione MP, Knoblich JA, Fuchs E (2014) Par3-mInsc and Galphai3 cooperate to promote oriented epidermal cell divisions through LGN. Nat Cell Biol 16(8):758–769
19. Konno D, Shioi G, Shitamukai A, Mori A, Kiyonari H, Miyata T, Matsuzaki F (2008) Neuroepithelial progenitors undergo LGN-dependent planar divisions to maintain self-renewability during mammalian neurogenesis. Nat Cell Biol 10(1):93–101
20. Morin X, Jaouen F, Durbec P (2007) Control of planar divisions by the G-protein regulator LGN maintains progenitors in the chick neuroepithelium. Nat Neurosci 10(11):1440–1448
21. Postiglione MP, Juschke C, Xie Y, Haas GA, Charalambous C, Knoblich JA (2011) Mouse inscuteable induces apical-basal spindle orientation to facilitate intermediate progenitor generation in the developing neocortex. Neuron 72(2):269–284
22. Zigman M, Cayouette M, Charalambous C, Schleiffer A, Hoeller O, Dunican D, McCudden CR, Firnberg N, Barres BA, Siderovski DP et al (2005) Mammalian inscuteable regulates spindle orientation and cell fate in the developing retina. Neuron 48(4):539–545
23. Gonczy P (2008) Mechanisms of asymmetric cell division: flies and worms pave the way. Nat Rev Mol Cell Biol 9(5):355–366
24. Wodarz A, Ramrath A, Grimm A, Knust E (2000) Drosophila atypical protein kinase C associates with Bazooka and controls polarity of epithelia and neuroblasts. J Cell Biol 150(6):1361–1374
25. Kuchinke U, Grawe F, Knust E (1998) Control of spindle orientation in Drosophila by the Par-3-related PDZ-domain protein Bazooka. Curr Biol 8(25):1357–1365
26. Goldstein B, Macara IG (2007) The PAR proteins: fundamental players in animal cell polarization. Dev Cell 13(5):609–622
27. Schober M, Schaefer M, Knoblich JA (1999) Bazooka recruits Inscuteable to orient asymmetric cell divisions in Drosophila neuroblasts. Nature 402(6761):548–551
28. Wodarz A, Ramrath A, Kuchinke U, Knust E (1999) Bazooka provides an apical cue for Inscuteable localization in Drosophila neuroblasts. Nature 402(6761):544–547
29. Kraut R, Chia W, Jan LY, Jan YN, Knoblich JA (1996) Role of inscuteable in orienting asymmetric cell divisions in Drosophila. Nature 383(6595):50–55
30. Parmentier ML, Woods D, Greig S, Phan PG, Radovic A, Bryant P, O'Kane CJ (2000) Rapsynoid/partner of inscuteable controls asymmetric division of larval neuroblasts in Drosophila. J Neurosci 20(14):RC84

31. Yu F, Morin X, Cai Y, Yang X, Chia W (2000) Analysis of partner of inscuteable, a novel player of Drosophila asymmetric divisions, reveals two distinct steps in inscuteable apical localization. Cell 100(4):399–409
32. Schaefer M, Shevchenko A, Shevchenko A, Knoblich JA (2000) A protein complex containing Inscuteable and the Galpha-binding protein Pins orients asymmetric cell divisions in Drosophila. Curr Biol 10(7):353–362
33. Schaefer M, Petronczki M, Dorner D, Forte M, Knoblich JA (2001) Heterotrimeric G proteins direct two modes of asymmetric cell division in the Drosophila nervous system. Cell 107(2):183–194
34. Du Q, Macara IG (2004) Mammalian Pins is a conformational switch that links NuMA to heterotrimeric G proteins. Cell 119(4):503–516
35. Du Q, Stukenberg PT, Macara IG (2001) A mammalian Partner of inscuteable binds NuMA and regulates mitotic spindle organization. Nat Cell Biol 3(12):1069–1075
36. Siller KH, Cabernard C, Doe CQ (2006) The NuMA-related Mud protein binds Pins and regulates spindle orientation in Drosophila neuroblasts. Nat Cell Biol 8(6):594–600
37. Bowman SK, Neumuller RA, Novatchkova M, Du Q, Knoblich JA (2006) The Drosophila NuMA Homolog Mud regulates spindle orientation in asymmetric cell division. Dev Cell 10(6):731–742
38. Izumi Y, Ohta N, Hisata K, Raabe T, Matsuzaki F (2006) Drosophila Pins-binding protein Mud regulates spindle-polarity coupling and centrosome organization. Nat Cell Biol 8(6):586–593
39. Zhu J, Wen W, Zheng Z, Shang Y, Wei Z, Xiao Z, Pan Z, Du Q, Wang W, Zhang M (2011) LGN/mInsc and LGN/NuMA complex structures suggest distinct functions in asymmetric cell division for the Par3/mInsc/LGN and Galphai/LGN/NuMA pathways. Mol Cell 43(3):418–431
40. Culurgioni S, Alfieri A, Pendolino V, Laddomada F, Mapelli M (2011) Inscuteable and NuMA proteins bind competitively to Leu-Gly-Asn repeat-enriched protein (LGN) during asymmetric cell divisions. PNAS 108(52):20998–21003
41. Yuzawa S, Kamakura S, Iwakiri Y, Hayase J, Sumimoto H (2011) Structural basis for interaction between the conserved cell polarity proteins Inscuteable and Leu-Gly-Asn repeat-enriched protein (LGN). PNAS 108(48):19210–19215
42. Kotak S, Busso C, Gonczy P (2012) Cortical dynein is critical for proper spindle positioning in human cells. J Cell Biol 199(1):97–110
43. Kaushik R, Yu F, Chia W, Yang X, Bahri S (2003) Subcellular localization of LGN during mitosis: evidence for its cortical localization in mitotic cell culture systems and its requirement for normal cell cycle progression. Mol Biol Cell 14(8):3144–3155
44. Natochin M, Gasimov KG, Artemyev NO (2001) Inhibition of GDP/GTP exchange on G alpha subunits by proteins containing G-protein regulatory motifs. Biochemistry 40(17):5322–5328
45. Blumer JB, Cismowski MJ, Sato M, Lanier SM (2005) AGS proteins: receptor-independent activators of G-protein signaling. Trends Pharmacol Sci 26(9):470–476
46. Yoshiura S, Ohta N, Matsuzaki F (2012) Tre1 GPCR signaling orients stem cell divisions in the Drosophila central nervous system. Dev Cell 22(1):79–91
47. Afshar K, Willard FS, Colombo K, Johnston CA, McCudden CR, Siderovski DP, Gonczy P (2004) RIC-8 is required for GPR-1/2-dependent Galpha function during asymmetric division of *C. elegans* embryos. Cell 119(2):219–230
48. Couwenbergs C, Spilker AC, Gotta M (2004) Control of embryonic spindle positioning and Galpha activity by *C. elegans* RIC-8. Curr Biol 14(20):1871–1876
49. David NB, Martin CA, Segalen M, Rosenfeld F, Schweisguth F, Bellaiche Y (2005) Drosophila Ric-8 regulates Galphai cortical localization to promote Galphai-dependent planar orientation of the mitotic spindle during asymmetric cell division. Nat Cell Biol 7(11):1083–1090

50. Hampoelz B, Hoeller O, Bowman SK, Dunican D, Knoblich JA (2005) Drosophila Ric-8 is essential for plasma-membrane localization of heterotrimeric G proteins. Nat Cell Biol 7(11):1099–1105
51. Wang H, Ng KH, Qian H, Siderovski DP, Chia W, Yu F (2005) Ric-8 controls Drosophila neural progenitor asymmetric division by regulating heterotrimeric G proteins. Nat Cell Biol 7(11):1091–1098
52. VanDongen AM (2009) Biology of the NMDA receptor. CRC Press, Boca Raton
53. Sans N, Wang PY, Du Q, Petralia RS, Wang YX, Nakka S, Blumer JB, Macara IG, Wenthold RJ (2005) mPins modulates PSD-95 and SAP102 trafficking and influences NMDA receptor surface expression. Nat Cell Biol 7(12):1179–1190
54. Oliva C, Escobedo P, Astorga C, Molina C, Sierralta J (2012) Role of the MAGUK protein family in synapse formation and function. Dev Neurobiol 72(1):57–72
55. Knoblich JA (2005) Pins for spines. Nat Cell Biol 7(12):1157–1158
56. Bourne JN, Harris KM (2008) Balancing structure and function at hippocampal dendritic spines. Annu Rev Neurosci 31:47–67
57. Hunt DL, Castillo PE (2012) Synaptic plasticity of NMDA receptors: mechanisms and functional implications. Curr Opin Neurobiol 22(3):496–508
58. Wiser O, Qian X, Ehlers M, Ja WW, Roberts RW, Reuveny E, Jan YN, Jan LY (2006) Modulation of basal and receptor-induced GIRK potassium channel activity and neuronal excitability by the mammalian PINS homolog LGN. Neuron 50(4):561–573
59. Sanada K, Tsai LH (2005) G protein betagamma subunits and AGS3 control spindle orientation and asymmetric cell fate of cerebral cortical progenitors. Cell 122(1):119–131
60. Bowers MS, McFarland K, Lake RW, Peterson YK, Lapish CC, Gregory ML, Lanier SM, Kalivas PW (2004) Activator of G protein signaling 3: a gatekeeper of cocaine sensitization and drug seeking. Neuron 42(2):269–281
61. Yao L, McFarland K, Fan P, Jiang Z, Inoue Y, Diamond I (2005) Activator of G protein signaling 3 regulates opiate activation of protein kinase A signaling and relapse of heroin-seeking behavior. Proc Natl Acad Sci USA 102(24):8746–8751
62. Groves B, Gong Q, Xu Z, Huntsman C, Nguyen C, Li D, Ma D (2007) A specific role of AGS3 in the surface expression of plasma membrane proteins. Proc Natl Acad Sci USA 104(46):18103–18108
63. Johns DC, Marx R, Mains RE, O'Rourke B, Marban E (1999) Inducible genetic suppression of neuronal excitability. J Neurosci Off J Soc Neurosci 19(5):1691–1697
64. Wang F (2009) The signaling mechanisms underlying cell polarity and chemotaxis. Cold Spring Harb Perspect Biol 1(4):a002980
65. Neptune ER, Bourne HR (1997) Receptors induce chemotaxis by releasing the betagamma subunit of Gi, not by activating Gq or Gs. Proc Natl Acad Sci USA 94(26):14489–14494
66. Stephens L, Milne L, Hawkins P (2008) Moving towards a better understanding of chemotaxis. Curr Biol 18(11):R485–R494
67. Kamakura S, Nomura M, Hayase J, Iwakiri Y, Nishikimi A, Takayanagi R, Fukui Y, Sumimoto H (2013) The cell polarity protein mInsc regulates neutrophil chemotaxis via a noncanonical G protein signaling pathway. Dev Cell 26(3):292–302
68. Wright CE, Kushner EJ, Du Q, Bautch VL (2015) LGN directs interphase endothelial cell behavior via the microtubule network. PLoS One 10(9):e0138763
69. Ezan J, Lasvaux L, Gezer A, Novakovic A, May-Simera H, Belotti E, Lhoumeau AC, Birnbaumer L, Beer-Hammer S, Borg JP et al (2013) Primary cilium migration depends on G-protein signalling control of subapical cytoskeleton. Nat Cell Biol 15(9):1107–1115
70. Tarchini B, Jolicoeur C, Cayouette M (2013) A molecular blueprint at the apical surface establishes planar asymmetry in cochlear hair cells. Dev Cell 27(1):88–102
71. Bhonker Y, Abu-Rayyan A, Ushakov K, Amir-Zilberstein L, Shivatzki S, Yizhar-Barnea O, Elkan-Miller T, Tayeb-Fligelman E, Kim SM, Landau M et al (2016) The GPSM2/LGN GoLoco motifs are essential for hearing. Mamm Genome 27(1–2):29–46

72. Goodrich LV, Strutt D (2011) Principles of planar polarity in animal development. Development 138(10):1877–1892
73. Shotwell SL, Jacobs R, Hudspeth AJ (1981) Directional sensitivity of individual vertebrate hair cells to controlled deflection of their hair bundles. Ann N Y Acad Sci 374:1–10
74. Jones C, Roper VC, Foucher I, Qian D, Banizs B, Petit C, Yoder BK, Chen P (2008) Ciliary proteins link basal body polarization to planar cell polarity regulation. Nat Genet 40(1):69–77
75. Montcouquiol M, Rachel RA, Lanford PJ, Copeland NG, Jenkins NA, Kelley MW (2003) Identification of Vangl2 and Scrb1 as planar polarity genes in mammals. Nature 423(6936):173–177
76. Almomani R, Sun Y, Aten E, Hilhorst-Hofstee Y, Peeters-Scholte CM, van Haeringen A, Hendriks YM, den Dunnen JT, Breuning MH, Kriek M et al (2013) GPSM2 and Chudley-McCullough syndrome: a Dutch founder variant brought to North America. Am J Med Genet A 161A(5):973–976
77. Diaz-Horta O, Sirmaci A, Doherty D, Nance W, Arnos K, Pandya A, Tekin M (2012) GPSM2 mutations in Chudley-McCullough syndrome. Am J Med Genet A 158A(11):2972–2973
78. Doherty D, Chudley AE, Coghlan G, Ishak GE, Innes AM, Lemire EG, Rogers RC, Mhanni AA, Phelps IG, Jones SJ et al (2012) GPSM2 mutations cause the brain malformations and hearing loss in Chudley-McCullough syndrome. Am J Hum Genet 90(6):1088–1093
79. Walsh T, Shahin H, Elkan-Miller T, Lee MK, Thornton AM, Roeb W, Abu Rayyan A, Loulus S, Avraham KB, King MC et al (2010) Whole exome sequencing and homozygosity mapping identify mutation in the cell polarity protein GPSM2 as the cause of nonsyndromic hearing loss DFNB82. Am J Hum Genet 87(1):90–94
80. Yariz KO, Walsh T, Akay H, Duman D, Akkaynak AC, King MC, Tekin M (2012) A truncating mutation in GPSM2 is associated with recessive non-syndromic hearing loss. Clin Genet 81(3):289–293
81. Chudley AE, McCullough C, McCullough DW (1997) Bilateral sensorineural deafness and hydrocephalus due to foramen of Monro obstruction in sibs: a newly described autosomal recessive disorder. Am J Med Genet 68(3):350–356
82. Alrashdi I, Barker R, Patton MA (2011) Chudley-McCullough syndrome: another report and a brief review of the literature. Clin Dysmorphol 20(2):107–110
83. Hilgert N, Smith RJ, Van Camp G (2009) Function and expression pattern of nonsyndromic deafness genes. Curr Mol Med 9(5):546–564
84. Speicher S, Fischer A, Knoblich J, Carmena A (2008) The PDZ protein Canoe regulates the asymmetric division of Drosophila neuroblasts and muscle progenitors. Curr Biol 18(11):831–837
85. Wee B, Johnston CA, Prehoda KE, Doe CQ (2011) Canoe binds RanGTP to promote Pins(TPR)/Mud-mediated spindle orientation. J Cell Biol 195(3):369–376
86. Carminati M, Cecatiello V, Mapelli M (2016) Crystallization and X-ray diffraction of LGN in complex with the actin-binding protein afadin. Acta Crystallogr Sect F Struct Biol Commun 72(Pt 2):145–151
87. Carminati M, Gallini S, Pirovano L, Alfieri A, Bisi S, Mapelli M (2016) Concomitant binding of Afadin to LGN and F-actin directs planar spindle orientation. Nat Struct Mol Biol 23(2):155–163
88. Mandai K, Rikitake Y, Shimono Y, Takai Y (2013) Afadin/AF-6 and canoe: roles in cell adhesion and beyond. Prog Mol Biol Transl Sci 116:433–454
89. Beaudoin GM 3rd, Schofield CM, Nuwal T, Zang K, Ullian EM, Huang B, Reichardt LF (2012) Afadin, a Ras/Rap effector that controls cadherin function, promotes spine and excitatory synapse density in the hippocampus. J Neurosci Off J Soc Neurosci 32(1):99–110
90. Xie Z, Huganir RL, Penzes P (2005) Activity-dependent dendritic spine structural plasticity is regulated by small GTPase Rap1 and its target AF-6. Neuron 48(4):605–618
91. Miyata M, Ogita H, Komura H, Nakata S, Okamoto R, Ozaki M, Majima T, Matsuzawa N, Kawano S, Minami A et al (2009) Localization of nectin-free afadin at the leading edge and

its involvement in directional cell movement induced by platelet-derived growth factor. J Cell Sci 122(Pt 23):4319–4329
92. Slovakova J, Speicher S, Sanchez-Soriano N, Prokop A, Carmena A (2012) The actin-binding protein Canoe/AF-6 forms a complex with Robo and is required for Slit-Robo signaling during axon pathfinding at the CNS midline. J Neurosci Off J Soc Neurosci 32(29): 10035–10044
93. Petritsch C, Tavosanis G, Turck CW, Jan LY, Jan YN (2003) The Drosophila myosin VI Jaguar is required for basal protein targeting and correct spindle orientation in mitotic neuroblasts. Dev Cell 4(2):273–281
94. Hasson T, Mooseker MS (1994) Porcine myosin-VI: characterization of a new mammalian unconventional myosin. J Cell Biol 127(2):425–440
95. Wells AL, Lin AW, Chen LQ, Safer D, Cain SM, Hasson T, Carragher BO, Milligan RA, Sweeney HL (1999) Myosin VI is an actin-based motor that moves backwards. Nature 401(6752):505–508
96. Avraham KB, Hasson T, Steel KP, Kingsley DM, Russell LB, Mooseker MS, Copeland NG, Jenkins NA (1995) The mouse Snell's waltzer deafness gene encodes an unconventional myosin required for structural integrity of inner ear hair cells. Nat Genet 11(4):369–375
97. Ahmed ZM, Morell RJ, Riazuddin S, Gropman A, Shaukat S, Ahmad MM, Mohiddin SA, Fananapazir L, Caruso RC, Husnain T et al (2003) Mutations of MYO6 are associated with recessive deafness, DFNB37. Am J Hum Genet 72(5):1315–1322
98. Melchionda S, Ahituv N, Bisceglia L, Sobe T, Glaser F, Rabionet R, Arbones ML, Notarangelo A, Di Iorio E, Carella M et al (2001) MYO6, the human homologue of the gene responsible for deafness in Snell's waltzer mice, is mutated in autosomal dominant nonsyndromic hearing loss. Am J Hum Genet 69(3):635–640
99. Self T, Sobe T, Copeland NG, Jenkins NA, Avraham KB, Steel KP (1999) Role of myosin VI in the differentiation of cochlear hair cells. Dev Biol 214(2):331–341
100. Roux I, Hosie S, Johnson SL, Bahloul A, Cayet N, Nouaille S, Kros CJ, Petit C, Safieddine S (2009) Myosin VI is required for the proper maturation and function of inner hair cell ribbon synapses. Hum Mol Genet 18(23):4615–4628
101. Osterweil E, Wells DG, Mooseker MS (2005) A role for myosin VI in postsynaptic structure and glutamate receptor endocytosis. J Cell Biol 168(2):329–338
102. Chibalina MV, Poliakov A, Kendrick-Jones J, Buss F (2010) Myosin VI and optineurin are required for polarized EGFR delivery and directed migration. Traffic 11(10):1290–1303
103. Buss F, Kendrick-Jones J, Lionne C, Knight AE, Cote GP, Paul Luzio J (1998) The localization of myosin VI at the golgi complex and leading edge of fibroblasts and its phosphorylation and recruitment into membrane ruffles of A431 cells after growth factor stimulation. J Cell Biol 143(6):1535–1545
104. Geisbrecht ER, Montell DJ (2002) Myosin VI is required for E-cadherin-mediated border cell migration. Nat Cell Biol 4(8):616–620
105. Woodard GE, Huang NN, Cho H, Miki T, Tall GG, Kehrl JH (2010) Ric-8A and Gi alpha recruit LGN, NuMA, and dynein to the cell cortex to help orient the mitotic spindle. Mol Cell Biol 30(14):3519–3530
106. Kataria R, Xu X, Fusetti F, Keizer-Gunnink I, Jin T, van Haastert PJ, Kortholt A (2013) Dictyostelium Ric8 is a nonreceptor guanine exchange factor for heterotrimeric G proteins and is important for development and chemotaxis. Proc Natl Acad Sci USA 110(16):6424–6429
107. Godin JD, Colombo K, Molina-Calavita M, Keryer G, Zala D, Charrin BC, Dietrich P, Volvert ML, Guillemot F, Dragatsis I et al (2010) Huntingtin is required for mitotic spindle orientation and mammalian neurogenesis. Neuron 67(3):392–406
108. Elias S, Thion MS, Yu H, Sousa CM, Lasgi C, Morin X, Humbert S (2014) Huntingtin regulates mammary stem cell division and differentiation. Stem Cell Rep 2(4):491–506
109. Elias S, McGuire JR, Yu H, Humbert S (2015) Huntingtin is required for epithelial polarity through RAB11A-mediated apical trafficking of PAR3-aPKC. PLoS Biol 13(5):e1002142

110. Gunawardena S, Her LS, Brusch RG, Laymon RA, Niesman IR, Gordesky-Gold B, Sintasath L, Bonini NM, Goldstein LS (2003) Disruption of axonal transport by loss of huntingtin or expression of pathogenic polyQ proteins in Drosophila. Neuron 40(1):25–40
111. Gauthier LR, Charrin BC, Borrell-Pages M, Dompierre JP, Rangone H, Cordelieres FP, De Mey J, MacDonald ME, Lessmann V, Humbert S et al (2004) Huntingtin controls neurotrophic support and survival of neurons by enhancing BDNF vesicular transport along microtubules. Cell 118(1):127–138
112. Colin E, Zala D, Liot G, Rangone H, Borrell-Pages M, Li XJ, Saudou F, Humbert S (2008) Huntingtin phosphorylation acts as a molecular switch for anterograde/retrograde transport in neurons. EMBO J 27(15):2124–2134

Editors' Biographies

Monica Gotta was born in Turin, Italy, where she also did her undergraduate studies. She received her PhD from the University of Lausanne, Switzerland, where she worked on chromatin organization in the laboratory of Susan Gasser at the Swiss Institute for Experimental Cancer Research (ISREC). In 1998, she joined the group of Dr Julie Ahringer at the Welcome CR/UK Gurdon Institute to study the mechanisms of cell polarization and mitotic spindle positioning during asymmetric cell division of the *C. elegans* embryo. In 2002, she went back to Switzerland as a Swiss National Science Foundation Assistant Professor at the Institute of Biochemistry, ETH, Zurich. In 2008, she moved to the Medical Faculty of the University of Geneva where she was first Associate Professor and then Full Professor in 2014. Research in the lab of Dr. Gotta focuses on the regulation of cell division processes during development and is funded by the Swiss National Science Foundation and the University of Geneva.

Patrick Meraldi was born in Zurich, Switzerland, where he studied Biology at the ETH Zurich. He joined the laboratory of Erich Nigg to study centrosome biology, first for his PhD at the University of Geneva, Switzerland, then as a post-doctoral researcher at the Max-Planck Institute of Biochemistry in Martinsried, Germany. As an EMBO post-doctoral fellow in the laboratory of Peter Sorger at MIT in Cambridge, USA, he studied the function and composition of human kinetochores. In 2005, he returned to Switzerland as a Swiss National Science Foundation Assistant Professor at the

M. Gotta, P. Meraldi (eds.), *Cell Division Machinery and Disease*, Advances in Experimental Medicine and Biology 1002, DOI 10.1007/978-3-319-57127-0

Institute of Biochemistry of the ETH Zurich, investigating chromosome segregation and the mitotic spindle. He was nominated in 2012 to an Associate Professorship of the Medical Faculty of the University of Geneva, where he coordinates since 2016 the Translational Research Centre for Oncohematolgy. Research in the Meraldi group is funded by the Swiss National Science Foundation, the Swiss Cancer league and the University of Geneva. It focuses on the fundamental mechanisms that govern human chromosome segregation, and investigates how a deregulation of these mechanisms contributes to erroneous cell divisions in cancer cells.

Index

M. Gotta, P. Meraldi (eds.), *Cell Division Machinery and Disease*, Advances in Experimental Medicine and Biology 1002, DOI 10.1007/978-3-319-57127-0

MIX
Papier aus verantwortungsvollen Quellen
Paper from responsible sources
FSC® C105338

If you have any concerns about our products,
you can contact us on
ProductSafety@springernature.com

In case Publisher is established outside the EU,
the EU authorized representative is:
Springer Nature Customer Service Center GmbH
Europaplatz 3, 69115 Heidelberg, Germany

Printed by Libri Plureos GmbH
in Hamburg, Germany